Marine Plastics Abatement

Marine Plastics Abatement

Technology, Management, Business and Future Trends (Volume 2)

Edited by

Thammarat Koottatep, Ekbordin Winijkul, Wenchao Xue, Atitaya Panuvatvanich, Chettiyappan Visvanathan, Tatchai Pussayanavin, Nantamol Limphitakphong and Chongrak Polprasert

Published by
IWA Publishing
Unit 104–105, Export Building
1 Clove Crescent
London E14 2BA, UK
Telephone: +44 (0)20 7654 5500
Fax: +44 (0)20 7654 5555
Email: publications@iwap.co.uk
Web: www.iwapublishing.com

First published 2023
© 2023 IWA Publishing

Disclaimer

British Library Cataloguing in Publication Data
A CIP catalogue record for this book is available from the British Library

ISBN: 9781789063431 (Paperback)
ISBN: 9781789063448 (eBook)
ISBN: 9781789063455 (ePub)

Doi: 10.2166/9781789063448

This eBook was made Open Access in January 2023.

© 2023 The Editors.

Contents

Foreword

Since 2015, the importance of marine ecosystems has been highlighted in Sustainable Development Goal (SDGs) No. 14, which emphasizes the need to 'Conserve and sustainably use the oceans, seas and marine resources for sustainable development.' A constant stream of alarming facts demonstrates that the sustainability of our oceans is under severe threat from acidification, ocean warming, eutrophication, fisheries collapses and, most notably, marine plastic pollution while over 3 billion people, or 42% of the global population, rely on oceans for their livelihoods. Marine plastic litter has become a serious global issue due to the fact that about 10 million metric tons of plastic waste generated on land enters the marine environment annually, contaminating major river basins and oceans. Plastics are also difficult to biodegrade and some types are non-degradable, resulting in accumulation rather than decomposition of plastics in the environment. One estimate predicts that by 2050, the weight of plastic waste in the ocean will be greater than the weight of fish. For this reason, in March 2022, the United Nations Environment Assembly adopted a resolution entitled, 'End Plastic Pollution' related to the marine environment, and negotiations for an internationally legally binding instrument will begin from the second half of 2022 onward.

In the last decade, several global/regional programs to develop innovative and practical solutions have been initiated by both the public and private sectors to tackle mismanaged plastic pollution. Among these initiatives, the 'Osaka Blue Ocean Vision' (OBOV) with the overarching aim of reducing additional pollution by marine plastic litter to zero by 2050 was shared at the G20 Osaka Summit in 2019, and the Government of Japan has launched the **MARINE Initiative** in order to realize OBOV. Japan's MARINE Initiative aims to advance effective actions to combat marine plastic litter on a global scale focusing on (1) management of waste, (2) recovery of marine litter, (3) innovation, and (4) empowerment.

One of the crucial factors in translating the initiative into action is to empower all stakeholders who play a significant role in marine plastic abatement, whether governmental offices, private companies, non-governmental organizations, reuse/recycling enterprises or small-scale waste pickers. The Ministry of Foreign Affairs (MOFA), Japan has thus supported the Asian Institute of Technology (AIT), Thailand in establishing and implementing an intensive empowerment program with an emphasis on marine plastic pollution. This initiative led to the very first one-year Master's in 'Marine Plastics Abatement (MPA)' program in the region, officially inaugurated in August 2020.

This unique program has recruited almost 100 young environmental leaders from more than 30 countries in Asia, Africa, and Latin America for training through comprehensive coursework and innovative research which will contribute immensely to realizing SDG14: Life Below Water and others such as SDG11: Sustainable Cities and Communities; SDG12: Responsible Consumption and Production; and SDG17: Partnerships for the Goals. The curriculum of the MPA program has drawn widely from up-to-date research findings, process innovations, technological advancement as well as social interventions/campaigns by experts and professionals from AIT and its partner institutions. To increase awareness and widen empowerment on this subject, it is essential to consolidate new areas of knowledge and expertise into a book which is accessible to other audiences from different sectors.

I am certain that readers of this book will come to understand not only the root causes and negative impacts on human and environmental health of the marine plastics issue, but also various means to reduce mismanaged plastics through innovative technology. They will also learn about the application of the circular economy and become familiar with innovative business models and lessons learnt from regional case studies around the world. I, therefore wish to acknowledge the authors and editors led by AIT and their respective partner universities, i.e., Thammasat University, Ramkhamhaeng University, Chulalongkorn University, Thailand for coordinating the edition and publication of this reference book. As the community of professionals grows, my personal expectation is for this book to be regularly updated to capture new evidence and scientific findings for new generations who might face and be affected by even more serious marine pollution.

H. E. Mr. NASHIDA Kazuya
Ambassador of Japan to Thailand

doi: 10.2166/9781789063448_0223

Chapter 6

Plastic-to-values: technologies and applications

Thammarat Koottatep

Table of Contents

Plastic debris and plastic pollution are the newly emerged pollutants in many regions affecting developed, developing, and undeveloped countries. Not only has the land been affected, but the marine system has also experienced the negative physical and chemical effects from plastic waste (PW). Plastic abatement has become one of the human challenges.

To get rid of and manage plastic debris both pre- and post-products, several opportunities have been suggested. One is the value enhancement of PW. This chapter discusses ways to increase the value of PW or upcycling technologies. It provides three possible technologies which are based on recycle, reuse and disposal. Moreover, the reduction of waste generation through prevention, reduction, reuse and recycling is one way to ensure sustainable consumption and production patterns (SDG12) targets. Therefore, this chapter will introduce the concepts to deal with PW.

Although plastic debris can be a valuable material, the upcycling processes also produce negative intermediate products. Thus, to complete plastic upcycling, pollution control devices are necessary for upcycling facilities. Moreover, in real situations, the waste management processes can generate many kinds of intermediate products and wastes (see Chapter 1). Therefore, the value enhancement of plastic using alternative technologies must be accompanied by proper pre-treatment technologies. Thus, Chapter 6 also discusses the pre-treatment principles and segregated machine suggestions (Section 6.2).

This chapter aims to provide knowledge on the entire process of PW upcycling. The chapter consists of two main sections: (1) pretreatment requirements and (2) upcycling technologies and application. Moreover, case studies of upcycling applications are presented in Section 6.4.

6.1 INTRODUCTION

There are several solutions to the problems related to PW. However, dumpsites and landfills are the most commonly selected by most people as a method to manage PW. In the case of developing countries such as Thailand, approximately 60% of PW is disposed of with improper sanitation in landfill and open dumping sites while only 14.5% is recycled and recovered within proper management systems (Bureecam *et al.*, 2018). On the other hand, Japan (a developed country) can manage more than 80% PW by recovery and recycling processes (Ministry of the Environment, Japan, 2018) while only 10% is sent to the landfill. Due to the lack of education and limited resources, waste management systems in developing and undeveloped countries tend to fall behind and are mostly improperly operated. PW upcycling can be used to motivate people to pay attention to proper waste management.

Figure 6.1 shows well-known management practices that are common in solid waste management, namely reducing, reusing, recycling, recovery and disposal. Reducing and reuse are considered onsite waste management or waste management at the source. Thus both reducing and reuse depend on human behavior and their education. On the contrary, recycling, recovery and disposal require other facilities or technologies. Therefore, this chapter only discusses upcycling technologies for recycling, recovery and disposal.

In Chapter 1, two main types of plastics, namely thermoset and thermoplastic are introduced. Thermoset plastic cannot be recycled or recycled as raw materials due to its properties while thermoplastic can be more easily recycled and thus can be reused as raw material or plastic resin. Furthermore, the popular methods within the thermoset plastic management are disposal and burning, which can generate a high volume of pollutants which are then released to the environment. Therefore, this chapter will provide discussions on the plastic recycling employing physical processes by concrete combination or plastics to concrete technologies (see Section 6.3.2).

To manage both thermoset and thermoplastic using upcycling technologies, the pre-treatment process and the selection of technology are crucial. Moreover, the pre-treatment selection for plastic upcycling depends on the purity

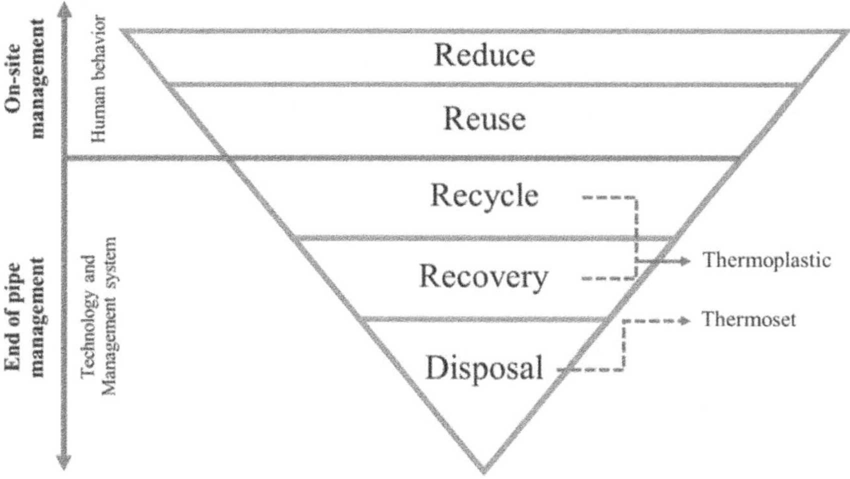

Figure 6.1 Waste management concepts.

requirement of the end product. For example, if PW is converted into plastic textile, the plastic requires high purification of plastic because textile needs high plastic properties such as a virgin plastic. Thus, the pre-treatment must be matched with the end-products processes.

6.1.1 General waste composition concept

Before discussing management facilities, it is necessary to examine the composition of general municipal solid waste (MSW) and industrial solid waste (ISW) composition, and the different characteristics of the two types of solid waste. This chapter also classifies solid waste using three main components which consist of inorganic (metals, aluminum cans, iron, sand, etc.), water/ or moisture, and organic waste. The organic waste is further classified into biological volatile solid (BVS) and residual volatile solid (RVS). BVS is a common organic compound which has a simple and non-complex chain structure (such as food waste, paper, cardboard, yard, etc.). Thus, BVS is easily broken down by microorganisms. On the contrary, RVS has a complex structure and a strong crosslink.

6.1.1.1 MSW composition

Several types of components can be found in MSW, especially in undeveloped and developing countries. Lack of knowledge on waste, lack of proper segregation systems and lack of financial support are the main underlying reasons for the problem. In undeveloped and developing countries, the major components of MSW are BVS and RVS. Table 6.1 shows MSW compositions in ASEAN (Association of South East Asian Nations) countries. It can be seen that organic waste accounts for 60% of MSW components in most of the ASEAN countries (UNEP, 2017). Therefore, organic waste, especially food waste is an important element of MSW and PW management. The high volume and the diverse types

Table 6.1 MSW composition in ASEAN countries.

Country	Organic Waste						Inorganic Waste			Others
	Food	Paper	Plastic	Textile	Rubber	Yard	Metal	Glass	Construction	
Brunei	36.0	18.0	16.0	0.0	0.0	0.0	4.0	3.0	0.0	23.0
Cambodia	60.0	9.0	15.0	1.0	1.0	0.0	0.0	3.0	0.0	11.0
Indonesia	60.0	9.0	14.0	3.5	6.0	0.0	4.3	1.7	0.0	1.5
Laos	64.0	7.0	12.0	5.0	3.0	0.0	1.0	7.0	0.0	1.0
Malaysia	45.0	8.2	13.2	0.0	0.0	0.0	0.0	3.3	0.0	30.3
Myanmar	73.0	2.2	17.8	1.1	0.0	0.0	0.0	0.5	0.0	5.4
Philippines	52.0	8.7	10.6	1.6	0.0	0.0	14.6	2.3	0.0	10.2
Singapore	10.5	16.5	11.6	2.1	0.0	8.6	20.8	1.1	17.0	11.8
Thailand	64.0	8.0	17.6	1.4	1.0	1.0	2.0	3.0	0.0	2.0
Vietnam	55.0	5.0	10.0	0.0	4.0	0.0	5.0	3.0	0.0	18.0
Average	52.0	9.2	13.8	1.6	1.5	1.0	5.2	2.8	1.7	11.4

Source: UNEP (2017).

of components found in MSW lead to the contamination of valuable waste such as plastic, aluminum, iron and glass.

6.1.1.2 ISW composition

ISW is the waste scrap or residual waste from manufacturing processes, which can be both liquid and solid phases. The ISW compositions normally correlate with the type of industry. According to Sharma *et al.* (2016) plastic components, especially foam and fluff, account for more than 65% followed by rubbers and metal at 20% and 9.5%, respectively. Moreover, the contamination in ISW is very low because such solid waste comes from manufacturing process, which is not contaminated with organic waste. Thus, ISW is suitable for recycling or upcycling processes.

6.1.1.3 Limitation of upcycling facilities

With the different characteristics of solid waste compositions, the management or upgrade processes are related to the quality of waste. This section discusses the consideration factors for upcycling facilities.

(1) **Contamination:** the contamination in plastic can occur both in the texture and on the surface of the materials. Contamination in the texture of plastics means contamination from components involved in the plastic production process such as additives used to improve chemical properties of the plastics while food, labels or oil may only contaminate the surfaces of the plastic materials. Both types of contamination bring about decreasing plastic quality.

(2) **Recycling loop:** the recycling loop is another limitation of plastic upcycling. Normally, thermoplastic can be recycled, but the thermal process or the reproduction process itself can affect plastic properties, thus potentially reducing its quality.

(3) **Environmental effect:** due to plastic reaction with the environment such as UV radiation or pH, the structure or properties of plastic will be changed (see Chapter 3 for more details).

Therefore, PW properties must be considered prior to the recycling processes. The plastic purifying and cleaning process can improve the properties of PW. Thus, pre-treatment and plastic cleaning facilities are crucial for the plastic upcycling.

6.1.2 Trends and current situations of plastic upcycling in the world

6.1.2.1 Asia region

6.1.2.1.1 ASEAN

In 2017, the Association of South East Asian Nations signed a memorandum of understanding on the PW combat in Phuket, Thailand. This agreement proposes four priorities which are used to combat marine debris. The four priorities are (1) policy support and planning; (2) research, innovation and capacity building; (3) public awareness, education and outreach; and (4) private sector engagement.

6.1.2.1.2 North Asia in Russia

In Russia, waste management has been hugely neglected over the years. MSW and ISW treatment can be best described as inefficient and is causing both negative environmental impact and suboptimal use of raw materials and energy. While the European Union Member States recover up to an average of 60% of MSW, Russia's waste recovery rate is almost zero. Recycling operations have not existed in Russia for a long time. Landfilling the waste was the only solution for waste disposal, but after 2000, the landfills increasingly reached their capacity limits. Many of the existing landfills are very poorly equipped and do not function as sanitary landfills but as dumpsites, and this has led to a number of challenges with the local residents suffering from odor nuisance, groundwater pollution and toxic gas emissions.

According to the Russian Ministry of Natural Resources and the Environment (RMNRE), about 3–4 billion tons of waste is generated in the country every year. Fifty-four comes from mostly the coal industry which extracts raw materials used to generate energy. About 17% is from nonferrous metallurgy, 16% is from the steel and iron industry, and 12% is from other waste sections, including housing and utilities. MSW accounts for 1–2% of total waste generated. The volume of MSW generated in Russia has been steadily increasing annually with more than 49 million metric tons generated in 2010 and an estimated volume of over 60 million metric tons in 2019 and thereafter. MSW generation per capita average is about 400 kg/year. Over 25 million tons of waste is produced each year in Moscow alone. Ninety-five percent of all MSW is sent for landfill disposal, and about 30% of waste disposal facilities in the country do not meet sanitary landfill requirements. This problem stems from the inability of the government and waste disposal firms to efficiently manage waste disposal.

In total, 65 million tons of waste was landfilled in 2019. Russia has the capacity to landfill merely 55 million tons a year. It is estimated that about 32 regions will reach their full landfilling capacity within 5 years. The country has a total of 1699 MSW landfills, 576 industrial waste disposal sites, 5500 authorized and 17 000 unauthorized landfills. Studies show that waste in a landfill generally is composed of 34% food, 19% paper, 14% polymers, 12% glass, 6% wood, 6% street waste, 4% metal, 3% textile and 2% other substances. Russia produces about 5.84 million tons of plastic every year. About 36.3 pieces of microplastic are found per kilogram of dry sediment in the beaches of the Baltic Sea in the Kaliningrad region. Local and governmental efforts have started in Russia to help fight plastic pollution, but these efforts are not enough as it needs to be scaled larger in an efficient manner due to Russia's size and population.

In 2018, the Russian government introduced the National Project 'Ecology' to protect the environment (Strategy 2018–2030). Currently, the plan is to introduce a solid waste management system for PW, ISW and MSW, and to close down all unauthorized landfills with 200 landfills in the countryside and 75 landfills within cities to be removed and renatured by 2025. In order to ensure a transition from landfilling to processing waste, the Russian government intends to build 36 waste separation- and processing plants (biofuels and composting), 7 new treatment facilities for hazardous waste and 154 plants for incineration and waste to energy recovery by 2024.

Russia utilizes the experience of the EU Member States, Germany in particular. For this to happen, a number of legal acts are being drafted to be used to reform the country.

Some new priorities in the waste management plan include the following:

- Building of eco-industrial parks.
- Reform of waste disposal regulation, where manufacturers and importers of packaging are obligated to dispose of the packaging they manufactured or imported.
- Heavy focus on recycling of waste (reduced landfilling) with about 50% of PW to be recycled by 2025.
- Setting up of waste collection and sorting facilities and building adequate infrastructure for other waste management activities.
- Reducing waste generation and raw resources use.
- Research and development on further technologies that can be used.
- Ensuring sound regulation of household waste in practice.
- Raising awareness among consumers and producers in Russia.
- Re-equipping and modernizing existing industries to reduce waste generation.

6.1.2.1.3 South East Asia in Thailand

In 2015, PW made up 76% of the marine debris found in the Gulf of Thailand (Figure 6.2). According to Marks *et al.* (2020), Thailand is one of the top 10 contributors of marine debris. Plastic ends up in the Gulf of Thailand due to mismanagement of MSW. More than 40% of MSW is not managed. Another cause of the high volume of marine debris found in the Gulf of Thailand is due to weak regulations. In 2015, 7777 local governmental organizations existed, but only 328 organizations could work on proper waste management systems. Therefore, more than half of solid waste is not properly collected and managed (<50%).

6.1.2.1.4 India

In India, around 60% of the generated PW is collected, and for its treatment, the hierarchy in the integrated waste management approach is followed. After recycling is explored, energy recovery and material recovery become the best options for the production of fuel and that of secondary raw materials, respectively. Commonly, this is through refuse derived fuel (RDF) production for cement and brick kilns.

Plastic Waste Management Rules (PWM Rules) (2016) suggest the use of non-recyclable PW for the construction of roads or for co-processing in cement kilns. Currently, about 10% of the 210 cement kilns in India is used in co-processing of PW. India views co-processing as an environmentally sound and sustainable method of PW management compared to incineration and landfilling, and recognizes its advantages in helping the country lower its reliance on conventional raw materials in cement kilns. In order to increase the amount of waste managed through upcycling techniques in India, there needs to be greater investment for the provision of capital to expand existing mechanical recycling facilities and also invest in pyrolysis facilities.

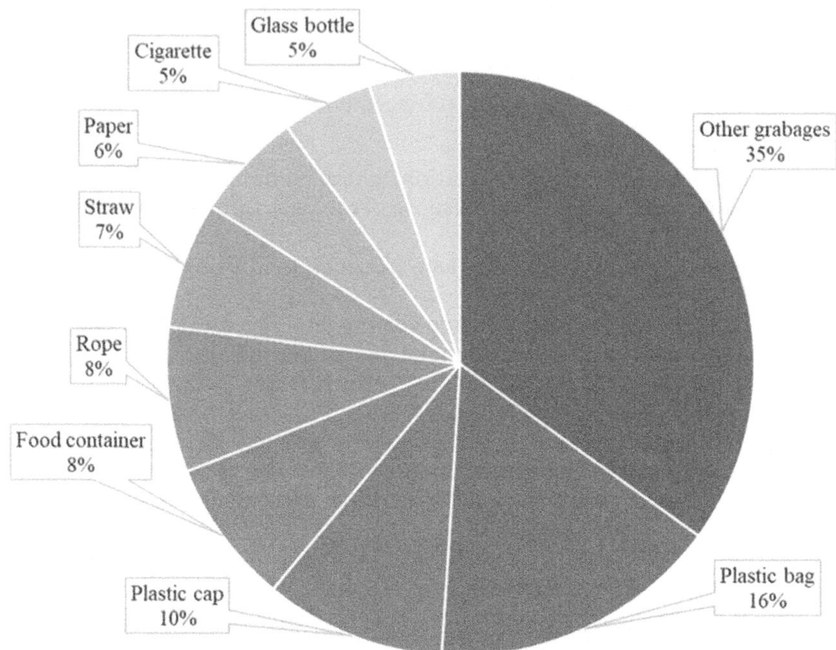

Figure 6.2 Type of debris in Thailand oceans (*Source*: Department of Marine and Coastal Resources, Thailand, 2015).

6.1.2.1.5 Sri Lanka

Sri Lanka generates around 7000 tons of PW per day with different waste compositions in urban and rural areas. In urban areas, plastic makes up around 10% of the total waste generated whereas in the rural areas it makes up around 5% of the total waste (JICA, 2018). Plastic consumption in Sri Lanka is increasing annually by 16% while the efficiency of PW collection stands at 33%, and that of recycling stands at 3%. Plastics processing is a thriving industry that has been present in Sri Lanka for over 45 years.

There are currently over 400 companies involved in plastics processing in Sri Lanka. Sri Lanka has invested a total amount of Rs. 15 billion in plastics processing. Almost 50% of this has been made through foreign direct investments. Of this, 66% of the total investment is exclusively in processing of plastic products for the export market (Samarasinghe *et al.*, 2021).

Regulations and Policies on PWM

- Prohibition on the manufacture of polyethylene (PE) or PE products 20 µm or less thick for country use.
- Prohibition on the sale, offer for sale, gratuitous offer, exhibition or use within the country of PE or PE products 20 µm or less in thickness.

- Use of PE or PE products with a thickness of 20 μm or less for lamination and medical or pharmaceutical purposes; written approval from the CEA required.
- Prohibition on the manufacture of expanded polystyrene (PS) food containers, plates, cups and spoons for use in the country.
- Prohibition on the sale, offer for sale, gratuitous offer, display or use of expanded PS food containers, plates, cups and spoons for use in the country.
- Prohibition of the use of all forms of PE, polypropylene (PP), and products as decoration at political, social, religious, national, cultural or other events or occasions.
- Prohibition on open burning of waste and other combustible materials including plastics.
- Prohibition on causing, permitting or allowing open burning of waste or other combustible materials, including plastics.
- Prohibition of the manufacture of bags (for the transport of products or goods, including food bags) made of high-density polyethylene (HDPE) as a raw material for domestic use.
- Prohibition on selling, offering for sale, offering for free, displaying or using of bags made of HDPE as raw material for domestic use (excluded articles: refuse sacks and textile sacks).
- Prohibition on the manufacture of food packaging (lunch sheets) made of PE (HDPE), low-density polyethylene (LDPE), PP) as raw material for use in the country.
- Prohibition on selling, offering for sale, offering for free, displaying or using of food packaging made of PE as a raw material in the country.

6.1.2.1.6 Pakistan

The plastic industry in Pakistan is expanding at a yearly average rate of 17% and further developing (Umer & Abid, 2017). The per capita consumption varies dramatically from rural to urban settings in the country and ranges between 0.283 and 0.612 kg/capita/day, and the waste generation growth rate is 2.4% per year (Khalid *et al.*, 2019). In all, 6000 plastic units in Pakistan are distributed which helps employ a number of people in various provinces (Pakplas, 2016). Karachi is the biggest and most (urban) populated city in the country, generating about 12,000 tons of waste every day. Sixty percent of solid waste is discarded in landfills, and the other 40% remains on the streets (Shahid *et al.*, 2014). Based on the 73.6 million urban population and per capita waste of 0.84 (kg/capita/day), the total MSW generation is roughly about 61.8 million (tons/day). It is difficult to determine the proportion of plastic in the total waste, but the proportion of organic waste all around the country is significant. In Pakistan, organic waste makes up 50% of MSW on a wet weight basis (Safar *et al.*, 2018).

In order to address this rising plastic pollution concern, the government of Pakistan has recently taken some measures during 2019–2021. These measures have nevertheless been significantly impacted by the global pandemic situation.

- 'Development of National Inventory of Plastic Waste in Pakistan'.
- Zero plastic in hiking trails and other areas of MHNP, 'Plastic-free National Park'.
- Environmental protection orders on violation of Environmental Laws including violators of Plastic Shopping Bags Regulations.
- A ban on polythene plastic bags, which also includes fines on manufacturers, shopkeepers and users of Rs. 100 000, Rs. 10 000 and Rs. 5000, respectively.
- Alternate and renewable energy target of at least 20% renewable energy generation by capacity by 2025 and at least 30% by 2030 in the revised Alternate and Renewable Energy Policy 2019. This policy supports the efforts for waste to energy.

6.1.2.2 *Australia region*

In 2016, municipal waste generation in Australia was at 54.5 million tons, which increased by about 30 million tons between 1996 and 2016. Moreover, the recycling rate of MSW increased from 7% to 58% in the same period (Figure 6.3). The reason for the increasing recycling rate during that time was China's National Sword import restriction.

According to O'Farrell (2019), plastic consumption in Australia was about 12.4 million tons between 2015 and 2018 while only 10.30% was recovered. During that time, the majority of recovery processes focused on energy recovery. The extent of energy recovery from plastics in Australia has grown and has been dominated by the manufacture of a C&D waste-based fuel which was manufactured, at the time of reporting, in South Australia and New South Wales, for combustion in local and overseas cement kilns. Timber is the main energy source for this fuel.

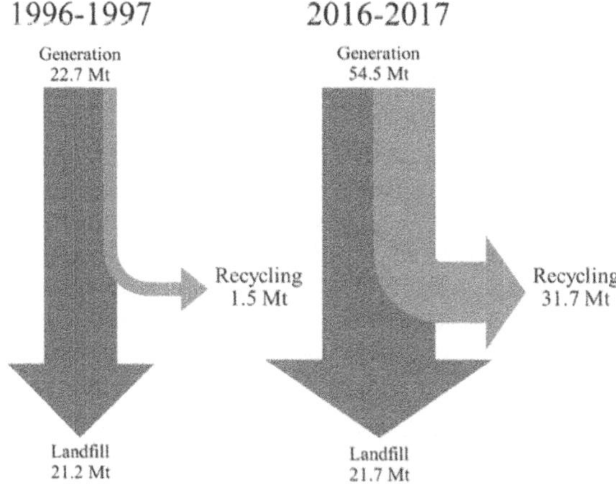

Figure 6.3 Solid waste generation and waste recycling in Australia (*Source*: modified from https://mraconsulting.com.au/mra-submission-to-the-inquiry-into-australias-waste-management-and-recycling-industries/).

6.1.2.3 European (EU) region

In 2018, 29.1 million tons of post-consumer PW was collected in the EU28 in order to be treated. PW exports outside the EU decreased by 39% from 2016 to 2018. About 5 million tons of plastic recycles are converted back to manufacture new products (80% of recycling). Figure 6.4 presents the composition of new products which use PW as raw materials. The majority of plastic management in EU is energy recovery, accounting for 42.6% while recycling and landfill account for 32.5 and 24.9%, respectively. Energy recovery and recycling processes showed an upward trend between 2006 and 2018 while landfill showed a downward trend as presented in Figure 6.4.

In 2019, EU Commission and EU industry committed to significant plastic recycling. By 2025, at least 10 million tons of PW can potentially be recycled if the commitment stands (Figure 6.5).

In 2020, the banning of landfills was pushed by the EU industry. However, more than 40% of waste in landfill contained PW. Due to the banning of

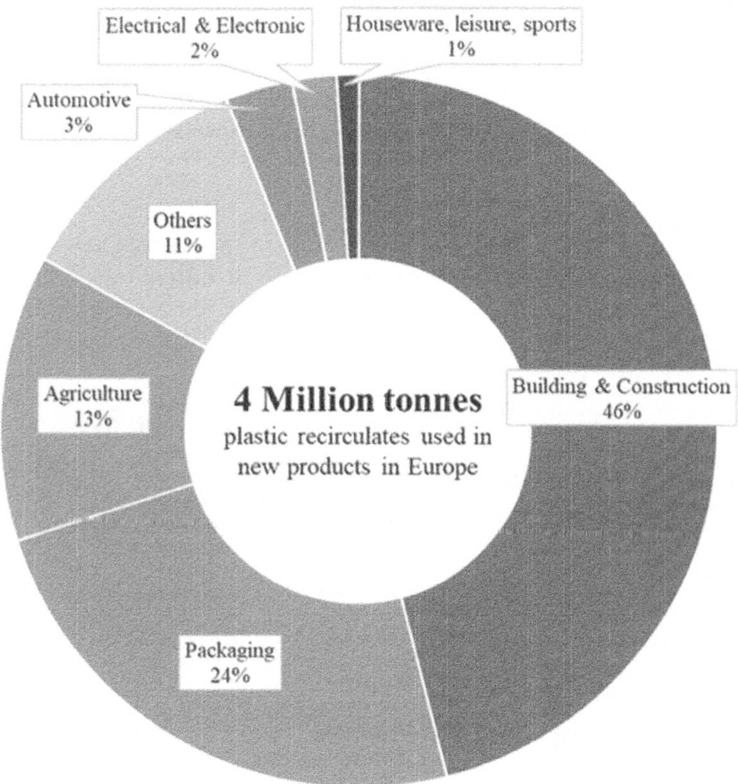

Figure 6.4 Plastic product in EU, 2018 (*Source*: https://plasticseurope.org/knowledge-hub/plastics-the-facts-2020/).

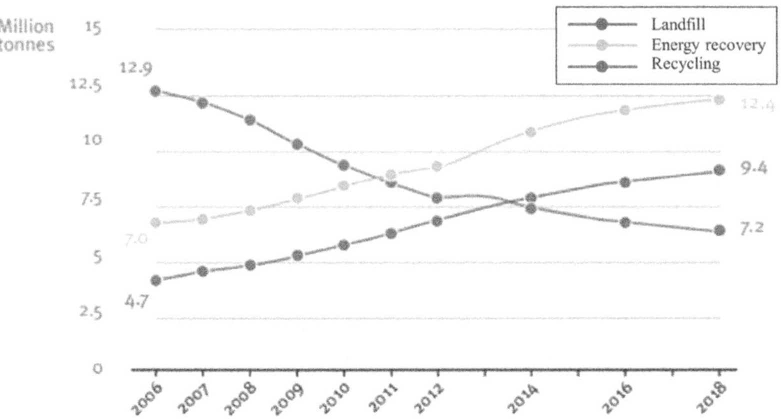

Figure 6.5 Trend of post-consumer plastic management during 2006–2018 (*Source*: https://plasticseurope.org/knowledge-hub/plastics-the-facts-2020/).

landfill, there was an upward trend in plastic upcycling. Ninety-six percent of recyclable plastic was recycled via material recovery methods (34%) and energy extraction (62%) (Figure 6.6). Most of energy recovery is created by combined heat and power recovery plants (CHP Plants) and solid recovered fuel (SRF).

6.1.2.4 USA region

In 2010, 37.83 million tons of PW was generated in the USA alone. In 2018, 292.4 million tons of MSW was generated. Of this, 9 million tons was recycled,

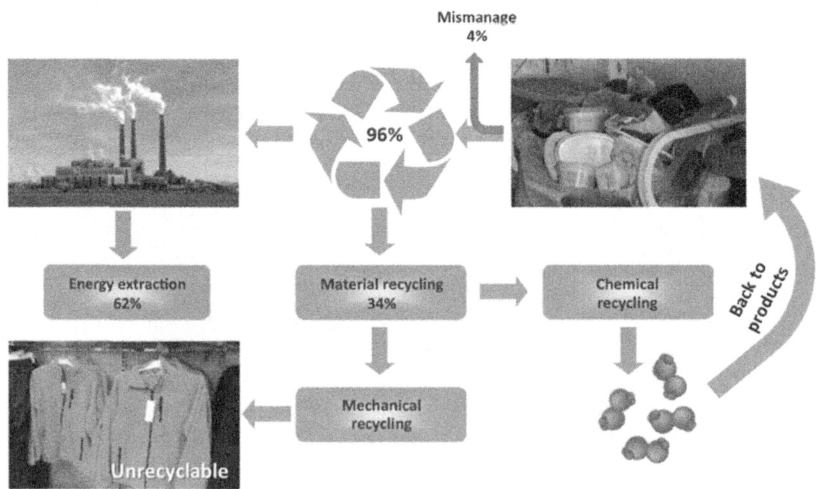

Figure 6.6 PW management in EU 2021 (*Source*: Bruder, 2019).

and 25 million tons was composted. Only 4.47% of the 9 million tons of recycled waste was from recycled plastics. In 2018, 3090 tons of plastic was recycled, and 5620 tons was utilized in waste to energy facilities. In 2010, it was estimated that as little as 0.9% of PW was mismanaged in the USA, and in 2019, data suggest that per capita, 0.81 kg of PW were mismanaged

6.1.2.5 *Goals*
In November 2020, the EPA set national recycling goals. The main goal was to increase the recycling rate in the country to 50% by 2030. Other goals included reducing contamination in recycling, making recycling processes more efficient, and helping to strengthen economic markets for recycles.

6.1.2.5 *Africa region*
6.1.2.5.1 South Africa
In 2010, South Africa generated 4.47 million tons of PW per year. In 2019, it was estimated that 12.1 kg of PW was mismanaged per capita. Due to this high volume of mismanaged waste, it is estimated that 11 000 tons of PW ends up as pollution in South Africa every year, an equivalent of 10 billion 20-g crisp packets.

Approximately 519 000 tons of plastic entered formal and informal waste management systems in 2017, 70% of which was disposed of in non-compliant landfills and open dumps, leading to a high risk of leakage into the natural environment. However, informal waste pickers collected between 20 and 30% of the dumped and landfilled PW. Leakage rates from non-compliant landfills and open dumps were estimated to be 30% and 80%, respectively.

Currently, there are no commercial or large-scale waste-to-energy facilities in South Africa. However, in 2019, it was estimated that 1000 tons of plastic was processed in this manner in informal and trial or small-scale private operations. Chemical plastic recycling makes use of gasification or pyrolysis to convert PW back into the primary constituent polymers intended for reprocessing. Globally, these technologies are still novel in the plastic recycling sector and are thus not fully established in many countries as yet. Polyolefin chemical recycling is being trialed in the Eastern Cape, and a few private pyrolysis plants are undergoing trials for diesel production.

As not all plastics are currently recycled in South Africa for various reasons, we need to find other solutions to reduce end-of-life PW. South Africa is currently trialing a variety of solutions, including plastic bricks, ecobricks and plastic roads. However, projects such as these often face logistical and organizational issues, and the longevity of these projects is not guaranteed. Plastic bricks are lightweight concrete blocks made from post-consumer expanded PS mixed with concrete. They are used as a low-cost construction material. Plastic bricks are produced in accordance with rigid building standards to ensure the flammability and strength of the bricks. They are used in loadbearing walls and concrete frames, among other structures. They resemble conventional bricks and can be cemented and plastered in the same way. Plastic bricks also provide improved noise absorption. However, there are concerns regarding microplastic leakage and the ability of the bricks to be reused after demolition of current structures. Ecobricks are made from 2-L cool drink bottles filled with dry, non-recycled plastics, such as cling film and crisps packets. They are used for filler material and insulation in various indoor and outdoor buildings and constructions; however, they are non-loadbearing. Many South Africans have turned to making ecobricks as a sink for their non-recycled PW. There are multiple concerns about the suitability of ecobricks used as building material, particularly the flammability risks, non-uniform weights and insulation properties. Thus, it is important that consumers follow the guidelines for size and weight when creating ecobricks from their home PW. Ecobricks unfortunately cannot be reused and thus are not a long-term solution to the plastic pollution problem. In December 2019, South Africa unveiled its first plastic road in Jeffery's Bay, in the Eastern Cape. The road is 1 km in length and comprises 1.5 tons of PW that was used as a partial replacement for bitumen. Another plastic road is currently under construction in KwaZulu Natal. There are concerns as to whether plastic roads can potentially leach microplastics over time.

In South Africa, 63% of all polyethylene terephthalate (PET) bottles get recycled. However, recycling is not the only way to achieve a circular plastics economy. Improving plastic recycling is only one solution of many. In order to achieve a circular PW economy, South Africa needs to remove unnecessary and problematic plastic items from circulation by stopping the production of these items in favor of using plastics that are more easily recycled, and products should be designed to be reused rather than disposed of.

South Africa already has a partial ban on plastic bags, with many stores opting to completely eliminate the use of plastic bags and others choosing to use plastic bags made partially out of recycled material. This has proven highly successful in the country, and after a small amount of push back from consumers, it has become the norm to carry reusable shopping bags. The success of the plastic bag ban should be used as an incentive to ban other plastic products, such as straws. Straws make up part of the Dirty Dozen™, a list of the 12 most commonly found items along the South African Coastline. Many establishments are already opting to use paper or even pasta straws, however, a complete ban can significantly reduce the volume of straws that end up polluting the environment.

Educating the new generations of South Africans on waste management is an important aspect of waste and PW management. The management of waste should be taught and practiced in all schools in South Africa in order to create mindfulness regarding the quantity of waste produced as well as to ensure that recycling becomes a habit.

South Africa is making great strides in improving waste management from landfilling around 90% of the general waste in 2011 to recycling nearly 40% in 2017. This improvement is due to recent changes in waste legislation, policies and plans, as well as a focus on waste beneficiation and the value of waste as a resource. There has also been a large number of media awareness campaigns in the last 10 years, and conscious consumerism is starting to take hold in many parts of the country.

6.1.2.5.2 West Africa

- Nigeria

Over 800 000 metric tons of PW is generated in Lagos every year, which has a population of over 21 million. One of the recent upcycling methods in Nigeria is the introduction of a waste to chemical plant by BASF Nigeria. PW is collected, sorted into types and colors, baled and transported to be crushed and converted to pellets.

PW is being collected by BASF and turned into chemicals in order to properly manage the waste. The waste is then processed into chemicals in the BASF waste to chemical project. This project was launched in 2018 with three objectives as follows.

(1) To build a scalable model to collect and aggregate the waste.
(2) To leverage on cost-effective pyrolysis technology to regenerate polycarbons from PP, PET and PS.
(3) To enable a circular economy by reengaging the monomers in the local chemical value chain.

Currently, there is a 1000 kg/day continuous pyrolysis unit, and one upstream recycling hub processes 1000 kg/day of PW. By 2022, they intend to set up two or three more recycling hubs to process up to 3–5 tons of PW per day. By 2025, they intend to set up 50 or 60 of these units in communities around Lagos. Another major upcycling project is the Lafarge Geocycle cement kiln project where PW is collected and used to power the cement kiln. About 100 000 tons of PW is diverted from landfills to their kilns per year since inception in 2017. Other upcycling techniques have been mechanical recycling by local conversion plants to plastic pellets for mostly exports or local production to other products. Although this has existed for quite some time, other upcycling techniques are creative artworks done by small businesses or entrepreneurs for educational, fun or sales purposes such as turning sorted plastics to ottomans, PET bottle caps to picture portraits, nylons bags to interior décor, or plastic bottles to vases and cups.

However, even with all these efforts, the end point of solid waste disposal is usually the landfill. However, waste landfills were set up to be the major means of plastic disposal before the chemical recycling technologies began. They are operated properly in order to minimize the risks posed to health and the environment.

Still, less than 15% of PW is being collected and upcycled by all the methods and technologies listed above as majority of waste is dumped in the landfills or leaks into the environment.

Policies for proper PW management are being deliberated and reviewed by policy makers to ensure nationwide enforcement.

- Rwanda

The Government of the Republic of Rwanda has made giant strides through its single-use plastic ban, Umuganda (Rwanda's compulsory monthly community cleanup work day) and a decade-long awareness campaign through publicizing the proposal of the ban since 2004, and have arguably made Kigali the cleanest city in Africa. Since the plastic ban in 2008, it is said that Rwanda has experienced an upward economic development in over 10 years, with the GDP growth currently sitting at 7.9%. The capital Kigali and other urban centers contribute roughly 50% of the GDP.

However, Kigali is still a city very much in development and is noted to have poor solid waste management practices. It has only one official dumpsite, in Nduba, Gasabo District. The city's recycling rate is between 2 and 12%. This variation is due to the informal way of handling PW collection and recycling.

According to Rwanda Environment Management Authority (REMA), total MSW generated in Kigali is about 800 tons a day, and about 5% of it is PW. Waste collection efficiency is about 300 tons daily with 450 tons of unsorted MSW dumped in the Nduba daily. This correlates with the fact that about 66% of rural households with no access to proper waste management collection facilities dispose of their waste in dumpsites, fields and bushes. However, the National Sanitation Implementation Strategy of 2016 has stated its desire to improve waste collection from households to about 60% collection by 2020 and 80% by 2030.

It is said that about 70–80% of the plastic in use is claimed to be recycled. There are about 14 recycling companies operating in business since the plastic ban. Though some of them are complaining there are not enough plastics disposed to remain in business, some companies are still existing to handle the recycling focus mostly on paper and plastics recycling. An example of the paper treatment and recycling plant called Trust Industries transforms paper waste into toilet paper. Other plastic recycling companies turn HDPE and/or LDPE plastics into furniture, household objects and industrial or agricultural materials. Currently, there is no recycling facility for PET bottles, meaning that the majority of the bottles generated, which is estimated to be about 100 000 bottles daily, are sorted, crushed and sold to recyclers in neighboring countries such as Uganda, Kenya, Tanzania and, until recently, China, though this is difficult as the selling price of their products were slightly higher compared to prices of virgin plastic pellets that were imported from abroad.

Almost all plastic materials that is recycled are collected from households, and very little recyclable material is obtained from the landfill. REMA's environmental reports indicate that only about 2% of PW is recovered from dumpsites. Plastic recyclers receive materials usually as industrial waste which is then separated, cleaned and recycled and then combined with virgin materials to create new products for the market. Similar trends are likely to exist for paper recyclers.

Some recycling companies have cited, as reasons for recycling difficulties, high cost and availability of inputs and difficulty in selling recycled products due to cheap plastic imports. Cost of electricity was also remarked to be high, accounting for about 25–30% of costs and this affects the profitability of the companies as electricity costs are expected to be about 5% for there to be good profitability. Significant resources are also deployed to sort and clean waste.

President Kagame has shown strong political will, though his government is mostly accused of being authoritarian, in enforcing the ban and is also challenging the industry to come up with plastic alternatives. A grace period of 3 years was given to businesses by the Rwandan government to explore and adapt to alternative packaging materials for their products. Some alternatives such as wax paper for bread products and bamboo cutlery and straws in shops and restaurants are some of the alternatives that have emerged. Some products such as biscuits and potato chips and other foods packed in plastic are allowed by the government and the company's manufacturing them are given approval after they have demonstrated and provided a detailed business plan that shows how they plan to collect and recycle their bags after customer consumption. In Rwanda, every school has an environment club where teachers are trained to teach the message of cleanliness and the importance of the plastic ban to students and through them to their parents.

6.2 PRE-TREATMENT TECHNOLOGY

Due to the common commingling of solids or MSW, the PW for upcycling processes especially, post-consumer should be purified or cleaned before upcycling. For example, Sumai's MSW consists of several waste compositions, and organic waste accounts for more than 60% (EPPIC, 2020). Therefore, to purify PW and get rid of the organic element, the biodegradation is necessary. Not only the organic waste, but other waste such as wood and glass must also be segregated to purify the PW. However, not only the contamination on the material surfaces, but also the added substances in the materials such as additives and improving compounds should be taken into consideration.

6.2.1 Biological process

The aim of biological processes is to convert the organic substances in the waste, especially BVS composition, to stabilization form (such as methane gas, water, carbon dioxide, etc.) with biological organism. Two main bioprocesses for organic waste removal are aerobic (or composting) and anaerobic (or digestion). However, biological processes can be man-made or naturally occurring. For example, the bottom of a sanitation landfill without oxygen gas can produce methane gas through anaerobic digestion, or a pile of leaves can also create a similar biological process. Although, most PW elements are organic, their structures are complex and consist of several polymer chain structures, which generates the non-biodegradation substances. Therefore, the PW cannot be decomposed by either the aerobic or anaerobic process.

6.2.1.1 Aerobic composting

Aerobic composting is the conversion of organic matter into compost or manure in the presence of air. This process is created by microbial oxidative reaction which occurs in aerobic condition. The composting process is shown in Figure 6.7. Moreover, anaerobic requires 4–6 weeks to decompose yard and organic matter from municipal waste on average.

Both biochemical and physical properties in waste such as nutrient balance (C/N ratio), pH, moisture content, temperature, oxygen requirement or particle size are the consideration factors of aerobic composting. To purify the PW from

$$\begin{matrix} \text{Proteins} \\ \text{Lipids} \\ \text{Carbon hydrates} \\ \text{Cellulose} \end{matrix} + O_2 + \text{Nutrients} + \text{Microorganisms} \rightarrow \text{Compost} + \text{New cells} + CO_2 + H_2O + NO_3 + SO_4 + \text{Heat}$$

Figure 6.7 Aerobic composting equation

MSW through aerobic composting, the air supply is necessary as it is related to energy consumption. For example, a high amount of organic compound in MSW requires a large volume of oxygen. This means a large amount of energy is required. Due to the high energy consumption, aerobic composting is suitable for the waste which requires short-time decomposition and which consists of a low amount of organic composition.

6.2.1.2 Anaerobic digestion

Anaerobic digestion is a biological process which is caused by anaerobic microorganisms. This process occurs in the absence of oxygen. As anaerobic digestion consists of several intrinsic processes such as hydrolysis, acidification and methanogenesis, it generates several end-products from the degraded organic waste as revealed in Figure 6.8. Apart from organic removal, anaerobic digestion also creates biogases such as methane gas (CH_4) which can be used as fuel. However, this process requires a greater deal of time for decomposition compared with the aerobic composting.

Due to the toxics in oxygen and odor issues, a close system is normally applied to anaerobic digestion process. Further consideration factors are nutrient balance (C/N ratio), pH, moisture content, temperature, particle size and mixing condition.

6.2.2 Physical process

The objective of the physical process is to remove unwanted materials by relying on physical properties of the materials such as visual, particle size, density and color. The target waste material which is segregated by this process is inorganic such as glass, stone or sand. This section introduces the physical segregation process consisting of manual separation, size (or screen) separation, density separation and metal separation. Moreover, the mechanism and basic principles are included in this section.

6.2.2.1 Manual and conveyer separation

Manual separation and sorting (see Figure 6.9) is usually necessary at the beginning of the recycling process for the preliminary removal of films, cardboard and bulky items, and is usually carried out by operators who

$$\begin{matrix} \text{Proteins} \\ \text{Lipids} \\ \text{Carbon hydrates} \\ \text{Cellulose} \end{matrix} + H_2O + \text{Nutrients} + \text{Microorganisms} \rightarrow \text{New cells} + CO_2 + CH_4 + NH_3 + H_2S + \text{Heat}$$

Figure 6.8 Anaerobic digestion.

Figure 6.9 Manual sorting (*Source*: Wongpanit International Co. Ltd, Phitsanulok, Thailand).

monitor the waste stream on a conveyor belt. Normally, conveyors used to transfer waste from one location to another are installed at the highest point of the waste recycling stream using gravity flow to reduce cost and energy. Moreover, conveyors may be classified as hinge, apron, bucket or pneumatic. Horizontal and inclined belt conveyors, where the material is carried above the belt, and drag conveyors, equipped with flights or crossbars to drag the material, are commonly used to handle solid waste (see Figure 6.9). Because solid waste is contained on the belt conveyor, other separation units such as magnetic separation or near-infrared radiation (NIR) sorting can be applied to additionally further remove such materials from the solid waste.

Belt conveyors are made from rubber, canvas or synthetic materials allowing them to handle relatively lightweight materials (see Figure 6.10) while the heavy-duty applications, such as unseparated MSW or industrial metal requires hinged steel belts (see Figure 6.11).

Because manual separation relies on workers to segregate waste, the performance of waste segregation depends on the experience and skills of the workers. Furthermore, the working area or sorting lines should be located in a well-lit and air-conditioned facility designed to meet Occupational Safety and Health Administration (OSHA) requirement. The consideration factors of manual and belt conveyer sorting are waste components the characteristic of the waste, the number of commingled recyclable items, and the throughput capacity of the facility. Critical factors in the design of the picking facility are the width, the belt speed, sorting or picking rates, and thickness of waste material.

6.2.2.2 *Size separation*
Size separation or screening is performed to remove small objects, especially yard and inorganic parts such as glass, sand and stones. The typical screening

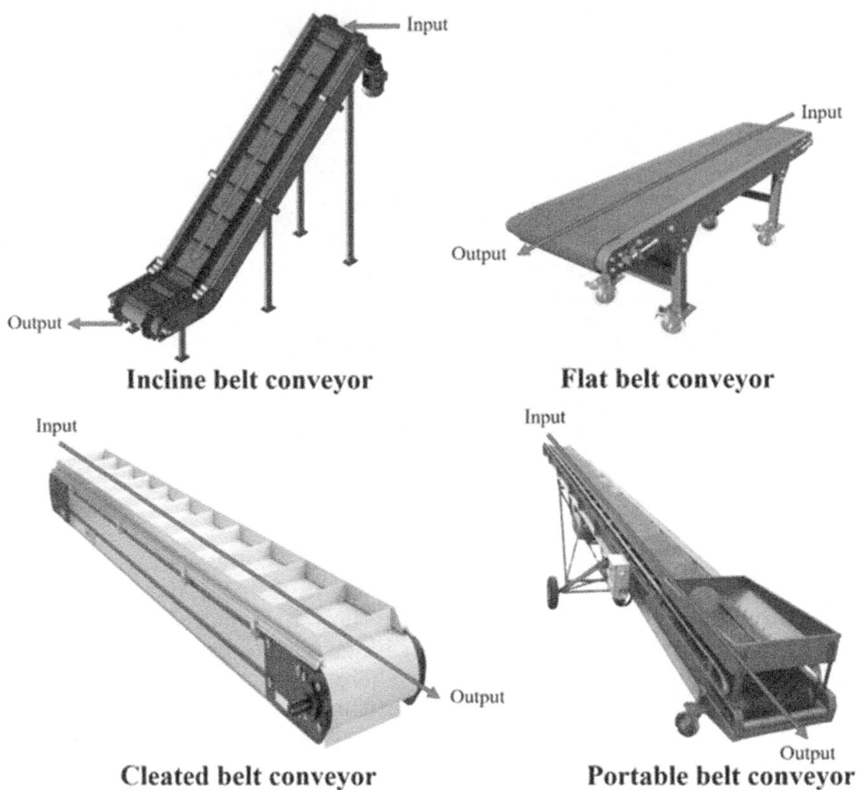

Figure 6.10 Conveyor belt used for lightweight materials (*Source*: modified from https://www.beidoou.com/construction/belt-conveyor-types.html).

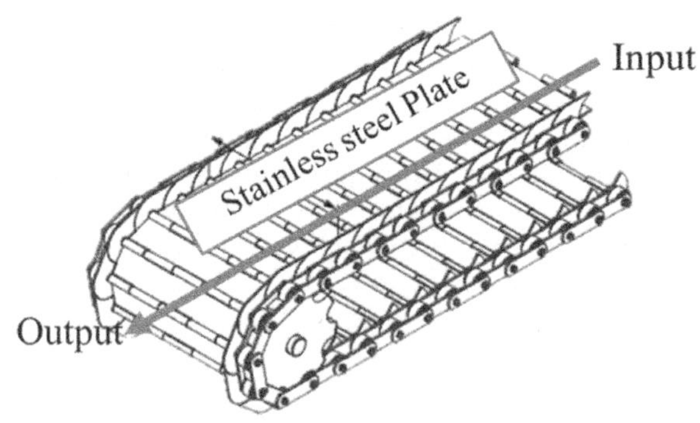

Figure 6.11 Hinged steel belts for heavy material (*Source*: OETIKER International; http://dktransportbaand.dk/upload_dir/pics/ALLERT_Hinged_Steel_Belts.pdf).

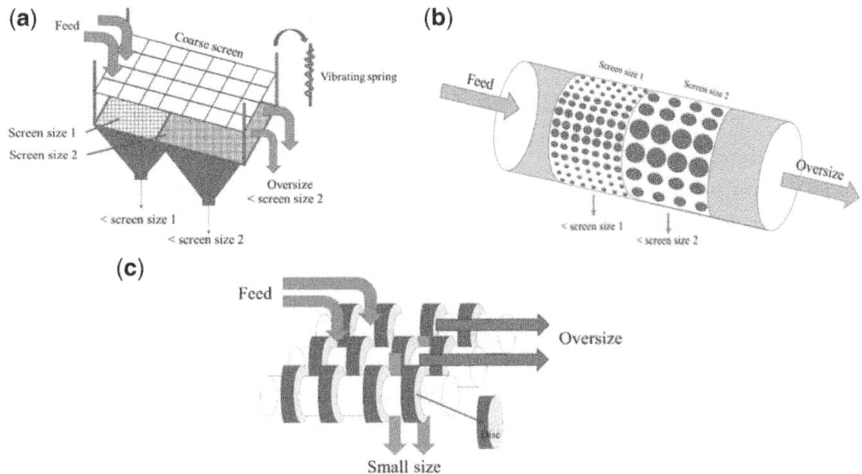

Figure 6.12 Typical screening: (a) vibrating screen; (b) rotary or trommel screen; (c) disc screen.

equipment for solid waste segregation consists of vibrating screens, drum or rotary screens, and dish screens (see Figure 6.12). Commonly, solid waste can be classified into three fractions, namely undersize (<50 mm), middle size (from 50 to 300 mm), and oversize (>300 mm) (Tchobanoglous *et al.*, 1993). Furthermore, most of the PW is classified as belonging to the middle size fraction. As such, screening separation is crucial for pre-treatment or plastic selection in plastic upcycling (Table 6.2).

6.2.2.3 Material sorting

Material or polymer sorting is used to eliminate unwanted materials using physical–chemical properties of waste material, such as density, color or chemical composition. Material sorting enhances the performance in the

Table 6.2 Typical screen for MSW.

Type	Detail
Vibrating screen (Figure 6.12a)	Used to segregate the unwanted material by inclined and vertical motion. Normally, this screen is used to classify the construction or demolition waste. Thus, vibrating screen could be installed at the source.
Rotary/ trommel screen (Figure 6.12b)	The trommel screen is the most popular for MSW separation. It can separate several size fractions. This machine is operated using vertical, inclined and rotary motion, which can enhance the performance and efficiency of waste separation.
Disc screen (Figure 6.12c)	The disc screen involves parallel interlocking lobed (or star-shaped) discs with horizontal shafts. The undesirable materials fall through the space between the discs. Moreover, the disc screen can adjust the size of the gap between the discs, which can help vary the target sizes.

removal of the unwanted contaminants such as pieces of metal, glass or paper from the PW stream. In this section, material or polymer sorting processes using density, electrostatic, magnetic density and sensor-based separation are explored.

6.2.2.3.1 Density separation

Density separation is a technique widely used to separate materials based on their density and hydro or aero dynamic characteristics. Physical properties such as weight and density are the functions of the density separation process. The target materials are classified into two groups, namely those considered as the light fraction (paper, plastic bags, films, etc.) and those considered as the heavy fraction (metals, aluminum cans, wood, etc.). Typically, the density separation includes hydrodynamic and aerodynamic processes. The hydrodynamic process is explained in Section 6.2.3. Meanwhile, the aerodynamic separation process, which is used to separate light plastic from other waste materials, is discussed in this part.

The aerodynamic separation process is applied to separate shredded wastes including two major components: (1) the light fraction materials such as paper, plastic, and organic; (2) the heavy fraction materials such as metals, wood, and other relatively dense inorganic materials. Air classifiers use the airflow to separate lighter materials from heavier materials. Waste enters the upper part of the height column while the air is supplied at the bottom part of the column (Figure 6.13). The light material is raised to the upper part and is eliminated from the waste stream while the heavy objects fall to the bottom part. Air classifiers not only come in a vertical shape, but also in a zigzag shape.

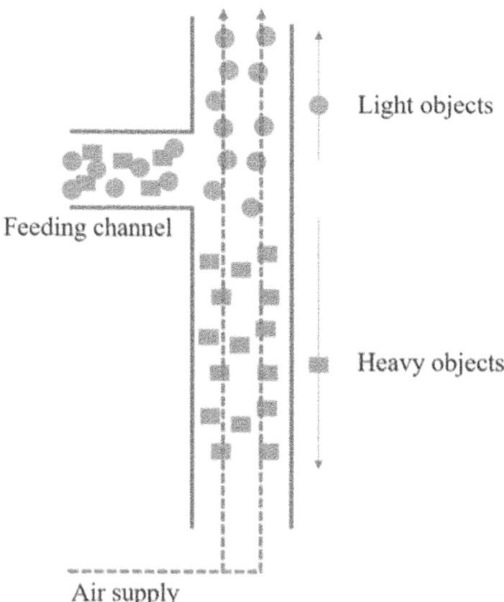

Figure 6.13 Schematic diagram of air classifier.

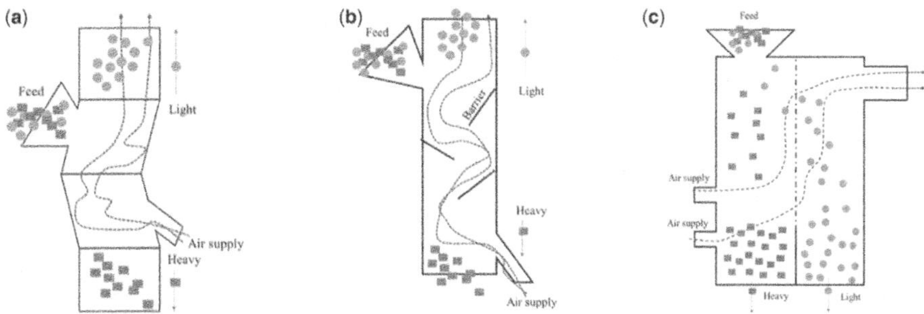

Figure 6.14 Cascade classifiers (*Source*: Shapiro & Galperin, 2005): (a) zigzag; (b) shelf classifier; (c) with horizontal scavenging.

Shelf classifiers and shelf classifiers with horizontal scavenging (Figure 6.14) are also used to improve the performance and efficiency of air classifiers.

6.2.2.3.2 Magnetic separation

The aim of the magnetic separation (MS) is to separate bulky metallic waste such as aluminum cans, keys or electronic cables from the waste stream using magnetic force. Normally, the commingled waste is fed on the open conveyor and travels along the conveyor under the magnetic field which then captures and thus removes metal materials from the waste stream. Figure 6.15 shows the modern MSs. Moreover, the developed magnetic separation can be used to segregate the different plastic types using the different magnetic densities, a technique referred to as magnetic density separation which is explained in Section 6.2.3.3c.

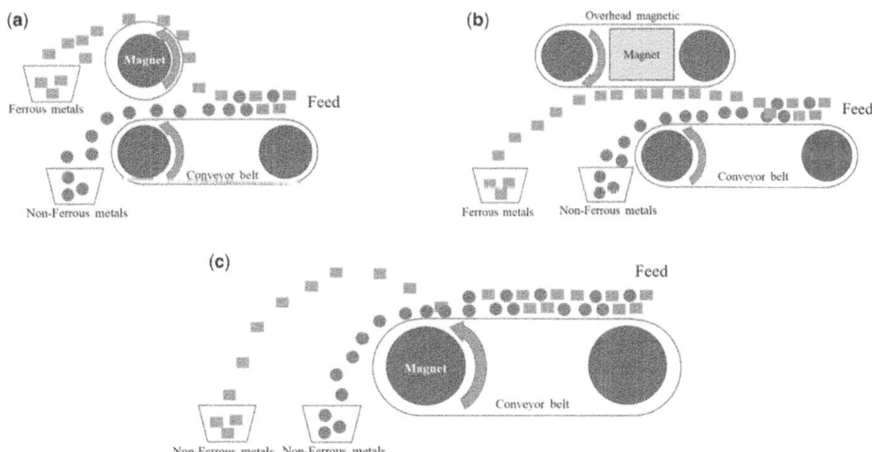

Figure 6.15 Magnetic separation facilities (*Source*: Gundupalli *et al.*, 2017): (a) magnetic dump sort; (b) magnetic overhead belt sort; (c) eddy current sort.

6.2.2.3.3 Sensor-based separation

Sensor-based separation selects target materials using a sensor. The sensor is a device or a system that applies light or signal to measure the physical characteristics of target materials. Once the sensor detects target materials, it sends a signal to the acceptor to segregate the target items. For example, light absorption is used to select bulky plastic, especially plastic bottles (see Figure 6.16). Other materials also can be selected using other sensors or signals. The segregation device might be an air gun or an air blower. Furthermore, many separating devices such as near-infrared spectroscopy (NIR) and X-ray fluorescence (XRF) apply sensor-based separation to classify waste (see Chapter 9 for more details).

6.2.3 Plastic preparation

Although plastic is selected from the waste stream, such plastic is still contaminated with other waste materials such as oil, sand or labels. Apart from the contamination, the bulkiness of plastic also presents a problem for upcycling. Each upcycling process requires a specific plastic type from the separation techniques. For example, the reprocessing process is specifically used for converting PET bottles to PS fibers. Therefore, this section provides the discussions on plastic preparation processes prior to the explanations of the upcycling facilities.

6.2.3.1 Size reduction

Size reduction is a process that is used to reduce the size of waste materials using cutting and shredding techniques. Moreover, size reduction can be used to reduce the size of not only PW, but also other waste materials such as yard and metal waste. The aim of this process is to produce a uniform waste material size. However, the size depends on the technique and equipment selected for the process. The commonly used size reduction devices for MSW and SW are the hammer mill shredder and the shear shredder.

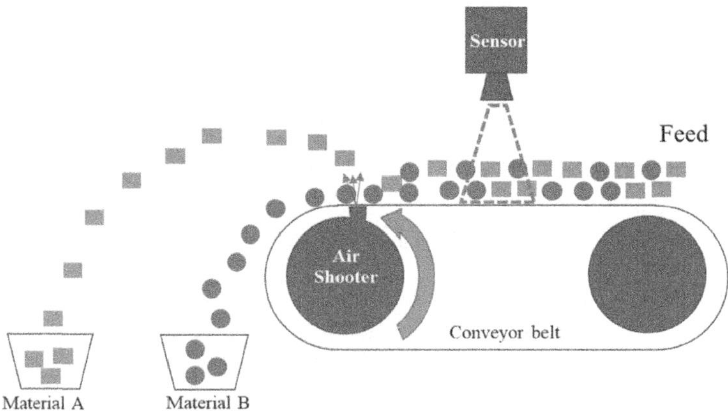

Figure 6.16 Symmetric diagram of sensor-based belt sorter (*Source*: Pacheco-Torgal *et al.*, 2018).

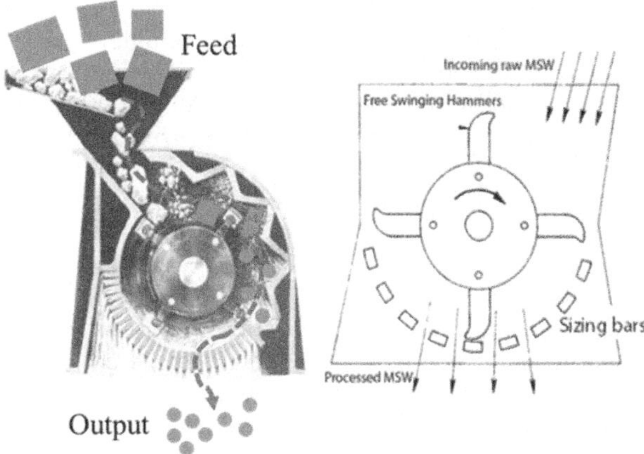

Figure 6.17 Hammer mill shredder (*Source*: Fitzgerald & Themelis, 2009).

The hammer mill shredder is a mill machine that is used to shred or crush waste materials into small fragments using impact force from a small hammer inside the machine. Materials are broken down into small pieces, and pass through the sizing bars (see Figure 6.17). Due to the impact force of hammer, the hammer mill shredder is suitable for brittle solid materials such as glass, stone construction waste.

The shear shredder is commonly used in the recycling industry as it works with both brittle and elastic materials. The principle mechanism of the shear shredder involves a row of hook discs, gaps between the discs and shear force (see Figure 6.18). Due to the shear force of the hook discs, the shear shredder can be used to reduce the size of highly elastic materials, especially tires and PW.

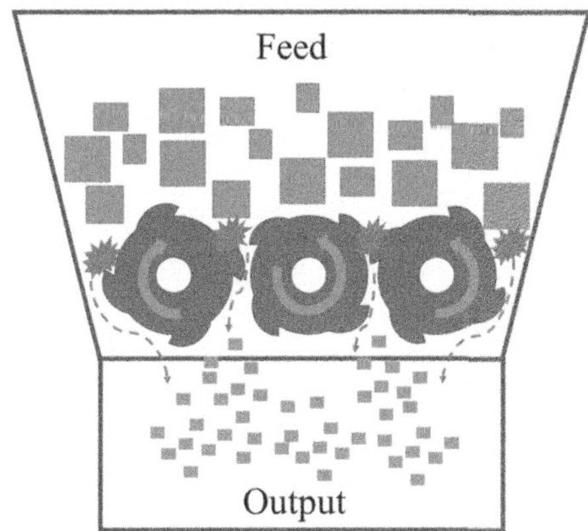

Figure 6.18 Shear shredder.

6.2.3.2 *Plastic washing and drying*

Although the plastics are separated from the commingle waste, they are still contaminated by oil, sand, soil, and so on. Such contaminations can affect the upcycling process. In order to reduce the impact, PW materials should undergo a thorough cleaning process. Water is commonly used as the washing agent which in turn can create moisture which can affect the subsequent upcycling processes. Thus, the drying process becomes an integral part of the cleaning process as well.

6.2.3.2.1 Plastic washing

The aim of plastic washing is to eliminate contamination on plastic surfaces which can be organic (such as oil, protein, food waste, etc.), inorganic (such as stone, sand, dust, etc.) and sticky impurities (such as glue, label, etc.). Plastic washing uses two main principles, namely water flow and the rubbing of flake plastic, to remove the impurity on plastic surfaces. Such impurities are transferred into the water, and thus the water from this process must be treated. Three common plastic washing facilities are floating washers, rotary washers and friction wash washers.

The floating washer is a washing process that relies on the relationship of density and mix condition to cleanse the impurities from the plastic flakes. The shredded plastic is fed to a water tank with flat wheels at the top of the tank, which are used to remove impurity from the floating plastic using mixing and rubbing actions. On the other hand, impurities that have high density, especially inorganic particles, sink down to the bottom of the tank. Because plastic density is generally lower than that of the water, the washed plastic floats on the water surface. After that, the washed plastic will be drained out from the tank (see Figure 6.19). This process is suitable for the PE and PP film washing.

The rotary washer allows flakes or shredded plastic to be fed into a rotary tunnel (see Figure 6.20) which contains water. The tunnel is connected with a wash and rinse section where a mixing plate is installed to create turbulence and remove impurities. Because plastic tends to have sticky impurities, especially labels, oil, and grease, the water flow and the rubbing action cannot completely remove those impurities. Thus, the corrosion agent and heat are applied to the

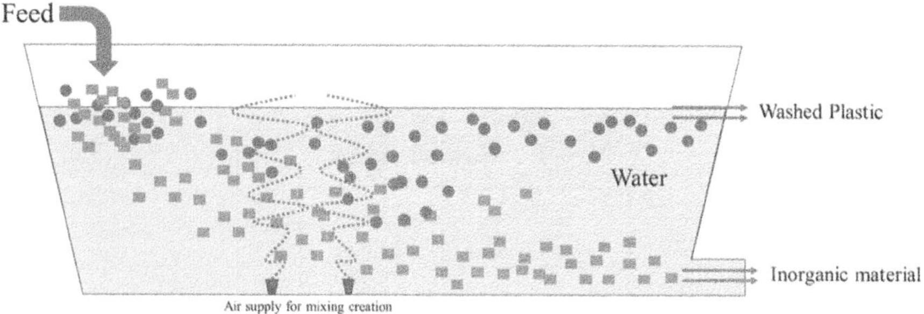

Figure 6.19 Floating washer.

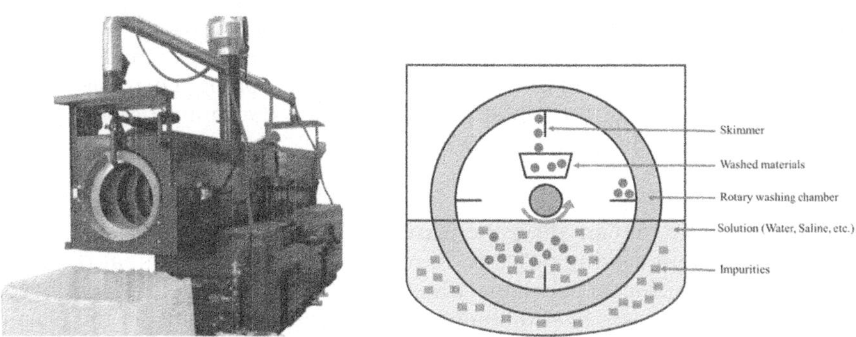

Figure 6.20 Rotary washer (*Source*: Rotajet Systems Ltd., 2020).

water to improve the performance of impurity removal. For example, Rotajet Systems Ltd. (2020) applies 1% of caustic solution and 65–70°C to the water inside their rotary washer to improve oil and grease removal efficiency. After the wash section, the plastic flakes are transferred to the rinsing section. The plastic is rinsed with water to remove the agent from the washing section.

The friction washer is used to wash the dirty and contaminated plastic flakes using rotary screws and water. The friction washer consists of three main components which are screws with multiple helical continual plates, dewatering screen, and water injection tunnel (see Figure 6.21). As its working principle, the

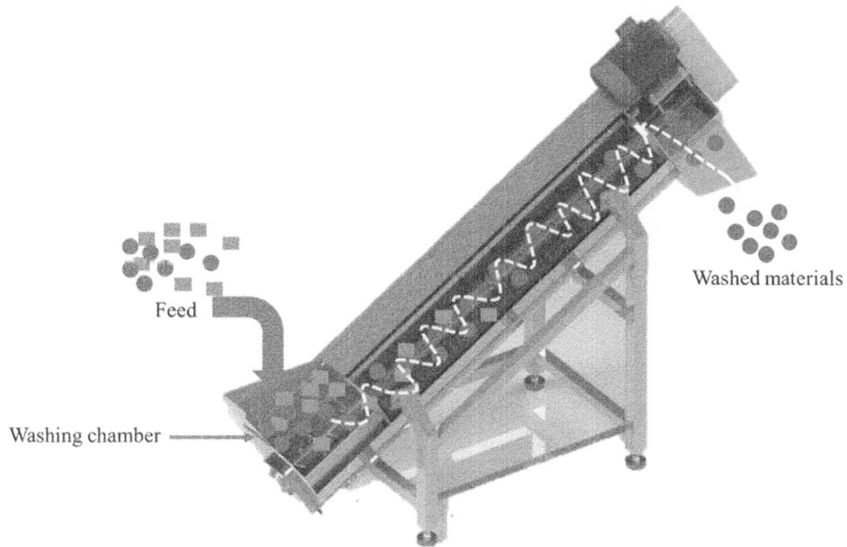

Figure 6.21 Friction washer (*Source*: https://www.ldmachinery.com/index.php?r=article/Content/index&content_id=28).

friction washer receives shredded plastics though the inlet hopper. The washing process is performed inside a tank which is under the hopper. As the screws rotate, the plastic flakes will be washed by water and rubbing. Then, the washed plastic is transferred to the outlet at the top while the impurities are settled down and discharged from the system.

6.2.3.2.2 Plastic drying

The moisture content can affect the plastic properties and upcycling facilities, especially the reprocessing process. To reduce the moisture content of plastic flakes, the drying process becomes a crucial part in the plastic cleaning process. The drying process uses heat to remove the moisture content from the plastic. However, the temperature has to be controlled as high temperatures can also affect plastic properties. Table 6.3 shows appropriate drying temperatures for plastic resin.

The dehumidifying dryer is used to remove moisture from plastic materials using dry air (see Figure 6.22). Air is blown through a desiccant bed to produce dry air. Moreover, the dry air will be heated to an exact temperature to enhance moisture removal performance. Then, the dry air is fed into the drying tank which contains dried materials. The dry air will draw the moisture out from the materials, after which, it will return to the desiccant bed and the heating device to be reused in the cycle.

The rotary dryer, also known as the rotary drum dryer or rotary drier (see Figure 6.23) is suitable for high humidity materials such as sludge, mud, animal manure or slag. The rotary dryer will directly apply heated air to the materials which are placed on a feed conveyor in a cylindrical tube. The rotation of the rotary dryer leads to an increase in the heated air contact with the materials. To increase the mixing condition, plates and turbines are installed inside the cylindrical tube. In order to convey the materials inside the dryer, the dryer is installed with a small slope to ensure that the discharge point is lower than the feeding point.

Table 6.3 Drying temperature for plastic resin.

Plastic Type	Drying Temperature (°C)
ASB resin	77–88
Acrylic	71–82
Nylon	71
Polycarbonate (PC)	121
LDPE	71–79
HDPE	71–104
PP	71–93
PS	71–82
PVC	60–88

Source: Mujumdar (2006)

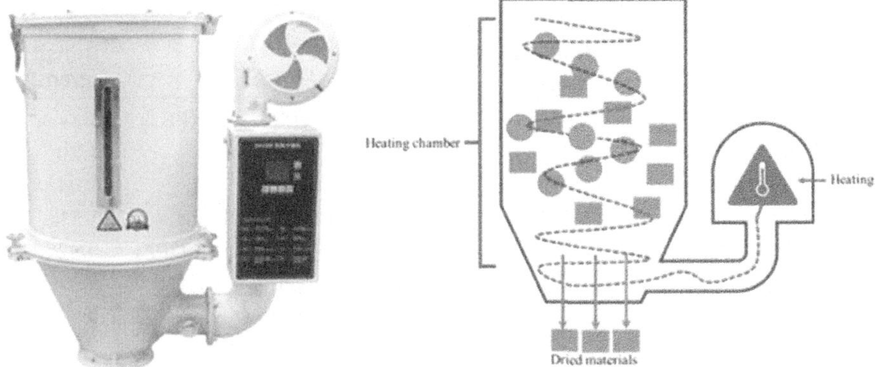

Figure 6.22 Dehumidifying dryers (*Source*: Zhangjiagang Lianda Machinery Co., Ltd., 2020).

6.2.3.3 *Plastic type segregation*

Various upcycling industries usually target specific plastic materials. For example, the textile recycling industry targets PET and PE waste, and thus they do not accept PP and HDPE waste. This section explores simple principal techniques used to segregate plastics, and also focuses on size reduction process prior to the plastic flake.

6.2.3.3.1 Sink-float separation

Sink-float separation is a common separation process used to separate types of plastic. As the density of plastic is quite similar (Table 6.4), air separation sometimes cannot effectively separate plastics. The sink-float separation uses a liquid with specific density to separate the materials. Lighter materials float, and heavier materials sink to the bottom. According to Pacheco-Torgal *et al.* (2018), this method can be used to separate polyolefins (PP, LDPE, HDPE) from heavier plastics such as PET and polyvinyl chloride (PVC) (see Figure 6.24). The process is also used to remove contamination.

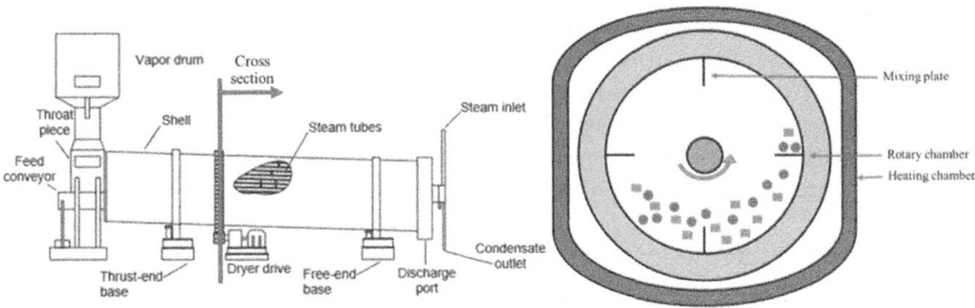

Figure 6.23 Rotary dryers (*Source*: Mujumdar, 1995).

Table 6.4 Density of plastics at room temperature.

Material	Density (g/cm³)	Material	Density (g/cm³)
ABS, extrusion grade	1.052	Polybutylene terephthalate (PBT)	1.34
ABS, high impact	1.024	PC	1.20
Acetal, 20% glass	1.550	Polyester (thermoset)	1.04–1.46
Acetal, copolymer	1.412	Polyetheretherketone (PEEK)	1.31
Acetal, homopolymer	1.412	Low-density polyethylene (LDPE)	0.925
Acrylic	1.190	High-density polyethylene (HDPE)	0.959
Butadiene – acrylonitrile (nitrile)	0.98	Ultrahigh molecular weight polyethylene (UHMWPE)	0.94
CPVC	1.550	PP	0.905
Epoxy	1.11–1.40	Polytetrafluoroethylene (PTFE)	2.17
Fiberglass sheet	1.855	Polyurethane	1.052
Styrene – butadiene (SBR)	0.94	PVC	1.384
Silicone	1.1–1.6	Polyvinylidene fluoride (PVDF)	1.772
Nylon 6/6 extruded	1.135	Phenolic	1.28

Source: Callister (2007) and Oberg *et al.* (2016).

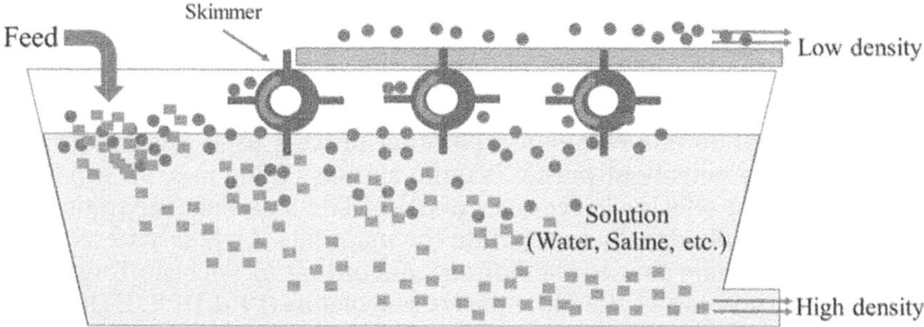

Figure 6.24 Sink-float separation.

6.2.3.3.2 Hydrocycloning separation

Hydrocycloning separation is a density sorting technology using the centrifugal or centripetal forces and fluid resistance of different particles. The liquid is fed into the clone using a high-pressure system to create a vortex which is the most important part of mixed plastic separation. Heavier elements fall to the bottom, while the lighter components will be lifted to the upper part (see Figure 6.25).

6.2.3.3.3 Electrostatic separation

Electrostatic separation uses the electrostatics principle to separate mixed plastics. The mixed plastic separation process can be completed by applying the different electric charges to the mixed plastic as shown below.

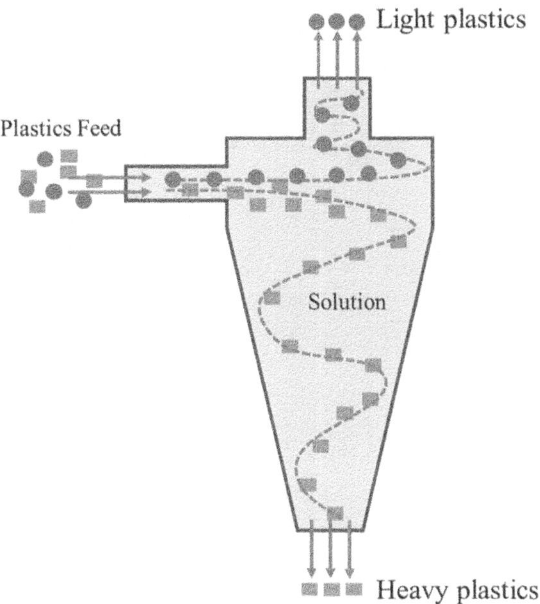

Figure 6.25 Hydrocyclone separation (*Source*: Pacheco-Torgal *et al.*, 2018).

(Positive charge; +)**ABS** > **PP** > **PC** > **PET** >
PS > **PE** > **PVC** > **PTFE** (Negative charge; −)

The electric charges can be created when mixed plastics rub each other. Commonly, the rotating tube or tribocharger is a popular technology used to create electric charges. Due to the different properties of mixed plastics, different charges may be used. After electric charges are applied, mixed plastics are fed though the magnetic field which then allows separation to take place (see Figure 6.26). For example, when PVC is rubbed with PET flakes PVC and PET can release a negative charge and a positive charge, respectively, while PET releases a positive charge when it is rubbed by PP flakes.

6.2.3.3.4 Magnetic density separation (MDS)

The aim of the MDS is to separate the mixed plastic that has similar density by using the relationship of liquid (such as water) and magnetic force. Due to similar density of the plastic, most plastics float during float–sink separation. Thus, to classify mixed plastics, the magnetic force is added to float–sink separation. The magnetic force then carries the mixed plastics to be separated in the liquid at a different level. For example, PP and PE flakes have lower density than that of water, and thus float. MDS then can be used to further separate the two plastics (see Figure 6.27).

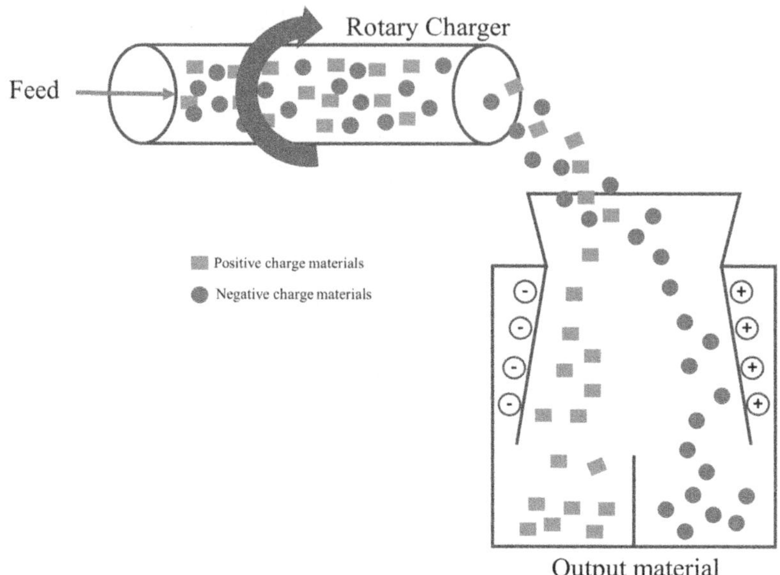

Figure 6.26 Electrostatic separation (*Source*: Pacheco-Torgal *et al.*, 2018).

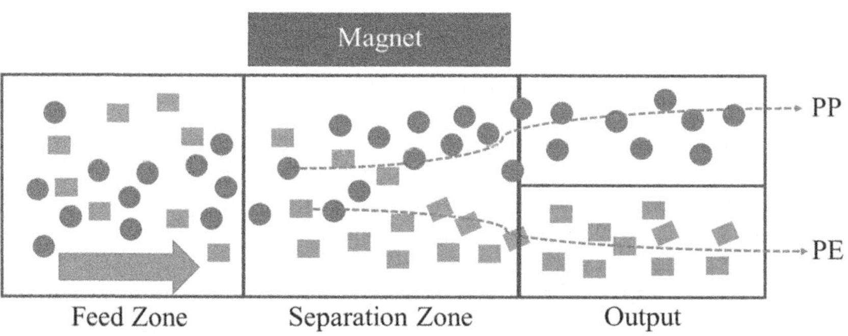

Figure 6.27 Magnetic density separation for PP and PE separation (*Source*: Pacheco-Torgal *et al.*, 2018).

6.2.4 Pre-treatment case studies
6.2.4.1 Pre-treatment case study of MSW

PW pre-treatment is generally done to remove contamination that may affect the PW that will be mechanically or chemically recycled. According to the EU Landfill Directive, biodegradable waste from MSW needs to be rid of its biodegradable content which can be further used in composting, recycling and anaerobic digestion. The local authorities are to implement this. Also, all waste, including industrial waste, should be treated in order to reduce its negative environmental impact before being transferred to the landfills.

PW is usually treated in a material recovery facility (MRF) before they are finally disposed of. An MRF may not be technologically advanced where most work is done by hand through sorting or by magnetic separators which makes this particular MRF low in capital but high in labor. Another MRF may be more technologically advanced where separation is done mechanically and an assortment of materials is done by machines. Such a MRF would require much lower labor cost, but much higher capital cost.

Some common issues in ISW and MSW are the presence of residues, odor, additives, inks, metals and adhesives. These contaminants can cause complications during recycling, cause health risks and affect the degradation of the recyclability of the plastic. Some pre-treatment methods commonly practiced in MRFs are listed below.

- Composting or the removal of the biodegradable aspects of the MSW
- Sorting – possibly done manually or by ballistic separation, screening, air separation, sensors, eddy current
- Washing
- Float–sink – identifying different plastic types according to their density
- Odor removal technology (deodorization) such as friction washers or water-based washing, solvent-based washing with the use of hot ethyl acetate usually at 65°C and polyethylene glycol
- Deinking technology, a patented technology called Nordenia extraction and cleaning which helps to remove a range of inks such as water-based inks and solvent-based inks. Another method is the use of surfactants such as cetyl trimethylammonium bromide which carries out the deinking process of both water and solvent-based inks by (a) adsorption to the plastic surface, (b) the binding compound in the surfactant solubilizing, (c) ink particle adhering to the binder and detaching from the surface of the plastic and (d) the surfactant stabilizing the particles from the ink.
- Delamination plastic pretreatment in the case of multilayer packaging through various ways such as the method of a series of dissolution of the multilayers in PW through heating the waste in a single solvent system. Another method for delamination is by decomposition of the plastic polymer layers in the packaging in organic solvent systems.

6.2.4.1.1 Pre-treatment of MSW for gasification process using refuse derived fuel (RDF)

MSW though inhomogeneous has an advantage of not having heavy metals contamination and high alkaline content. Most of the time, RDF, used for combustion, is derived from it. MSW is pre-crushed using a shredder to generate a more homogeneous size distribution and to increase the bulk density. Heavy contents such as glass, stones and ceramics are then removed using a wind sifter. During this step, the lower heating value increases. The ferrous and non-ferrous metal components are removed in the third and fourth steps by an over-belt magnetic separator and an eddy current separator. Secondary crushing is done

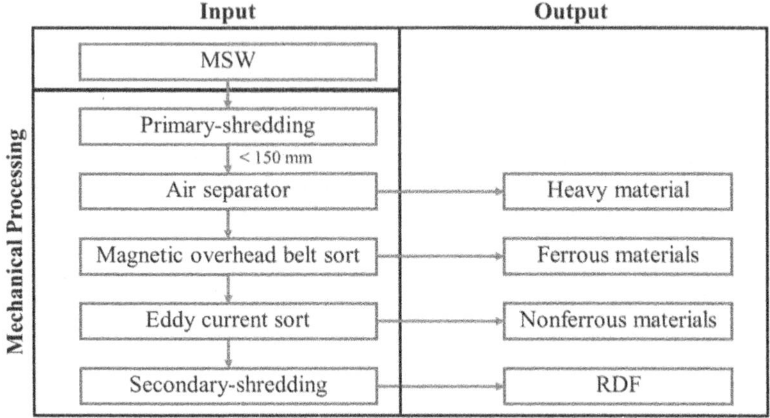

Figure 6.28 Pre-treatment process of gasification by using MSW (*Source*: Staph *et al.*, 2019).

to the remaining fractions using a fast-running shredder. The overall diagram of the MSW pre-treatment process for gasification is shown in Figure 6.28.

6.2.4.1.2 Pre-treatment of MSW for RDF process at Campania, Italy

In Campania Region in Southern Italy, an MSW management system is planned and designed for 35% of the total solid waste generated, in order to promote composting and materials recycling. The treatment of the rest of the waste is done to obtain RDF, stabilized organic fractions and ferrous materials. The pre-treatment process includes the following phases.

(1) First-stage preliminary shredding of the MSW.
(2) Obtaining a primary over-sieve and under-sieve by feeding the shredded MSW into the first trommel screen.
(3) Magnetic separation and hand sorting of the primary over-sieve.
(4) Primary under-sieve fed into the second trommel screen to get both secondary over-sieve and under-sieve.
(5) Magnetic and ballistic separation of the secondary over sieve, to obtain fractions for RDF production, composting and landfill disposal.

The overall concept of MSW pre-treatment process for RDF is shown in Figure 6.29 while Figure 6.30 shows the amount of materials recovered from each unit during the pre-treatment process (Belgiorno & Panza, 2008).

6.3 UPCYCLING TECHNOLOGIES

Due to the special properties of PW, especially thermoset plastic, there are several ways to enhance the value of PW. This section explores two possible choices which are used to add values to PW. Two value-added processes are recycling and reusing.

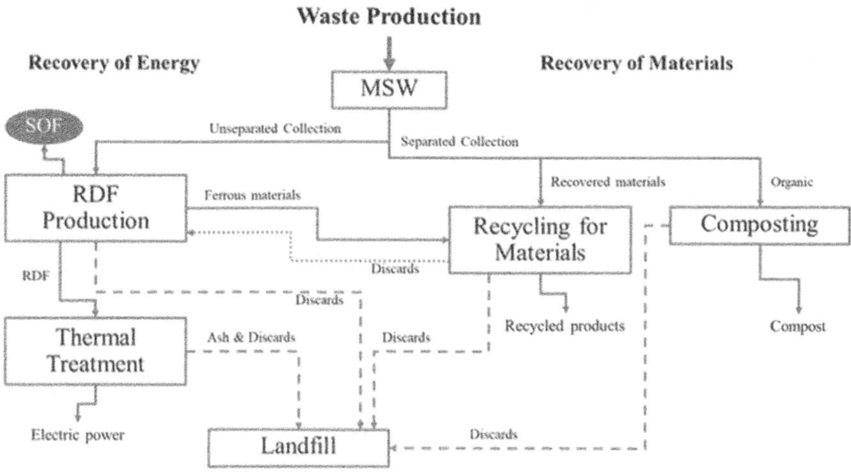

Figure 6.29 Overall concept of MSW pre-treatment process for RDF (*Source*: Belgiorno & Panza, 2008).

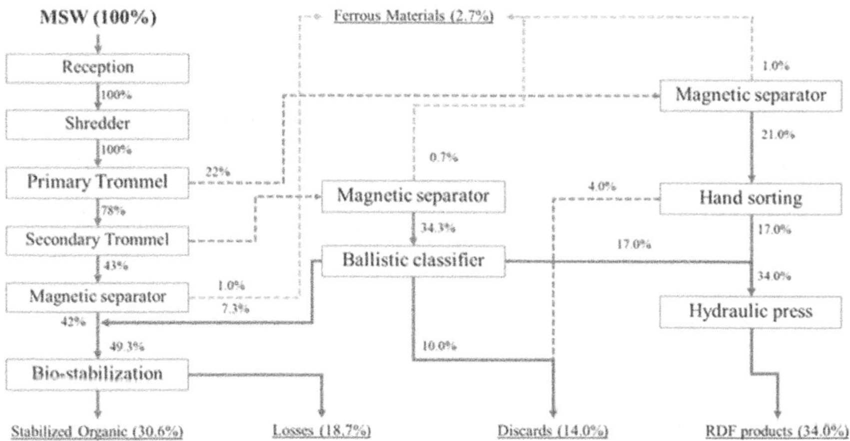

Figure 6.30 Amount of material recovery from each unit during Pre-treatment process (*Source*: Belgiorno & Panza, 2008).

6.3.1 Plastic recycling technologies

The recycling process is a process that converts materials to reusable products, or turn waste materials to raw material as a cyclic process. According to ASTM (2000), plastics recycling can be classified into four groups: (1) primary recycling (or closed-loop recycling), (2) secondary recycling (or downgrading recycling), (3) tertiary recycling (or feedstock recycling), and (4) quaternary recycling (or calorization). Table 6.5 shows the details of each type.

Table 6.5 Terminology of type of plastics recycling and recovery.

Terminology	Definition	End Product (Example)
Primary recycling (closed-loop recycling)	Mechanical reprocessing into original properties	• Virgin plastic resin • PET bottle to PET bottle
Secondary recycling (downgrading recycling)	Mechanical reprocessing into lower properties	• Wood plastics composite (WPC)
Tertiary recycling (feedstock recycling)	Recovery to chemical elements (both chemical and biological processes)	• PET bottle to polyester (PS) • By depolymerization • HDPE bag to ethylene glycol • By organism
Quaternary recycling (calorization)	Recovery to energy	• Heat • Electrical

6.3.1.1 Basic fundamental of recycling technology

6.3.1.1.1 Basic technology for plastic recycling

The extruder is used in the conventional plastic production process, converting polymer to plastic. The extruder is typically applied to produce recycling plastic/plastic resins. In the extruder, the plastic forming relies on three main factors for creating the proper conditions for re-melting process and transforming polymers into plastic materials. With re-melting, the extruder is normally selected for recycling high-quality PW to primary material. To enhance the recycled plastic properties during the extruding processes, several improving agents and ingredients are added in different zones in the extruder (see Figure 6.31).

Mold injection is the machine that melts plastic or polymer substances inside a heating channel and injects melted plastic into a mold. The melted plastic is then solidified inside the mold. The injection mechanism consists of a mold clamping part controlling the opening and closing of the mold, and the injection

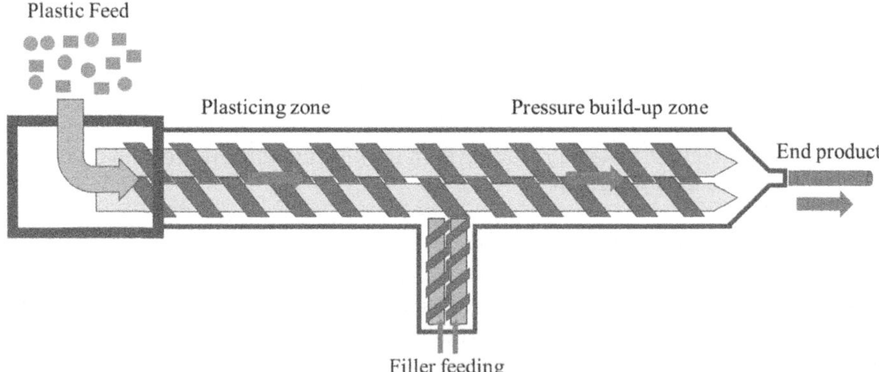

Figure 6.31 Process zones of a twin-screw extruder (*Source*: Bonten, 2019).

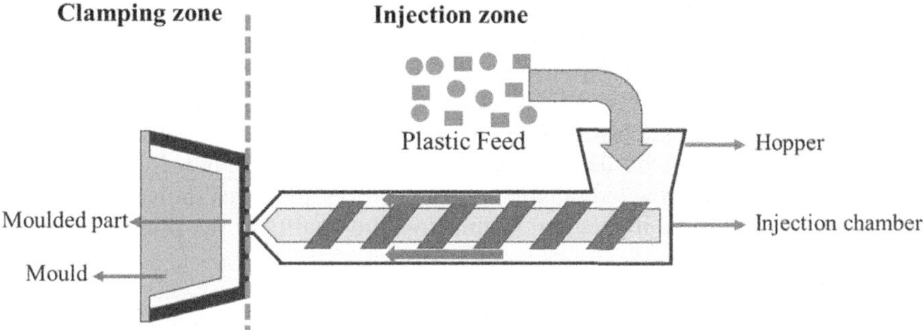

Figure 6.32 Mold injection.

part including the melting and injecting units (see Figure 6.32). Normally, the injection process requires an extruder, and injection speed, injection time, and the injection pressure must be determined. Mold injection also allows flexibility in how the end products are shaped. However, one limitation of this method is the low quantity production.

Compression molding is a process that produces plastic products using the plastic melting process. Heating and pressurizing are the cores of the compression molding process. The machine consists of two molds, one at the top, and the other at the bottom. The top mold applies heat while the bottom mold is fixed with the base of the machine. Moreover, compression molding does not require a separate heating section as in the case of the injection mold. This machine melts plastic flakes by applying heat to the mold itself (see Figure 6.33). Without a separate heating section, the compression molding machine is cheaper compared to the injection mold. On the other hand, compression molding consumes a greater amount of time compared to extrusion or injection. The consideration factors of mold compression are material quality, temperature, pressure and cure time.

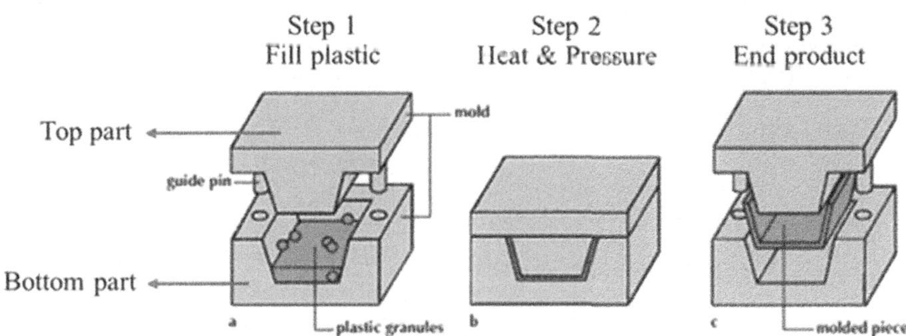

Figure 6.33 Mold compression (*Source*: https://kids.britannica.com/students/assembly/view/53835)

6.3.1.1.2 Plastic and resin production

Plastic resins are an important material in the production of plastic products. They are made with the extrusion process which involves re-melting (see Figure 6.34). Furthermore, plastic production usually involves a thermal process which can destroy the plastic properties. This creates an issue as not 100% of recycled plastic can be used to produce plastic resins. Therefore, the resin production in which recycled plastic is used requires additives such as the virgin plastic resin or additive ingredient to improve the quality of the plastic resin.

6.3.1.1.3 Consideration factors for plastic reprocessing

During the re-melting process: Because thermoplastic consists of crystalline properties, the heating process can change the state of plastic. Therefore, crystalline melting temperature (T_m) is the key factor for the plastic re-melting process. Table 6.6 shows suitable temperatures for the re-melting process for commercial thermoplastic products. Apart from the applied temperature, the melting behavior of plastics such as the flow property is also a consideration factor. Moreover, the melting behavior of melted plastic has a direct relationship with the applied force

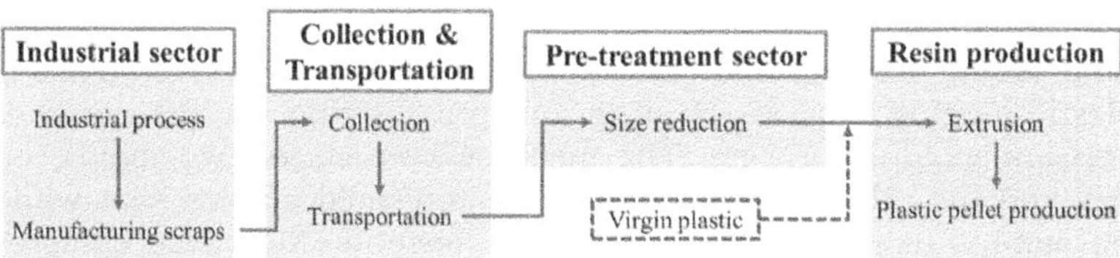

Figure 6.34 Plastic and resin recycling process for industrial sector. (*Source*: modified from Kosior *et al.*, 2020).

Table 6.6 Crystalline melting temperature (T_m) of thermal plastic products.

Structure unit	Polymer	T_m (°C)
$-CH_2-CH_2-$	PE	140–203
$-CH_2-CH-$ $\quad\ \ CH_3$	PP	170–297
$-CH_2-CH-$ $\quad\ \ C_6H_5$	PS	115–196
$-CH_2-CH-$ $\quad\ \ Cl$	PVC	100–150
$-\overset{O}{\overset{\|}{C}}-\langle\bigcirc\rangle-\overset{O}{\overset{\|}{C}}-O-CH_2-CH_2-O-$	PET	274–342

in the extrusion process. Therefore, T_m, melting behavior, and the extrusion force are three important factors affecting the re-melting process.

After the re-melting process or product requirements: Because plastic productions have special purposes, after the re-melting process, the properties of recycled plastic should meet the requirements. Such requirements can be physical, chemical or biological. The important physical properties of plastic products are strength, stiffness, toughness and stress–strain diagram. The important chemical and biological properties are moisture resistance, heat resistance, electric resistance and antimicrobial resistance. Therefore, additive agents such as fillers and plasticizers are important ingredients in the plastic production processes.

6.3.1.2 *Primary recycling technology*

Primary recycling is the re-extrusion or reproduction process converting the plastic scraps to raw materials again. This recycling technique does not involve any changes in the polymer structure. Thus, the end-products of this process have a similar structure with the original materials. As the process involves converting plastic scraps into original raw materials, the primary recycling technology requires high-quality raw materials. Therefore, the pre-consumer and manufacturing scraps are the suitable materials for the primary recycling process. For example, PET bottles which are eliminated by quality control (QC) process can be converted into PET resin again. As materials fed to the primary recycling process are manufacturing scraps, this recycling process does not require a complex pre-treatment process (see Figure 6.35). Moreover, the primary recycling process is referred to as 'closed-loop recycling'. Therefore, the purification of recycled plastic is the crucial step of the primary recycling process.

Three main factors affecting the re-melting process are polymer structures, temperature and mechanical properties. First, the plastic or polymer that is recyclable must contain crystallinity in their structure as the crystallinity is important for the re-melting process. This is because the crystallinity property represents the reversible state of plastic materials. Second, melting temperature is the factor which bring about the changing state (both solid-to-liquid phase and liquid-to-solid phase) of materials. The temperature will cause the atoms inside polymer chains to vibrate. When the vibration of atom is higher than intermolecular force, the attachment force between molecules will break,

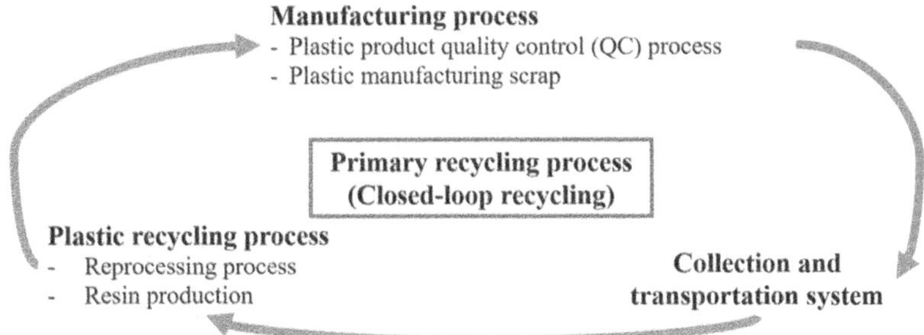

Figure 6.35 Concept of primary recycling process.

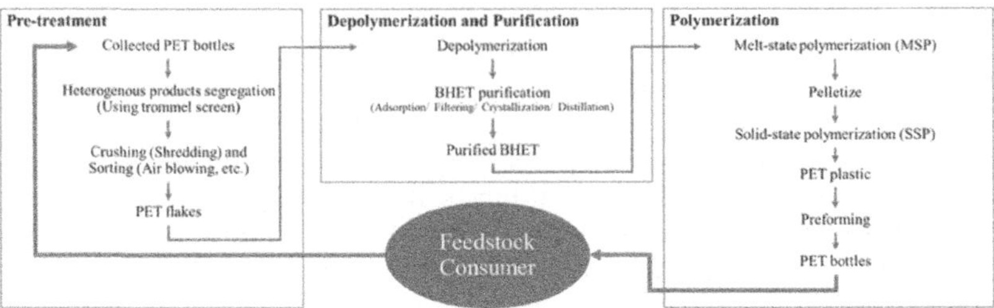

Figure 6.36 PET bottle to PET bottle (*Source*: https://www.jeplan.co.jp/en/technology/bottle/)

resulting in melting of plastic materials. Finally, mechanical properties can affect the re-melting process as they introduce deformation and flow of melted plastic during extrusion or the resin production process. The plastic scarps commonly fed into the primary recycling process are acetals, nylons, PE, PP, and polyesters or PET.

For examples of technology for recycling the PET bottle to PET bottles in Japan, the objective of the technology is to demonstrate that PET bottles can be a continually circulating resource to enable repeated PET bottle to bottle fabrication. This recycling is essential as there is a high global reliance on PET bottles. Japan has one of the highest collection rates of PET bottles standing at 85%. Figure 6.36 shows the overall picture of PET-to-PET bottle projects in Japan.

The collected PET bottles that are the feedstock to be recycled undergo a pre-treatment process of screening, crushing and sorting to form the PET flakes. The flakes are then depolymerized by a BHET purification process to form purified BHET. Then, the polymerization process takes place to form PET bottles.

6.3.1.3 Secondary recycling technology

According to ASTM (2000), the quality of plastic properties after secondary recycling will be diminished. Therefore, the quality requirement of raw materials for secondary recycling is lower than that for the primary recycling process. Secondary recycling (Figure 6.37) refers to material recovery by

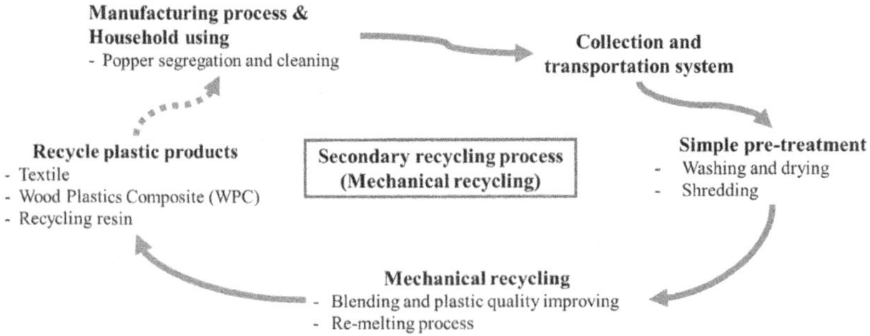

Figure 6.37 Concept of secondary recycling process.

mechanical recycling. The secondary recycling can be applied to various types of plastic, especially thermoplastic. The secondary process first requires the pre-treatment processes such as washing and shredding of the plastics. The prepared plastics are further processed by re-extrusion, which is a common process mainly used for the transformation process in the mechanical recycling. The heat in the re-extrusion process can physically change the plastic structure. The re-extruded virgin plastics can then be mixed with other materials such as wood fiber or wood waste to produce composite plastic products during the mechanical recycling. To improve the plastic properties, virgin resin and special additive (blending) agents are added during this mechanical recycling. During the mechanical recycling, the physical forms and structures of plastics are changed while the chemical structures of the plastics do not undergo a great deal of changes. Therefore, those composite plastic products cannot be returned to their previous cycle. Hence, secondary recycling process can be applied to upcycle the used materials namely, plastic waste although the properties of the recycled plastic from the secondary process have lower quality than those in the original or virgin plastic products.

6.3.1.3.1 Precious Plastic Bangkok, Thailand

Founded by Dave Hakkens in 2012 Precious Plastic is a company that attempts to tackle the plastic problem by tackling PW and by becoming a global alternative recycling system. With such aims, the company invented technology and equipment to convert PW to valuable products such as plant pots, chairs, and floors. Moreover, the company also provides courses about their technology and equipment for anyone interested (Figure 6.38).

Figure 6.38 Precious Plastic Bangkok workshop at AIT.

In 2018, Dominic Puwasawat Chakrabongse launched Precious Plastic Bangkok. The company has been gathering volunteers who collect recyclable PW such as bottle caps. After collection, the PW is shredded and turned into new products using extruders and compressive machines (Figure 6.39). Precious Plastic Bangkok has recycled around 100 kilograms of bottle caps, sourced from

Figure 6.39 Precious Plastic Bangkok equipments and their products: (a) shredder machine; (b) extruder machine; (c) Precious Plastic Bangkok's products (*Source*: https://www.khaosodenglish.com/featured/2019/05/13/precious-plastic-recycling-bangkok-one-bottle-cap-at-a-time/).

donations and from trash-collecting volunteer groups such as Trash Hero. The company even put together a Bangkok Design Week display in January 2022. Dominic recently received one of his biggest orders yet, from a corporation who has ordered 250 plant pots which will require 20 000 bottle caps.

6.3.1.3.2 Upcycling at Inle Rose; Nyaung Shwe, Myanmar

Nyaung Shwe is the main gateway to the tourism destination Inle Lake in Shan State and an important hub of hotels for travelers visiting Inle Lake. In 2018, the municipal government reorganized and formalized the waste management of the town and granted the license for municipal waste management for 30 years to the private SME Inle Rose Company Limited based in Nyaung Shwe.

The income generated by waste management fees was not sufficient to be financially sustainable in order to operate the waste management of Nyaung Shwe Thus, the company needed to generate additional income from waste management, mainly from recycling of the main tradeable wastes such as metal and tradeable plastic. There had not been implementation of value-added processing of waste, and only 20% was recycled. Within the GIZ project, Capacity Strengthening for Private Sector Development (PSD)-Phase III, of which PEM implements several components, Inle Rose started collaboration with GIZ, RecyGlo and Evergreen Social Ventures in Vietnam in December 2019 in order to increase the volume of plastic to be recycled. The developed simple but efficient technologies to process various value-added products from non-tradeable PW, for example by using heat presses. The project transferred an innovative technology from Vietnam to Nyaung Shwe to turn PW into products such as boards and sheets which can be sold to local construction and furniture industries which had traditionally been using local wood and imported non-wooden boards. Inle Rose has been equipped with all necessary tools and training to successfully operate its project and generate additional revenue streams. Although COVID19 has imposed significant challenges and slowed down the implementation, Inle Rose has successfully proven to be able to produce quality boards and create marketable products.

The company expects higher competitiveness through increased production and sales, new products and the development of new markets for the manufactured building materials and products. The business model is also open to other companies for application in Myanmar (Figure 6.40).

6.3.1.3.3 Parley for the Oceans and Adidas

Globally, there have been multiple companies that have explored the use of recycled polyester made from end-of-life PET products, and one of the largest is Adidas. Adidas and Parley originally partnered in 2015 to create the world's first running shoes made from recycled plastic. Adidas is aiming to reduce its environmental impacts by incorporating less virgin polyester into their products. As of 2020, more than half of the polyester used to manufacture their products was recycled polyester. The company aims to cease the use of all virgin polyester by 2024, and thus far is on track to doing so. Each Adidas and Parley for the ocean product is made from at least 75% recycled materials recovered from coastal areas.

Figure 6.40 Upcycling at Inle Rose; Nyaung Shwe, Myanmar (*Source*: https://www.pem-consult.de/news/myanmar-upcycling-at-inle-rose-nyaung-shwe.html).

First, Parley and its partner organizations collect PW from coastal regions, such as the Maldives. The collected plastic is then sent to Taiwan where it is upcycled into yarn. The plastic is then shipped to Adidas x Parley's supplier in Taiwan, where it is upcycled and transformed into yarn fibers (Parlay Ocean Plastic™). First, the waste is separated, and PET is shredded, washed and dried. The shredded plastics is then heated to melt the plastic, and then processed into small resin pellets. The pellets are then stretched to create plastic yarn.

This recycled yarn is used to make clothing items like shirts and the upper parts of shoes. The other plastic, such as HDPE, that was collected is sent to recycling facilities, and some flagship Parley for the Ocean stores used these recycled plastics to make items for their store, such as hangers and even mannequins (Figures 6.41 and 6.42).

Figure 6.41 Process of making a Parley for the Ocean Adidas product (*Source*: https://www.adidas.co.th/en/blog/639412-how-we-turn-plastic-bottles-into-shoes-our-partnership-with-parley-for-the-oceans).

Figure 6.42 Parley for the Oceans and Adidas shoes (*Source*: https://www.adidas.co.th/en/blog/639412-how-we-turn-plastic-bottles-into-shoes-our-partnership-with-parley-for-the-oceans).

6.3.1.3.4 DA.AI Technology Co. Ltd – recycled-to-recycled, Taiwan

DA.AI Technology is a recycling platform that is driven by the efforts of people from all walks of life ranging from volunteers and supporters to business partners. They believe that humans should 'coexist with the Earth,' and, therefore they use recycled PET bottles as raw material to manufacture recycled eco-products. This activates a new life cycle for the PET bottles without increasing the total PET volume and reduces the consumption of natural resources. DA.AI Tech develops textile fabric from recycled raw materials, such as recycled poly chips, recycled polyester fibers, and recycled fabrics. These textile materials are then used to produce typical everyday necessities, eco-products such as clothing, bedding, scarves, backpacks, furniture, blankets, quilt covers and other everyday textile products. Their products are 100% recycled materials. The products are customized to fit the customers' needs. The creation of these new eco-products consumes fewer resources and less energy than conventional means of production and are categorized into a series of products which are marketed. These include Recycle to Recycle (R2R) Series, Compassion Technology Series, and Low Carbon Lifestyle Series (Figure 6.43).

The patented DA.AI's Recycle-To-Recycle® (R2R®) Series of the company officially shifted from a 'waste reduction' green brand to a 'zero waste' green global role model using the Cradle to Cradle® certification. DA.AI products are made from recycled PET bottles that have been collected, treated, and sorted, and recycled into non-polluting (never piece-dyed), high-quality, eco-friendly rPET textiles without increasing the total PET volume. This production process has received a Global Recycled Standard (GRS) certification from the Netherlands' Peterson Control Union for rigorous quality control.

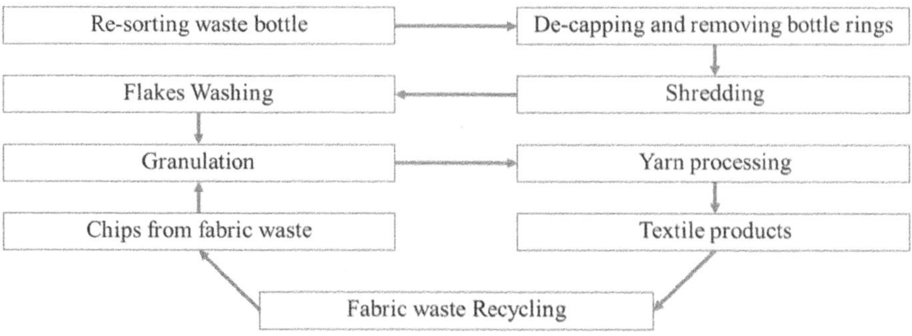

Figure 6.43 DA.AI Technology Co. Ltd – Recycled-to-Recycled, Taiwan (*Source*: https://daaitechnology.com/pages/brand-story).

6.3.1.3.5 Plaxtil, France

Among the many social and economic impacts of the COVID-19 pandemic, there have also been large environmental impacts owing to the increased use of single-use plastics and disposable PPE equipment, such as face masks. A study found that, as of 2021, 3.4 billion face masks and face shields are discarded every day. Asia alone is estimated to throw away 1.8 billion face masks per day. Each disposable mask contains 5 g of PP, and a single face mask can shed up to 173 000 microfibers/day. It is estimated that a disposable PP face mask can take upwards of 450 years to degrade. Plaxtil, a French start-up has converted equipment used in textile recycling to recycle PP face masks, turning the potentially hazardous waste into useful products. Plaxtil says the masks could be turned into a vast array of different objects, but for the moment the start-up is turning them into products that can be used in the fight against COVID-19, such as plastic visors. In six months Plaxtil says it recycled more than 50 000 masks, producing between 2000 and 3000 recycled products.

First, the masks are collected and placed in 'quarantine' for four days. They are then ground down into small pieces and subjected to ultraviolet light to ensure they are fully sterilized before the recycling process begins. Next, a binding agent is mixed with the shredded masks, and they are melted and transformed into a material called Plaxtil, which can be used in industry and molded like normal plastic (Figures 6.44–6.46).

6.3.1.4 *Tertiary recycling technology*

Tertiary recycling refers to an upcycling process which uses chemical mechanism. The main result of this process is the chemical form which can be used as the raw material for other manufacturing. Therefore, the term 'feedstock recycling', 'chemical recycling', and 'advanced recycling technologies' all represent tertiary recycling process. The complex structure of plastic is converted into smaller molecules as intermediate elements using heat/chemical principles. The products of tertiary recycling process can be liquid and gas phase, as well as solid phase, such as waxes. According to Kumar *et al.* (2011), tertiary recycling

Figure 6.44 Shredded PP facemasks (*Source*: https://recyclinginternational.com/gallery/plaxtil-puts-single-use-face-masks-back-in-the-loop/32591/).

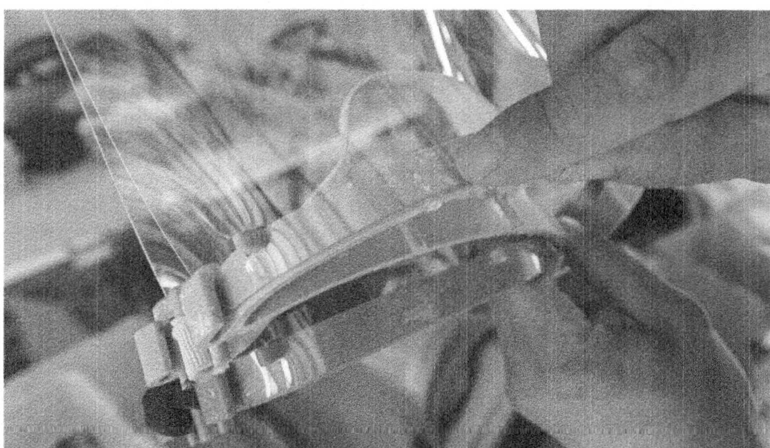

Figure 6.45 Disposable face mask (*Source*: https://www.cnnindonesia.com/teknologi/20200902111631-201-541868/foto-sampah-masker-bisa-diubah-jadi-plastik-daur-ulang).

can be classified into three main principles which are chemolysis, gasification and cracking. Moreover, the target material or raw material of the chemical recycling is segregated into mix PW or heterogeneous waste, and single PW or homogeneous waste (see Figure 6.47).

As chemical components are the products of the tertiary recycling process, the product processing is quite sensitive. Therefore, the raw material input should be cleaned by several units in the pre-treatment process to eliminate any impurity.

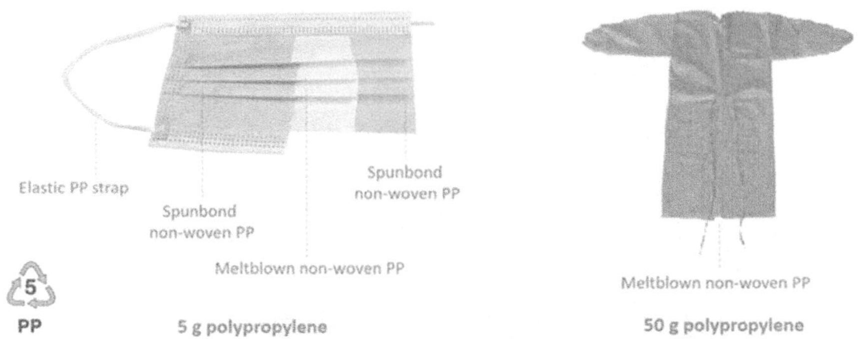

Figure 6.46 Visor made from recycled PP face masks (*Source*: Nghiem *et al.*, 2021).

Because the chemical process is the key function, the tertiary process should be properly controlled and monitored. Although tertiary recycling process is created by humans or synthesis reactions, there are several organisms such as fungi, bacterial and worms that can create an enzyme that breaks down the polymer to monomer elements, the process of which is referred to as the 'biochemical process'.

6.3.1.4.1 Chemolysis

Chemolysis is the chemical process that depolymerizes or turns plastic polymer into small elements, called monomers. To depolymerize plastic to monomers, special chemical reagents are used as catalysts. There are three common catalyst agents, namely hydrolysis, alcoholysis and glycolysis (Table 6.7). As specific monomer requirements are related to the polymer input, chemolysis is used with homogeneous plastic or individual plastics.

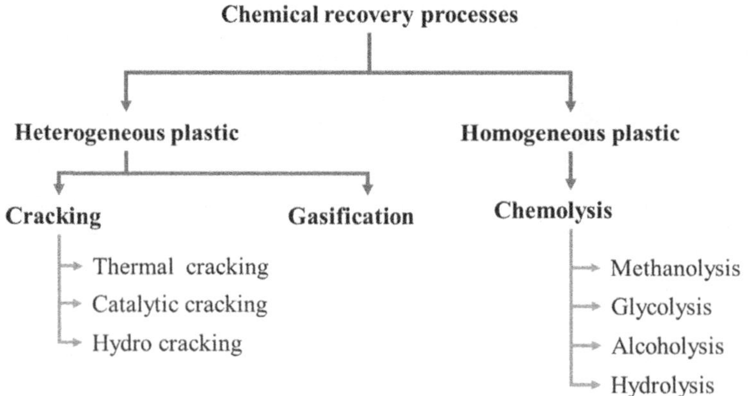

Figure 6.47 Tertiary recycling process diagram. (*Source*: modified from Kumar *et al.*, 2011).

Table 6.7 Chemolysis reactions.

Reaction	Catalyst Agent
Hydrolysis	Water
Alcoholysis	Methanol
Glycolysis	Ethylene glycol

Hydrolysis is an ancient Greek combination between 'Hydro or water' and 'lysis or unbind'. Therefore, hydrolysis reaction is the chemical reaction that converts the target substance to original form using water as the cutting agent. Water is absorbed by the plastic structure and breaks down the plastic crosslink. However, most plastics are designed to resist water under common conditions. Thus, the hydrolyzing performance of plastic is very low. To produce monomer using hydrolysis reaction, the plastic input has to increase the hydrophilic properties.

Alcoholysis is the chemical converting process which uses alcohol as the cutting agent. This reaction is the common chemical reaction that is used to convert polyurethanes to polyhydroxy alcohol and small urethane fragments. Meanwhile, the **glycolysis** takes place for regenerating monomer under the condition of glycol (catalyst) existence.

For a case study of dimethyl terephthalate production using hydrolysis reaction, Matsushita Electric Works, Ltd., Japan has been developing a technology for the depolymerization of flame retardant polymers (FRP) (thermosetting plastics comprising polyesters and their crosslinking moieties) using hydrolysis in subcritical water. In their technology, thermosetting resin in FRP can be recycled into basic materials with a material recycling rate of 70%. After subcritical water hydrolysis, the resin is dissolved into a liquid. The recovered components such as glycols and fumaric acid can then be separated from the aqueous solution and polymerized into polyester with fresh resin material to produce recycled resin. Also, a styrene–fumaric acid copolymer (SFC) can be separated from the aqueous phase. The concept of the sub-critical water hydrolysis in the recycling process is illustrated in Figure 6.48.

6.3.1.4.2 Gasification

Gasification is a direct combustion process that transforms energy or calorific value of PW or solid waste to chemical energy in gaseous form which is called syngas. Syngas consists of several compositions from the combustion process which are the mixture of hydrogen, carbon monoxide, carbon dioxide, water, hydrocarbons and methane. The important gaseous fuel is hydrogen (H_2) which can be used to serve several functions such as an engine for heat and electricity generation. Because gasification is a combustion process, it can produce pollutants when the raw material contains high amount of contamination such as heavy metal and PVC scrap. The important pollutants which can be released into the environment are tar, hydrogen sulfide (H_2S) and hydrogen chloride (HCl). Therefore, the gasification must have proper environmental control devices such as the scrubbing process which is explained in Section 6.3.1.5a.

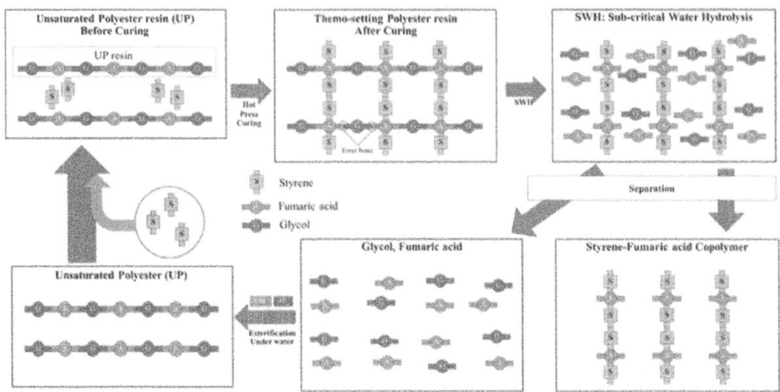

Figure 6.48 Polyester depolymerization process (*Source*: Matsushita Electric Works, Ltd.)

6.3.1.4.3 Cracking

Cracking or thermal cracking is known as pyrolysis reaction. The pyrolysis process is a combustion system which is created under the absence of oxygen. The cracking process is usually confused with the gasification process because the solid, liquid and gaseous fuels are the result of both processes. Pyrolysis is the thermal decomposition of substance in a non-reactive atmosphere at high temperatures. Therefore, the external heat source is applied to maintain a temperature which is in the range of 300–850°C for MSW. The main results of pyrolysis combustion are a carbonized char and a volatile gas fraction that can be converted into condensable hydrocarbon oil, while non-condensable is a gas phase that has high calorific value. Pyrolysis generally occurs at temperatures exceeding 430°C. It is an irreversible process which changes both the physical and chemical compositions of material. There are three types of pyrolysis, namely slow, fast and flash pyrolysis (see Table 6.8); 90% of the plastics is

Table 6.8 Types of cracking/pyrolysis process.

Type of Cracking/Pyrolysis	Details
Slow pyrolysis:	• Resident time 5–30 min • Low heating rate (<1°C/s) • Maximum temperature 600°C • Produces char, pyrolysis oil, and gas
Fast pyrolysis:	• Resident time 0.5–5 s • High heating rate (1–200°C/s) • Maximum temperature 650°C • Produces pyrolysis oil
Flash pyrolysis	• Resident time <1 s • High heating rate (>1000°C/s) • Maximum temperature >650°C • Produces pyrolysis oil and gas

Figure 6.49 Plastic bag sleeping mats (*Source*: https://www.sapeople.com/2019/07/13/port-elizabeth-residents-recycle-plastic-bags-as-sleeping-mats-for-the-homeless/).

produced using petroleum. Thus, the pyrolysis of PW is a feasible method of energy recovery and an alternative to the more commonly used plastic disposal methods such as landfilling and mechanical recycling. Generally, plastics with high volatile matter and low ash content (such as PS and PP) are preferred for the production of pyrolysis liquid. Unfortunately, PVC and PET are considered unsuitable for pyrolysis due to the presence of chlorine and benzoic acid, respectively (Figure 6.49).

6.3.1.5 Quaternary recycling technology

Quaternary recycling is the energy recovery process which uses the energy content of PW called 'Incineration process'. This incineration process typically employs the combustion to chemically converse waste to heat. The combustion is a chemical oxidation process that applies thermal processing with excess amount of air to create final products which are hot gas stream and noncombustible residue or ashes. The heat generated from the combustion can be recovered as energy. The quaternary recycling is a proper process to manage PW together with energy recovery because PW has higher calorific value (CV) than other wastes as presented in Tables 6.9 and 6.10. However, in some situations, management of the commingled waste with the improper quaternary recycling process can subsequently create drawbacks such as incomplete combustion which affects the environment and health. Incomplete combustion of the plastics can hence emit unwanted gases such as volatile organic carbons (VOCs), HCl, polynuclear aromatics hydrocarbons (PAHs), PM_{10} and $PM_{2.5}$. Therefore, incineration is not suitable for community areas.

Table 6.9 Common waste CV and energy content.

Type of Waste	Waste (%)	Moisture (%)	Solids (%)	Ash (%)	Combustible (%)	Energy Content (MJ/kg)
Food and food productions	45	66	34	13.3	20.7	1.9–17
Plastics	23.1	29	71	7.8	63.2	20.1–33.0
Textiles	3.5	33	67	40	63	11.7–20.0
Paper production	12	47	53	5.6	47.4	6.4–16.0
Leather and rubber	1.4	11	89	25.8	63.2	14.3–23.0
Wood	8	35	65	5.2	59	9.3–17.0
Metals	4.1	6	94	94	0	0
Glass	1.3	3	97	97	0	0
Inerts	1	10	90	90	0	0
Fines	0.6	32	68	45.6	22.4	2.6–15
Total	100					

Source: Al-Salem (2018) and Tchobanoglous *et al.* (1993).

Table 6.10 Common plastics CV.

Item	Energy Content (MJ/kg)	Item	Energy Content (MJ/kg)
PE	43.30–46.50	ELTs pyrolysis oil	42.66
PS	46.50	Rubber tires	64.50
PP	45.0	Kerosene	46.50
Polyurethane (PU)	23.4	Gas oil (GO)	45.20
Polyvinyl chloride (PVC)	20.4	Heavy oil (HO)	42.50
Petroleum	42.30	MSW	20.00

Source: Al-Salem (2018).

The challenge posed by quaternary recycling is the controlling of pollution and emissions of hazardous compounds. Therefore, incineration plants have to be designed and operated to produce the least amount of pollution.

6.3.1.5.1 Pollution control system

The pollutants that can be emitted from the combustion process are particulate matters (PM), NO_x, acid gases, carbon monoxide (CO), hydrocarbons (HC), dioxins and furans. These pollutants can lead to negative health effects. Apart from air emissions, wastewater and ash are also the unwanted by-products from the combustion process. Therefore, proper treatment and management for both generated wastewater and ash should be applied. The pollutants and their control equipment are shown in Table 6.11.

Table 6.11 Pollutants and their control equipment.

Pollutants	Control Equipment
Air pollution	
Particulate matters less than 10 μm	Electrostatic precipitators (ESP) Fabric filters Electrostatic gravel bed filters
NO_x	Selective catalytic reduction (SCR) Selective non-catalytic reduction (SCNR)
Acid gases (e.g. HCl, H_2SO_4 and HNO_3)	Wet scrubbers Dry scrubbers
Carbon monoxide (CO) and hydrocarbons (HC)	Balance air supply and combustible material
Dioxins and furans	Fabric filters
Wastewater	
Colloids	Neutralization, precipitation and settling
Sludge	Dewatering and landfill
Acid agent	Neutralization, precipitation and settling
Total dissolved solids (TDS)	Reverse osmosis (RO)
Ash	Sanitation landfill

6.3.1.5.2 A case study: RDF in Okhla, India

In India, the Ministry of New and Renewable Energy (MNRE) runs a programme on Energy Recovery from Municipal Waste providing grants to five waste-to-energy plants in India currently. Looking at the Okhla waste-to-energy (WtE) power plant that was commissioned in January 2012 gives us a good understanding of the different environmental and socio-political aspects involved in incineration plants.

The plant was foreseen to process 1300 TPD of MSW and projected to produce 450 TPD of RDF to generate 16 MW power. It also received subsidies to meet international standards on Dioxins and Furans and other standards as specified by the Central and State Pollution Control Boards. (CPCB/SPCB).

However, the project has been controversial with the National Green Tribunal (NGT) hearing the Okhla WtE case presenting the contention between the owner of the plant, The Jindals and the residents of Sukhdev Vihar, where the plant is located. The main concerns raised are regarding public health and environmental risks and environmental clearance violations. The CPCB in 2013 also uncovered that dioxins, furans and other particulate matter were higher than the permissible limit. The Okhla incinerator plants also violate the Delhi Master Plan 2021, which mandated that a compost plant was only permissible where the incinerator was built instead. Social uproar resulted in marches and constant public interest litigations (PILs). In the end, although the NGT held the plant owner guilty of acts of omission and other violation, in its final verdict the NGT allowed the plant to operate.

The main issues in this case study highlight that while incineration is beneficial in waste management, to be successful it must be accompanied by the following.

(1) Data transparency and quality
(2) Strengthening of monitoring and regulatory capacity
(3) Proper documentation of data, approaches and lessons learnt so that the process is improved

Consultation of stakeholders before setting up the process and following due process and all the stakeholders (Governments, States, Cities, Academia, NGOs and private sector) who must be involved in developing a joint agenda.

6.3.1.5.3 A case study: co-processing

Co-processing is the use of waste material as alternative fuel or raw material in industrial cement kilns to recover both energy and material. Due to the high temperature in the cement kiln (~1300–1450°C), various types of waste can be disposed of without the production of harmful emissions or fly ash. Many countries are starting to invest heavily in the use of PW for co-processing in the cement industry (e.g., China and India). In co-processing, the waste is converted to heat, and the energy from the heat is then directly utilized. RDF is fed in cement kilns where incineration occurs at very high temperatures and the heat generated from incineration is directly transferred to the raw material for cement production to produce clinker.

Co-processing in cement kilns results in almost 100% energy recovery and 100% material recovery. Co-processing in cement kilns is highly efficient (74% efficacy compared to 13% in WtE) and cheaper than WtE facilities. Almost any waste type can be co-processed on cement kilns. This means that the volume of waste in the dumpsite can be substantially reduced, making way for new waste or for the remediation of the land. Co-processing in cement plants is highly efficient and cheaper than WtE facilities, it also reduces the GHG and other emissions as well as the production of fly ash compared to WtE facilities. Co-processing does not generate any solid waste, and therefore all the material that is sent from the dumpsite mining operation to the cement plant will be completely recovered as heat, and no waste that will need to be disposed of in the dumpsites will be generated.

6.3.2 Plastic reusing technologies

Although the recycling process has multiple functions to manage the PW, those systems also are equipped with multi-control systems. Moreover, a great amount of pollution can occur if the recycling process is not controlled properly. Therefore, the reusing technology is the other recommendation to deal with PW. Reuse is the use of materials to serve the same or a different function. Thus, plastic reusing technology is the process that repurposes plastic without using heavy mechanic and chemical processes. Thus, plastic reusing process can include handicraft products. Reusing generally leads to higher values of reused products. More than 50% of plastic entering the ocean system would be eliminated if about 10–20% of reusable plastic was reused (World Economic

Forum, 2021). Therefore, the reusing technology is an effective and practical process to manage PW.

6.3.2.1 Reuse
Due to their specific plastic properties, plastic products can be reused easily. Although plastic products can be used to serve their original purpose again and again, they can be repurposed.

6.3.2.1.1 Plastic bag sleeping mats, South Africa
Suzette Hendricks began an initiative in 2019 to use plastic bags to crochet sleeping mats for the homeless in Gqeberha, South Africa. It takes around 250 plastic bags to make a full-size sleeping mat, which is lightweight and warm. Businesses and schools donated plastic shopping bags, and Suzette as well as residents from Lake farm Center (Charity organization and home for the physically and mentally disabled) doing good for the environment and helping fellow citizens out.

The first step is to make the plastic yarn which they call 'plarn.' (plastic yarn). Straighten out the bags and fold them vertically in half. Fold up as many times as needed for the size of the bag to make a small vertical pile.

Using scissors, cut off the top handles and bottom seams to create open-ended tubes.

Fold the tube in half with one closed end and start to cut strips open end. The width of the strips determines the thickness of the mat. Strips of two inches and larger are recommended to make the best mats.

Next, the strips should be rolled to get a large ball of 'plarn'. Hold two strips at one round end each and loop one through another. Pull the top of the looped strip from underneath itself to create a tight, perfect joint. Loop as many as you can together so you have a steady supply of 'plarn'. Next, roll the long-jointed strop into a ball so it's easier to carry around.

Then start to crochet the 'plarn'. It is recommended to make the mats 32–26 inches so that they are wide enough for most people to be able to sleep on them.

A long handle at one end of the mat is also crocheted as well as two roll-up strips at the bottom so that the mat can be rolled up, slung over the shoulder, and carried everywhere.

6.3.2.1.2 Turning trash into light – liter of light
Liter of light is an example of plastic bottles reusing which was discovered by Alfredo Moser in 2002, a Brazilian mechanic. He invented a light bulb which was created by filling recycled plastic bottles with water and chlorine and installing the small light kit (see Figure 6.50). Plastic bottles have high transparency and durability. Thus, Moser applies these properties to create the light diffuser or light bulbs. The recycled plastics can be modified to enhance the diffusing of the sunlight during day time (Figure 6.51). Also, they can be installed to improve diffusing light bulbs used in the night time. Moreover, the invention can reduce by 40%.

With all its benefits, the product is now widespread around the world, especially areas with the lack of electric supply such as India, Bangladesh, Tanzania, Kenya, Colombia and Fiji.

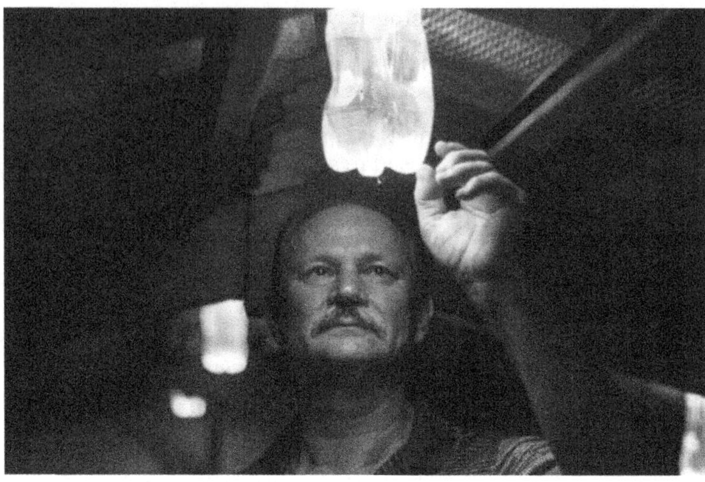

Figure 6.50 Alfredo Moser (*Source*: https://believe.earth/en/alfredo-moser-genie-of-the-bottle/).

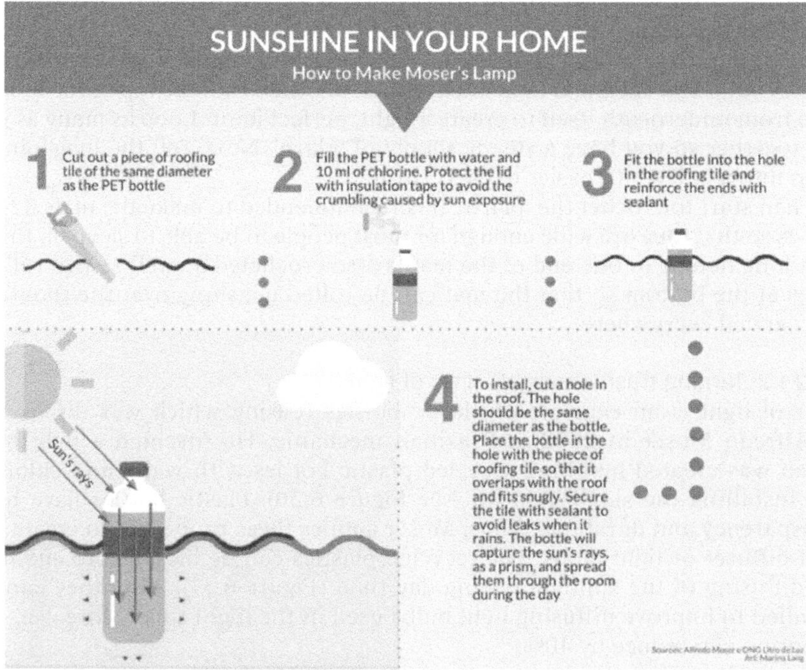

Figure 6.51 Liter of light and their production (*Source*: https://believe.earth/en/turn-a-plastic-bottle-into-a-lamp/).

6.3.2.1.3 Lily Loompa, South Africa

Lizl Naude is the CEO and Founder of Lilly Loompa, a company that upcycles waste into homeware. The company was founded in 2016 with the idea of changing people's perception about waste and reducing the waste that is produced through upcycling. The company has referred to their form of upcycling as 'hipcycling', they create a wide range of products including storage containers, platters, lamps and vases. The waste materials used to create the products are collected from waste from residential estates and waste management companies. The company aims to empower communities through employment as well as instilling a circular economic mindset, encouraging the use of products long after their initial end-of-life by transforming them from waste back into beautiful, useable products. South Africa is one of the world's leading wine producers, and as such wine has become a part of South African culture. Lilly Loompa uses old wine bottles to make platters, toothpick holders and snack caddies (used for anything from crackers to olives to condiments) (Figure 6.52). Some of the companies most popular upcycled products are their 'toona can' storage containers, which are primarily aimed at the tourism market and are sold in tourist shops around the country (Figure 6.53).

Figure 6.52 Upcycled products from wine bottles (*Source*: https://impakter.com/lilly-loompa/).

Figure 6.53 Upcycled storage containers made from tuna cans (*Source*: https://impakter.com/lilly-loompa/).

6.3.2.1.4 Bottle House Project, Nigeria

An estimated 1.6 billion people do not have adequate housing, globally. The bottle House project aims to provide low-cost housing while also reducing the large amount of waste plastic bottles seen in Nigeria. Upcycled PET plastic bottles for use as building blocks, sometimes termed Eco-bricks, has been explored in various countries as a possible solution to the global housing crisis while also reducing PW and promoting a circular economy. The bottle house constructed an initial prototype one bedroom house using upcycled PET bottles for the walls of the house and some parts of the foundation, zinc roofing and cement flooring.

For the foundation, instead of the hollow concrete blocks that are used in certain sections of the foundations, plastic bottles are used. This may prevent termites from climbing through the hollow blocks and damaging roofing structures which are often made from wood and thatch.

For the construction of the walls, plastic bottles are filled with sand and laid down horizontally, they are then tied together using rope to increase the structural integrity. Thereafter, a mortar mixture made using a combination of cement and mud is used to hold the bottles together and provide strength. As water shows better thermal conductivity than sand, the bottles surrounding the lintels for the windows and doors are filled with water instead of sand, and steel rods are included in the structure near lintels to provide increased strength and prevent sagging.

A one-bedroom house made using this technique costs around 3500 pounds, this is estimated to be only 35% of the cost of building using conventional bricks in Nigeria. Up to six PET bottles are required to cover the same wall space as concrete bricks. Nonetheless, a life cycle analysis determined that this process using PET bottles has a lower environmental impact than using conventional concrete bricks (Figure 6.54).

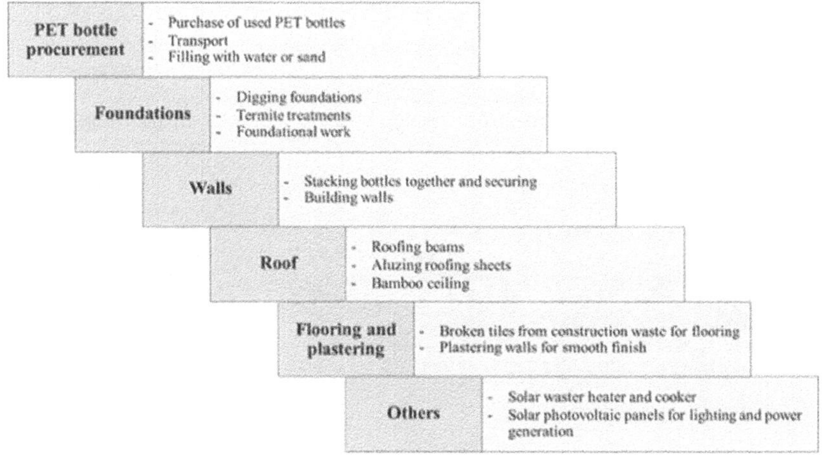

Figure 6.54 Bottle house creation process (*Source*: Oyinlola *et al.*, 2018).

Figure 6.55 Demonstration of Bottle House Project, Nigeria (*Source*: Oyinlola *et al.*, 2018).

6.3.2.2 Refill

Refill refers to the action of repetitively filling a container or collector with items. The containers for the refill can be made of several types of materials such as plastic, glass or aluminum. Refill is a common process that can be applied and integrated with the reusing process. Refill can directly reduce plastic consumption at source. Moreover, refill is the management concept which can be used to reduce single-use plastic consumption effectively. Thus, there has been an upward trend of plastic reduction by applying the concept of refill in many regions, especially Europe.

A case study: Loop platform is a shopping system that applies the refilling concept. Several investors which include P&G, Nestlé, SUEZ, Aptar, Sky Ocean Ventures, ImpactAssets and Quadia have joined this platform. Loop platform provides the closed-loop system for containers by creating durable packaging which can be reused, cleaned, refilled and used again (see Figure 6.55). Moreover, loop production covers 100 brands around the world which offer about 400 products. In 2021, six countries (United States, the UK, France, Canada, Japan, and Australia) participated in the Loop Project (Figures 6.56).

6.3.2.3 Replace

Given plastic properties, many raw materials, especially construction material such as sand and stone can be replaced by PW. Typically, thermosets or thermoplastics can be used to replace other materials for a specific purpose. For example, plastic replacement is commonly used in building construction and decoration processes. Some construction or décor parts such as floors, walls and light bricks used the processes can be made of recycled plastics. The quality of the replaced materials can vary based on the applied types of plastics. However, environmental impact must be considered as replacement requires a large amount of PW and degradation of the replaced plastics, especially bioplastics, there is a possibility of micro- and macro-plastic being released from the materials used for the replacement. However, increasing plastic consumption, the complexity of bioplastic production, and the high cost of bioplastic are the reasons why bioplastic production is not very common. However, petroleum-based plastic can be replaced if bioplastic is completely developed.

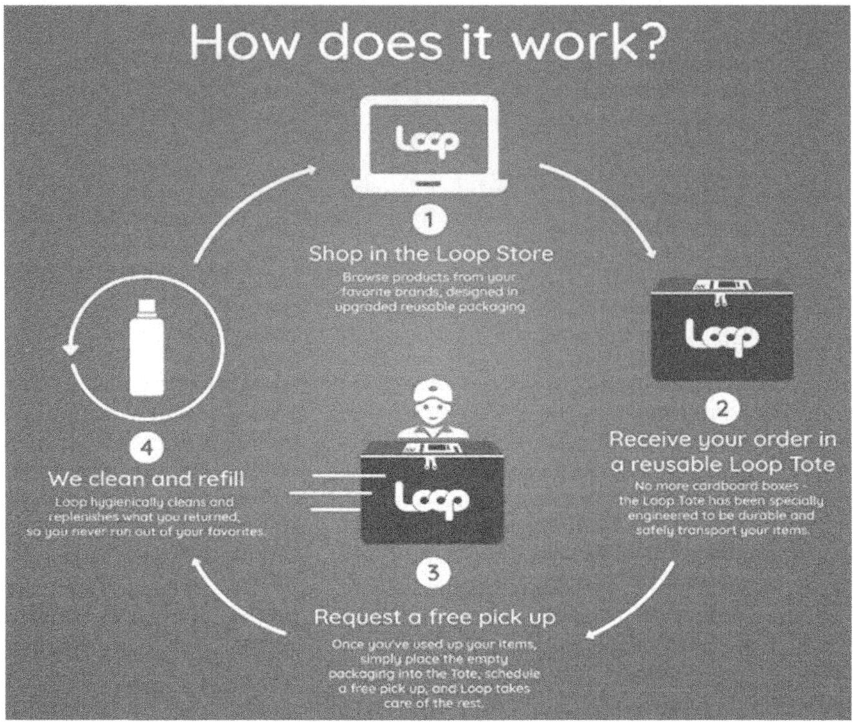

Figure 6.56 Concept of loop system (*Source*: TerraCycle, 2019).

6.3.2.3.1 A case study: Bottle 2 Build in Gauteng, South Africa

Bottle 2 Build was launched with the aim of finding a way to utilize the plethora of plastic bottles which would otherwise end up in landfills in Gauteng, South Africa. Bottle 2 Build manufactures 'brick shaped' bottles that at end-of-life can be returned to the company and used as building material to build schools. The company mainly sells their bottles to schools and business at wholesale. In all, 90 000 bottles are required to build one 54 m2 freestanding classroom, once the bottles are collected the company also raises R200 000 (400 000 Thai Baht) for the additional materials required to build the school including, labor, roofing materials, insulation, steel and so on. The bottles provide good insulation as they are filled with air. Once the steel frame is built, it is filled with the bottles, and each side is measured and plastered to make a completely watertight structure. As of 2017, the company had built 45 classrooms for schools across the Gauteng province. 'What is really nice about the initiative is that, not only is it about keeping plastic out of our landfills, because plastic is very harmful for the environment, but also uplifting local disadvantaged communities,' said the company's brand development manager, Kevin Petitt. The company has however faced many challenges. They say that changing people's mindset regarding using waste bottles as a building material has been difficult (Figure 6.57).

Figure 6.57 Bottle 2 Build Project in South Africa (*Source*: https://www.kirstywattsfoundation. org.za/fundraising/bottle-to-build). (a) Bottle production; (b) building created by Bottle

6.3.2.3.2 A case study: Pradhan Mantri Gram Sadhak Yojana (PMGSY) (upcycling PW for rural road construction in India)

The Pradhan Mantri Gram Sadhak Yojana (PMGSY) or the Prime Minister's Rural Road Program is an initiative by the Ministry of Rural Development, Government of India with an objective of poverty reduction (through efficient rural road development). The program also hopes to contribute to PW abatement (by diverting PW from incineration and landfill). The initiative pilots the use of PW as an alternative to typically use road construction material to lay roads in rural India as an innovative solution to address previous concerns in PMGSY with regard to cost, material availability, technology readiness and specifications. Through the innovative approach, the initiative aims to reduce construction costs, conserve natural resources and adopt a 'green approach', upscale to newer technologies to enhance road quality and serviceability and provide long-lasting, durable roads in rural India. Additionally, reducing the bitumen content can provide an overall economic advantage. As of January 2019, 46.1% of all roads constructed under the initiative has been through the innovative approach of utilizing PW.

Figure 6.58 Incorporating PW into a road asphalt mixture using the dry process (*Source*: Heriawan, 2020).

- Dry-processed PW in asphalt mix

A simple process called the 'dry-processed' approach is used where the PW is added to bitumen and subsequently mixed into the aggregate. Records of this approach indicate that it appears to be an effective way to add PW shred to asphalt mix. The source, dimensions, impurity level, melt flow value of the PW shred must comply with the Indian Road Congress (IRC) guidelines. Calibration is undertaken to ensure that the correct ratio of PW is utilized. A consistent and uniform distribution of the PW shred is required throughout the aggregate. Riding quality, texture depth and visual condition surveys are some assessments that are carried out to evaluate performance of these PW incorporated roads.

Figure 6.59 Plastic cells filled with cement concrete (*Source*: Heriawan, 2020).

- Plastic cell-filled concrete block pavement:

In this innovation, the PW used is the plastic cell-filled concrete block pavement (PCCBP) where interconnected diamond-shaped plastic cells are filled with cement concrete. The plastic cells are made from LDPE or HDPE with a thickness of 0.2–0.5 mm and welded together. The plastic cells are used to encase concrete blocks measuring 150 mm × 150 mm having a depth of 100–150 mm. Compaction then deforms the cell walls and provide block interlocking which is followed by curing process. Plastic cells allow for small movement and flexibility giving this the nomenclature of 'flexible-rigid' pavement. This makes it suitable for low-volume village roads or as overlay over existing pavements (Figures 6.58 and 6.59).

Limitation:

- While this innovation diverts PW from incinerators and landfills, it has potential environmental and human health impacts during production and use of PW in road construction.
- Recyclability of these roads build using PW must be further studied as this is a potential hurdle.
- PCCBP is unsuitable for heavy load of traffic and the surface may not be as good as asphalt.

REFERENCES

Al-Salem S. (2018). Plastics to Energy: Fuel, Chemicals, and Sustainability Implications. William Andrew, Elsevier, Netherlands.
ASTM (2000). Standard Guide for Development of ASTM Standards Relating to Recyling and Use of Recycled Plastics. Society for Testing and Materials, Pennsylvania, USA.
Belgiorno V. and Panza D. (2008). Solid waste management system: an impressive case study. *WIT Transactions on Ecology and the Environment*, **117**, 715–724, https://doi.org/10.2495/SC080671

Bonten C. (2019). Plastics Technology Introduction and Fundamentals. Hanser Publications, Munich.

Bruder U. (2019). User's Guide to Plastic. Carl Hanser Verlag GmbH Co KG, Munich, Germany.

Bureecam C., Chaisomphob T. and Sungsomboon P. Y. (2018). Material flows analysis of plastic in Thailand. *Thermal Science*, **22**(6 Part A), 2379–2388, https://doi.org/10.2298/TSCI160525005B

Callister W. D., Jr. (2007). Materials Science and Engineering: An Introduction, 7th edn. John Wiley & Sons, Inc., Hoboken, New Jersey, USA

EPPIC (Ending Plastic Pollution Innovation Challenge) (2020). EPPIC 2020 LOCATION II KOH SAMUI, THAILAND. United Nations Development Programme (UNDP).

Fitzgerald G. C. and Themelis N. J. (2009). Technical and Economic Analysis of Pre-shredding Municipal Solid Wastes Prior to Disposal. Master thesis, Department of Earth and Environmental Engineering, Columbia University, USA.

Gundupalli S. P., Hait S. and Thakur A. (2017). A review on automated sorting of source-separated municipal solid waste for recycling. *Waste Management*, **60**, 56–74, https://doi.org/10.1016/j.wasman.2016.09.015

Heriawan A. (2020). Upcycling Plastic Waste for Rural Road Construction in India: An Alternative Solution to Technical Challenges. https://www.adb.org/publications/upcycling-plastic-waste-rural-road-india

JICA (Japan International Cooperation Agency) (2018). Democratic Socialist Republic of Sri Lanka – Achieving sustainable waste management that is compatible with local characteristics – Passing on a Rich Natural Environment to Future Generations by Reducing Environmental Impact. https://www.jica.go.jp/english/publications/j-world/1810_02.html (accessed 31 August 2021).

Khalid S., Naseer A., Shahid M., Shah G. M., Ullah M. I., Waqar A., Abbas T., Imran M. and Rehman F. (2019). Assessment of nutritional loss with food waste and factors governing this waste at household level in Pakistan. *Journal of Cleaner Production*, **206**, 1015–1024, https://doi.org/10.1016/j.jclepro.2018.09.138

Kosior E. and Mitchell J. (2020). Current industry position on plastic production and recycling. In Plastic Waste and Recycling (Letcher, T.M. ed.). (pp. 133–162). Academic Press, Cambridge, Massachusetts, USA.

Kumar S., Panda A. K. and Singh R. K. (2011). A review on tertiary recycling of high-density polyethylene to fuel. *Resources, Conservation and Recycling*, **55**(11), 893–910, https://doi.org/10.1016/j.resconrec.2011.05.005

Marks D., Miller M. A. and Vassanadumrongdee S. (2020). The geopolitical economy of Thailand's marine plastic pollution crisis. *Asia Pacific Viewpoint*, **61**(2), 266–282, https://doi.org/10.1111/apv.12255

Ministry of the Environment, Japan (2018). Japan's Resource Circulation Policy for Plastics, https://ec.europa.eu/environment/international_issues/pdf/S2-02-Yusuke%20Inoue.pdf (accessed 15 September 2021).

Nghiem L. D., Iqbal H. M. and Zdarta J. (2021). The shadow pandemic of single use personal protective equipment plastic waste: a blue print for suppression and eradication. *Case Studies in Chemical and Environmental Engineering*, **4**, 100125.

Mujumdar A. S. (ed.). (1995). Handbook of Industrial Drying, Revised and Expanded. CRC Press, Vol. **2**.

Mujumdar A. S. (2006). Handbook of Industrial Drying. CRC press. Boca Raton, Florida, USA. https://doi.org/10.1201/9781420017618

Oberg E., Jones F. D., Horton H. L. and Ryffel H. H. (2016). Machinery's Handbook, 30th edn, Industrial Press Inc., South Norwalk, USA.

O'Farrell K. (2019). 2017–18 Australian plastics recycling survey: national report - final report. https://www.environment.gov.au/system/files/resources/3f275bb3-218f-4a3d-ae1d-424ff4cc52cd/files/australian-plastics-recycling-survey-report-2017-18.pdf

Oyinlola M., Whitehead T., Abuzeinab A., Adefila A., Akinola Y., Anafi F., Farukh F., Jegede O., Kandan K., Kim B. and Mosugu E. (2018). Bottle house: A case study of transdisciplinary research for tackling global challenges. *Habitat International*, **79**, 18–29, https://doi.org/10.1016/j.habitatint.2018.07.007

Pacheco-Torgal F., Khatib J., Colangelo F. and Tuladhar R. (eds.). (2018). Use of Recycled Plastics in Eco-Efficient Concrete. Woodhead Publishing, Sawston, Cambridge, UK.

Pakplas (2016). Plastic Industry of Pakistan. Pakistan Plastic Manufacturer Association, Karachi, Pakistan.

Rotajet Systems Ltd. (2020). Rotary Washers. https://plasticwashing.co.uk/rotary-washer/

Safar K. M., Bux M. R., Aslam U. M., Shankar B. A. and Goel R. K. (2018). The feasibility of putrescible components of municipal solid waste for biomethane production at Hyderabad, Pakistan. *Waste Management & Research*, **36**(2), 169–182, https://doi.org/10.1177/0734242X17748363

Samarasinghe K., Pawan Kumar S. and Visvanathan C. (2021). Evaluation of circular economy potential of plastic waste in Sri Lanka. *Environmental Quality Management*, **31**(1), 99–107, https://doi.org/10.1002/tqem.21732

Shahid M., Nergis Y., Siddiqui S. A. and Choudhry A. F. (2014). Environmental impact of municipal solid waste in Karachi city. *World Applied Sciences Journal*, **29**(12), 1516–1526.

Shapiro M. and Galperin V. (2005). Air classification of solid particles: a review. *Chemical Engineering and Processing: Process Intensification*, **44**(2), 279–285, https://doi.org/10.1016/j.cep.2004.02.022

Sharma P., Sharma A., Sharma A. and Srivastava P. (2016). Automobile waste and its management. *Research Journal of Chemical and Environmental Sciences*, **4**(2), 1–7.

Standard Guide for Development of ASTM Standards Relating to Recyling and Use of Recycled Plastics (2000). American Society for Testing and Materials, Pennsylvania, USA.

Staph D., Ciceri G., Johansson I. and Whitty K. (2019). Biomass pre-treatment for bioenergy: Case study 3-Pretreatment of municipal solid waste (MSW) for gasification. https://www.ieabioenergy.com/wp-content/uploads/2019/02/CS3-MSW-pretreatment-for-gasification.pdf (accessed 31 August 2021).

Tchobanoglous G., Theisen H. and Vigil S. (1993). Integrated Solid Waste Management: Engineering Principles and Management Lssues. McGraw-Hill, USA.

TerraCycle (2019). CPGs and TerraCycle Launch Zero-Waste Packaging Platform. https://www.packagingstrategies.com/articles/90949-cpgs-and-terracycle-launch-zero-waste-packaging-platform (accessed 31 August 2021).

Umer M. and Abid M. (2017). Economic practices in plastic industry from raw material to waste in Pakistan: a case study. *Asian Journal of Water, Environment and Pollution*, **14**(2), 81–90, https://doi.org/10.3233/AJW-170018

United Nations Environment Program (UNEP) (2017). Waste Management in ASEAN Countries: Summary Report. UNEP, Bangkok.

World Economic Forum (2021). Future of Reusable Consumption Models Platform for Shaping the Future of Consumption. https://www.weforum.org/agenda/2021/07/reusing-plastic-waste-pollution-economy-value/ (accessed 1 September 2021).

Zhangjiagang Lianda Machinery Co., Ltd. (2020). Hopper Dryer. https://www.shenzhoumac.com/Hopper-Dryer-pd42541380.html (accessed 20 December 2021).

doi: 10.2166/9781789063448_0289

Chapter 7

Non-recyclable plastics: management practices and implications

Thammarat Koottatep

Table of Contents

7.1 DEFINITIONS AND COMPOSITIONS OF NON-RECYCLABLE PLASTICS

7.1.1 Types of plastics

Plastics are the most widely used products in several applications such as packaging films, wrapping materials, shopping and garbage bags, fluid containers, clothing, toys, household and industrial products, medicinal electrical applications and building materials. Typically, they are organic polymers of high molecular mass and produced by conversion of natural products or by synthesis from primary chemicals generally coming from petrochemicals (SRI India, 2016).

In general, plastic products used in daily life can be classified into thermoplastics and thermosetting plastics, depending on how they react when exposed to heat. The primary difference between these two plastics is the ability to be remolded after initial forming. Thermoplastics can be reheated, remolded and cooled as required without losing their chemical properties. Thermosetting plastics, on the other hand, are the opposite since they cannot be remolded once heated. As a result of these physical and chemical properties, thermoplastics have low melting points, while thermosetting products can withstand higher temperatures without loss of structural integrity (Thomas Publishing Company, 2020).

In addition, plastics are also divided into recyclable and non-recyclable. Recyclable plastics are plastic waste that can be introduced into a manufacturing process to make new materials and convert them into new products, while non-recyclables are plastics that can no longer be used as raw materials in the manufacturing process. The majority of non-recyclables are disposed of through landfills, which has a significant risk of polluting the environment. The summary of types of plastics is presented in Figure 7.1.

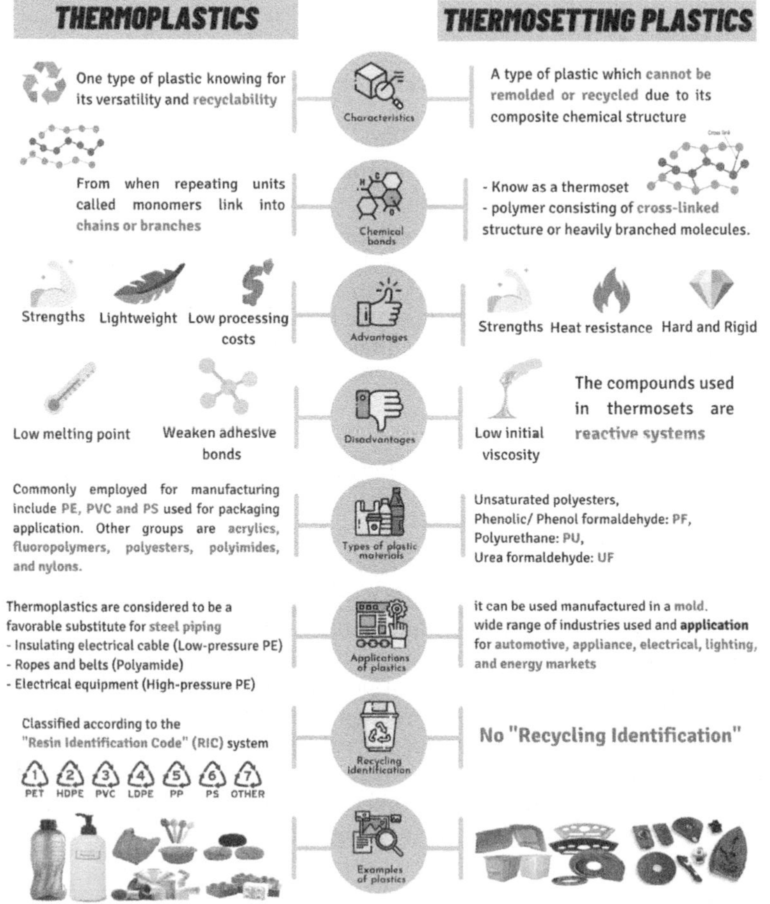

Figure 7.1 Plastic types: thermoplastics vs thermosetting plastics.

7.1.2 Definition of non-recyclable plastics

In terms of recyclability, non-recyclable plastics are plastics that cannot be reprocessed as raw materials. Non-recyclable plastics are often contaminated or for which no current recycling system exists (Bottone, 2019). The definition of non-recyclable plastics is based on different aspects, which can be considered as follows.

7.1.2.1 Original properties

Given the thermal behavior of plastics, thermoplastics are known for its versatility and recyclability. The polymers form when repeating units called *monomers link into chains or branches*, as shown in Figure 7.2. The chains are held together by Van der Waal forces, weak attractions or interactions between molecules. This characteristic allows thermoplastic material to be repeatedly melted, remolded and recycled without negatively affecting the material's physical properties.

Several kinds of thermoplastic resins offer various performance benefits. However, most of them have high strength, shrink-resistance, and high flexibility. Depending on the resin, thermoplastics can serve low-stress applications such as plastic bags or can be used in high-stress mechanical parts. Examples of thermoplastic polymers include plastics that are listed as numbers 1–6 in the Resin Identification Code (RIC) system such as polyethylene (PE), polyvinylchloride (PVC), and polypropylene (PP), and other plastics (number 7 of RIC) such as acrylonitrile butadiene styrene (ABS) and nylon.

Thermoset polymer is a material that cannot be remolded or recycled because of its composite chemical composition. They set in their physical and chemical characteristics after the formation process, as implied by their name, and are no longer affected by subsequent heat exposure. The chemical change in thermosetting plastics that occurs during molding, which is triggered by temperature and pressure, is a result of a curing process. Once heated, thermosets form permanent chemical bonds or crosslinks with each

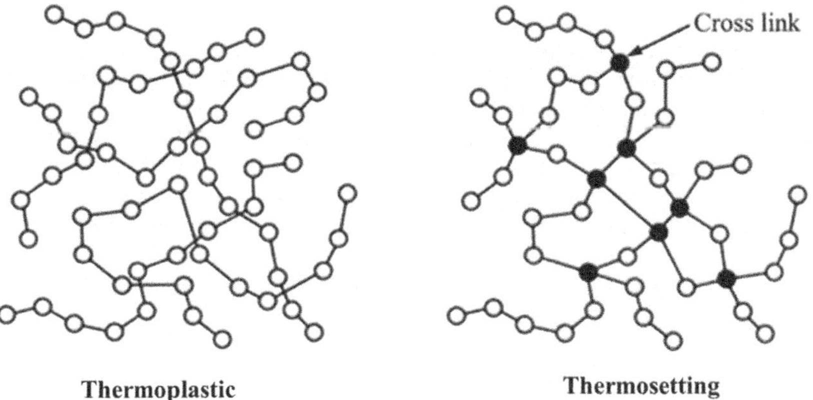

Figure 7.2 Polymer structure of thermoplastic and thermoset plastics (*Source*: Karuppiah, 2016).

other (Figure 7.2). A high crosslink density will result in a very stiff, hard, and sometimes brittle material. Thermosetting plastics are non-recyclable because they cannot be remolded into new products.

After initial heat treatment, thermoset materials can withstand heat, corrosion and mechanical creep. This makes them perfectly suitable for use in components that require tight tolerances and excellent strength-to-weight characteristics while being exposed to elevated temperatures (Central Pollution Control Board, 2018). Examples of thermoset plastics are epoxy, phenolic, urea-formaldehyde and unsaturated polyesters.

Box1: Curing of Resin

Curing is a chemical process applied in polymer chemistry and process engineering that produces the toughening or hardening of a polymer material by cross-linking of polymer chains. This process is dependent on the types of resin. It can be originated by heat, radiation, electron beams, or chemical additives (Chambon & Winter, 1987).

1. Room temperature curing

The goal of curing polymers is to make the materials structurally stable. There are three main components, namely catalyst, cross-linking agent and accelerator, all of which are used together with the resin. The most used catalyst is methyl ethyl ketone peroxide (MEKP) while the common accelerator is cobalt naphthenate or cobalt octoate. For cross-linking agents, styrene is the most common agent used in this process. Mostly, 24 hours of curing time is required for room temperature curing. After this, the material can be removed from the mold and sent for further processing, such as post-curing at elevated temperatures (Biswas *et al.*, 2019).

2. High-temperature curing

High-temperature curing is used for compression, injection, and pultrusion molding techniques. In this process, free radical chain growth co-polymerization reactions occur between the cross-linking agent and the polymeric chains. The initiator first breaks and generates free radicals. These free radicals link with, for example, unsaturated polyester resins (UPR) chain molecules through the styrene oligomer. Many researchers have observed the effects of temperature on the curing process of UPRs at 30–50°C throughout the whole conversion range. It was found that, at the start of the reaction, co-polymerization occurs. Then, styrene conversion takes place, followed by the opening of $C = C$ bonds. Eventually, the reaction of styrene and the side chain takes place (Biswas *et al.*, 2019).

Example: UV curing of thermosets by photopolymerization

- Curing by polymerization rather than vaporization
- Photopolymer: Light-sensitive resin that cures when light hits it
- UV light of 320–400 nm

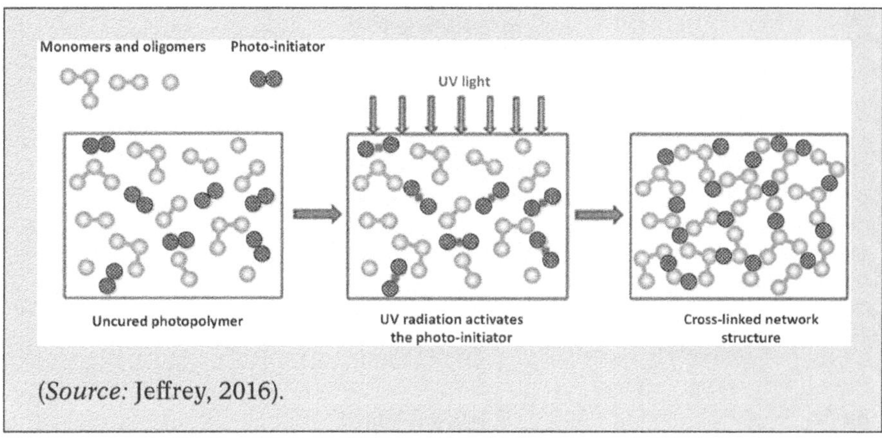

(*Source:* Jeffrey, 2016).

7.1.2.2 Composite materials

Composite materials are made by combining two or more materials that have significantly different physical or chemical properties. The components of composite materials include resin matrix, fiber reinforcements, core materials, additives and surface finishes (American Composites Manufacturers Association, 2022). The first modern composite material is fiberglass or fiber-reinforced polymer (FRP), and the most common FRP manufacturing process is hand lay-up or wet lay-up. Sheet molding compound (SMC) and bulk molding compound (BMC) are the later developed form of FRP composites which are widely processed by compression or injection molding (Forenge Technologies, N.D.).

Composite materials are high quality and have a long lifespan. Their higher strength, lower weight and less maintenance have led to many engineering applications, especially in the transport sector, for significantly reduced energy consumption and impact on the environment (Yang *et al.*, 2012). Three types of composite materials are developed and widely used in numerous kinds of engineering applications including polymer–matrix composites (PMC), metal–matrix composites (MMC) and ceramic–matrix composites (CMC). Regarding the reinforcement types, composite materials can be classified into particulate composites, fiber-reinforced composites, and structural composites. The classification of composite materials is presented in Figure 7.3.

For all types of composite materials, polymer–matrix is holding the largest share in the market, in which thermoset composites accounted for 72.0% of the overall revenue share in 2019. This is due to the growing demand for transportation and aerospace, and defense applications (Grand View Research, 2020). However, thermoplastic composites are also growing more rapidly in recent years.

The major application sector is automotive and transportation which accounted for 21.1% of the global market revenue in 2019. The aerospace and defense industry is also one of the most important to the composite market. In addition, composite materials are used in sports and recreation facilities,

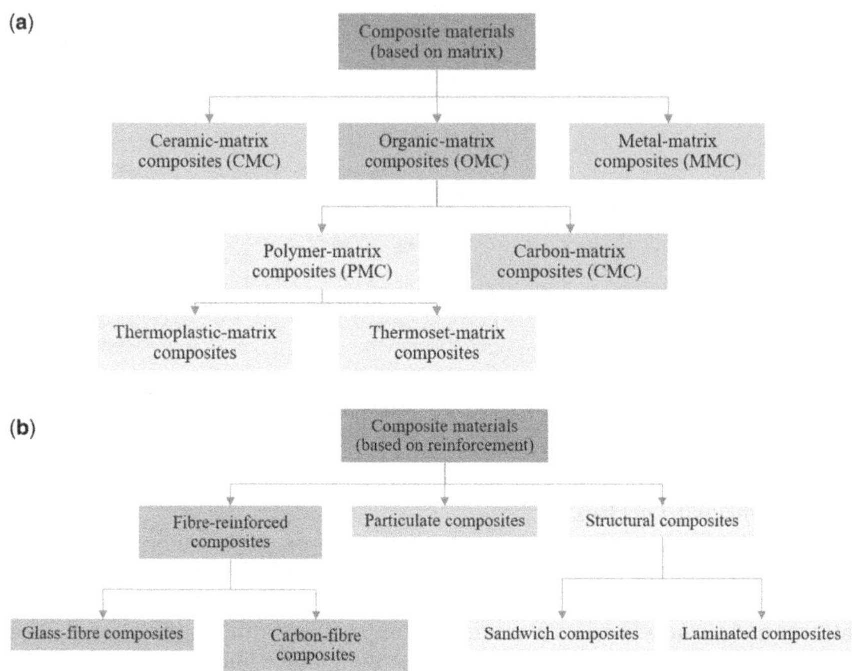

Figure 7.3 Classification of composite materials. (a) Based on matrix materials and (b) based on reinforcement materials (*Source*: modified from Ibrahim *et al.*, 2015).

boat and shipbuilding, wind energy generation for wind turbine blades, as well as in oil and gas offshore exploration. The examples of composite materials in different applications are shown in Figure 7.4.

Western European countries have the highest demand for composite materials, with Germany taking the largest market in 2018, and it is expected to continue its dominance over the forecast period (Grand View Research, 2019). In addition, the well-established automotive, aerospace and defence, construction, and electrical and electronics industries are anticipated to continue the growing trend in the European market. Increasing demand for cost-effective and high-performance composites in the renewable energy industry is also expected to drive the regional market. Figure 7.5 shows the statistics of the market value of composite materials worldwide in 2015, 2017 and 2019, with a forecast figure for 2027. It is estimated that the global market size for composites will increase upto 160.54 billion U.S. dollars in 2027.

Commercial recycling of composite materials remains limited due to technological and economic constraints. The basic problem is the difficulty in separating homogeneous components from composite materials. The presence of fibers and other types of reinforcement, as well as binders in composites, especially thermosets, is an obstacle to the recycling of this type of material. Because of these limitations, most of the recycling activities for composites are

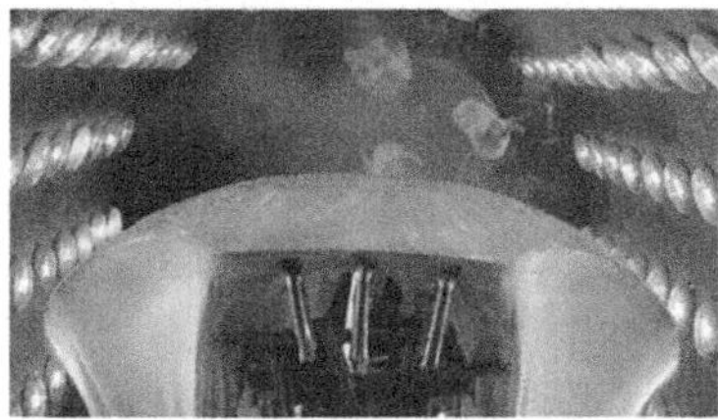

Aerospace: light-sport aircraft made from CFRP

Automotive: CFRP internal structure and body

Sports & Recreation: Motorcycle racing helmet made from CFRP

Marine: Recreational boat made from GFRP

Transportaion: Walmart's green concept truck made from CFRP

Energy: Wind turbines made from GFRP

Figure 7.4 Applications of composite materials, CFRP is carbon fiber-reinforced polymer and GFRP is glass fiber-reinforced polymer (*Source*: http://compositeslab.com/where-are-composites-used/).

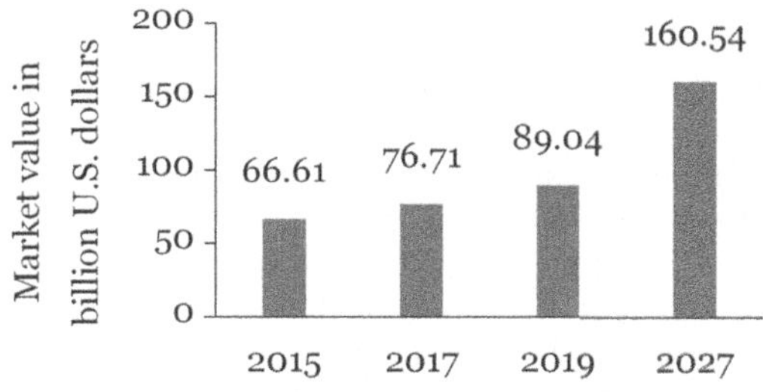

Figure 7.5 Market value of composite materials worldwide from 2015 to 2027 (*Source*: https://www.statista.com/statistics/944471/global-market-value-of-composites/).

limited to down recycling such as energy or fuel recovery with little materials recovery such as reinforcement fibers. Additionally, the lack of markets, high recycling cost, and lower quality of the recyclates compared to the virgin materials also impede the use of recycled composite materials in many applications such as automotive, aerospace, and other engineering products. Environmental regulations, as well as long-term technology improvements, are essential to support recycling (Yang *et al.*, 2012).

(a) Resins
Resins are used to transfer stress between reinforcing fibers. They are also used as glue to hold fibers together and protect the fibers from mechanical and environmental damage. Resins employed in reinforced polymer composites can be thermoplastic or thermoset.

(b) Reinforcements
Various reinforcements can be used to give the composite materials the desired properties according to their application. Natural and man-made materials, such as cellulose in wood, can be used as reinforcements. However, in commercial applications, most reinforcements are man-made. There are many commercially available reinforcement forms to meet the design requirements of the user. The ability to customize the fiber architecture allows for optimized performance of a product which translates to weight and cost savings.

Many forms of fiber are used as reinforcement in composite laminates, such as glass fibers, carbon fibers, aramid fibers (polyaramids), and polyester and nylon thermoplastic fibers. However, glass fibers are the most commonly used, which account for more than 90% of the fibers used in reinforced plastics. They are generally inexpensive to produce and have an excellent strength-to-weight ratio.

(c) Additives and fillers
Additives and modifiers are used to expand the usability of polymers by increasing their processability or extending product durability. They are generally used in relatively low quantity by weight compared to other components such as resins, reinforcements, and fillers. However, they perform critical functions. Although additives and modifiers often increase the cost of the basic material system, they always improve cost and performance.

- Additives
  There are several additives used to modify and enhance resin properties.
  - *Thixotropes*: Fumed silica or certain clays can be used as thixotropic agents in the hand lay-up or spray-up. In a standstill state, resins containing these agents have a high viscosity, reducing the liquid resin's tendency to flow or drain from vertical surfaces. The viscosity of the resin is reduced when it is sheared, making it easier to spray or brush on the mold.
  - *Pigments and colorants*: They can be added to resin or gel coat for cosmetic purposes or to enhance weatherability of products.

- *Fire retardants*: Most thermoset resins are combustible and produce toxic fumes when burned. Combustion resistance can be improved by selecting an appropriate resin type and using fillers or flame-retardant additives.
- *Suppressants*: In open mold applications, styrene emission suppressants are used to prevent evaporation for air quality compliance. These wax-based materials form a film on the resin surface and reduce styrene emissions during the curing process.

- Fillers

Fillers are used to improve mechanical properties of materials such as fire and smoke performance. The resins combined with fillers will shrink less than the unfilled resins, thus improving the dimensional control of molded parts. The composite properties can be improved for various applications by selecting the appropriate fillers. The important properties include water resistance, weathering, surface smoothness, stiffness, dimensional stability and temperature resistance.

The use of inorganic fillers in composite applications is increasing. When used in composite laminates, this type of fillers can account for 40–65% by weight. Several inorganic filler materials are used with composites.

- *Calcium carbonate*: It is the most widely used inorganic filler due to its low cost and is available in a wide range of particle sizes.
- *Kaolin (hydrous aluminium silicate)*: It is the second most commonly used filler with its common name, clay. This filler is also available in different sizes.
- *Alumina trihydrate*: This filler is usually used to reduce fire and smoke problems. When exposed to high temperatures, the filler gives off water (hydration), thereby reducing the flame spread and development of smoke.
- *Calcium sulfate*: It is a low-cost filler used for flame and smoke retarding purposes, often used in the tub and shower industry.

- Core

The core materials are used to produce rigid and lightweight composite products. The properties such as thermal conductivity, sound insulation and fire resistance can also be improved by use of the proper core material. The use of the core is called *sandwich construction* which consists of a face-skin laminate, the core material, and the back-skin laminate. The use of the core makes the laminate thicker, which results in a more rigid material.

- Surface finishes

Mostly, surface finishes are used for UV protection, corrosion resistance and aesthetics. They can be critical to the long-term appearance of composite products. The examples of surface finishes are listed below.

- *Gel coat*: It is used to improve weathering, filter out ultraviolet radiation, add flame resistance, provide a thermal barrier, improve chemical resistance, improve abrasion resistance, and provide a moisture barrier.

- ○ *Surface veils*: They are designed to improve the surface appearance and ensure the presence of a corrosion resistance barrier for typical composite products such as pipes, tanks and other chemical process equipment. Other benefits of the surface veils include increased resistance to abrasion, UV and other weathering forces.
- ○ *Adhesives*: They are used to bond the composites to themselves and other surfaces. The three most common adhesives used in various applications include acrylics, epoxies and urethanes.
- ○ *Ultraviolet protection*: Some materials are sensitive to UV degradation; thus, they should be protected from UV by an opaque gel-coated surface or by painting the exposed surfaces.
- ○ *Painting*: Painting systems are widely used in both architectural and marine fields. Applying paints on the surface of composites requires proper abrasion and removal of residual mold release agents without special surface preparation.

7.1.2.3 *Properties for non-recyclable plastics adjustment*

From the viewpoint of recycling, waste can be classified into three groups (Sabău, 2018).

- Most of the manufacturing waste is in the form of a *single material*. As it is not contaminated with other materials, recycling is then easier. Usually, these wastes are reintroduced into production lines.
- *Easily separated waste* is made up of one to two polymers or contaminated materials like fillers (on a macroscopic scale). These materials are separable, at least in theory.
- *Microscopic mixtures* or intimately connected (soldered, interpenetration) are considered as the most difficult case for treatment since the separation of components is very difficult or even impossible, requiring complicated operations. The organic matrix composites are the sample materials that are classified in this category. The most representative example is the waste from the automotive industry. In this case, the blend will find materials (resins) thermoplastic, polymer mix, fibers, fillers and multilayer composite materials.

However, the first two types of plastic waste are currently common in waste recycling. For the composite materials, landfilling and incineration are the common methods for disposal of this kind of materials. Although there are initiatives for mechanical recycling, organizing a practicable collection/transportation/processing system for the waste and finding markets for the recyclates are two main barriers to this approach (Jacob, 2011).

In the case of the microscopic mixtures, multilayered plastics are commonly found in some packaging industries. The multilayered packaging is any material used for packaging which has at least one layer of plastic as the main ingredients, and it is combined with one or more layers of other materials such as paper, polymeric materials, or aluminum foil. The combination can be in the form of a laminate or co-extruded structure (Central Pollution Control Board,

Table 7.1 Sources and uses of non-recyclable plastic waste.

No.	Sources	Examples
1	Food packaging	Multilayer films are used for packing of biscuits, chips, and juices
2	Pharmaceutical and cosmetic products	Multilayered packing is used for packing of medicines, tablets and cosmetics
3	Electrical and electronic goods	Multilayer films such as bubble raps, laminates are used for packing of electrical and electronic items
4	Items used for food storage and serving	Thermocol products such as plates or cups are used for serving food, tea, or coffee. They are also used as fillers in packing of goods and items

Source: Adapted from Central Pollution Control Board (2018).

2018). It is marked with the symbol '7' and 'other' in the resin identification code (RIC) system.

Multilayered plastics are generally used in the packaging industry since they provide good barrier properties which are required for the products such as food, pharmaceuticals and electronic products. Several goods are made of multilayered plastics as shown in Table 7.1 and Figure 7.6. Although this type of plastic is produced from thermoset or thermoplastic material, due to its complex structure, it is very difficult to separate from each other, thus considered as *non-recyclable*.

The packaging industry is the largest sector of plastic consumption, accounting for 36% of total global plastic production in 2015 (Malhotra, 2020). In addition, more than 45% of plastic waste generated in the same year was from packaging materials. Furthermore, among packaging films produced, the

Figure 7.6 Examples of multilayer plastics in daily life (*Source*: https://packagingsouthasia.com/she-safety-health-and-environment/sustainability-health/why-multilayered-plastic-structures-are-indispensable/).

multilayer materials account for approximately 17% (Lahtela *et al.*, 2020). The increasing demand for this type of plastic is also expected to be about 7%, and during 2018–2026, the global market value is predicted to expand by 4.4% with CAGR (compound annual growth rate). PE and PP films are the common manufactured films used for the production of multilayered packaging, either for use as a package or as a coating material. Both PE and PP account for 60–65% of the total quantity of the produced films.

The growing demand for packaging materials due to packed products' safety and quality requirements has resulted in the significant development of advanced packaging techniques (Tartakowski, 2010). Economic and environmental considerations have led to a replacement of single-layer films by multilayer films. Compared to common single-layer films, the multilayers provide special properties, including high barrier for water vapor, gases and aromatics, as well as high mechanical strength, good sealability and resistance at low temperatures. These properties result from the synergy of basic properties of the materials used such as PE, PP, polyethylene terephthalate (PET) and polystyrene (PS) and the use of ethylene vinyl alcohol (EVOH) and ethylene vinyl acetate (EVA) as a barrier and adhesion layer. An example of using multilayers instead of single-layer plastics for resource-saving purposes is EVOH in packaging products. The EVOH barrier layer with a thickness of 4–20 μm has the same barrier properties as polyamide (PA) with a thickness of 400 μm. An overview of commonly used materials and their particular functions are shown in Table 7.2 (Kaiser *et al.*, 2018).

Table 7.2 Overview of common functional layers.

Mechanical Stability	Oxygen Barrier	Moisture Barrier	Light Barrier	Tie Layers	Sealant
HDPE	EVOH	PE (LD, LLD, HD)	Aluminum	Polyurethanes	LLDPE
PP, OPP	PVDC	PP, OPP	TiO_2 filled polymers	Acid/anhydride grafted polyolefins	LDPE
OPET	Polyamides (nylon, BOPA)	EVA	–	–	EVA
PS	Polyesters, OPET	Ionomers	–	–	Ionomers
Paper	Coatings (SiO_x, Al_2O_3, PVOH, nano particles)	PVDC	–	–	PP, OPP
–	Aluminum	–	–	–	PA, OPA
–		–	–	–	PET, OPET

Source: Kaiser et al. (2018).
Note: Abbreviations: BOPA, biaxially oriented polyamide, EVA, polyethylene-vinyl acetate, OPA, oriented polyamide, OPET, oriented polyethylene terephthalate, OPP, oriented polypropene, PVDC, polyvinylidene chloride, PVOH, polyvinyl alcohol, PE, polyethene, LD, ow-density, LLD, linear low-density, HD, high-density.

The recycling of multilayer packaging is available for certain products such as liquid packaging boards (LPBs), in which their end-of-life recycling option is already established on the market. The components of LPBs include 21% of PE, 4% of aluminum, and about 75% of cardboard. The cardboard layer is used to provide stability of the product, while the PE layers provide protection against moisture from the outer and the inner sides of the packaging. For aluminum, it acts as a gas barrier.

Furthermore, in order to recycle multilayer post-industrial waste, a delamination process may be required. Nevertheless, for polymer–polymer multilayer materials, no industrial solution has been available until now (Kaiser *et al.*, 2018). The recycling of multilayer films is still an important challenge due to several reasons.

- Various materials used for each layer
- Large differences in the processing properties of the materials used for multilayer films
- Lack of systems for identification of multilayer film (e.g., PS/PE, PET/ PE, etc.)
- Lack of system solutions for the collection of these materials, which results in contamination with other materials with similar visible characteristics (color, thickness, stiffness)
- Lack of standard research of the properties, processing and applications of composites based on recycled multicomponent materials
- Lack of economically viable systems of segregation of the various materials

Due to the similar appearance of multilayers and single-layer films, it is possible that the target films will be processed with other films as a result of inappropriate segregation and false characterization of properties. This contamination leads to the production of material with undefined properties, which prevents the proper processing of the material and its application as a desired device (Tartakowski, 2010).

The multilayered plastics hold a large percentage in the packaging market, and mostly they are treated by incineration or landfilling due to the difficulty in recycling. However, in order to recycle these materials, the replacement of multilayer packaging by monomaterials can be a potential option for some multilayer systems. Nevertheless, it is expected that the monomaterials can substitute not all kinds of multilayers. In addition, functionality, costs and marketing are still the main factors of the packaging market. Therefore, it may be worthwhile to have a closer look at the recycling strategies of this type of material (Kaiser *et al.*, 2018).

7.2 FACTORS AFFECTING RECYCLABILITY OF THE PLASTICS

7.2.1 Degree of contamination

Currently, many recycling plants are optimizing their systems so that they can efficiently sort and process recycled materials. However, in some situations, it has been found that recyclable products are mixed with non-recyclable waste. This is because the recyclables are put in the incorrect recycling bins, causing the recyclable items to be contaminated with other unwanted materials such as food, grease, or liquid. Hence, the recyclable materials cannot be processed or recycled effectively. In addition, the recyclable plastics can become dirty non-recyclable waste through the contamination by their own product such as dirty food box, or the contamination from storage, collection, and transportation processes. Also, partly recyclable items such as children's car seats, where certain components are recyclable and others are not, can contaminate the recycling system as well (Rabbit Waste Management, 2020). The contamination of recycling streams by other non-recyclable waste or different recyclable waste results in the rejection of recyclable items, causing the entire waste to be disposed of by landfills.

The average recycling contamination rate is about 25%, meaning that out of every four items there is one contaminant presented (Cleanriver Recycling Solutions, N.D.). The high concentrations of non-recyclable items or contamination of other forms such as food waste in the recyclable stream may happen for several reasons (Stephenson, 2018).

- People are confused about what goes in which bin.
- People are not always very careful about what they put in.
- In areas where all types of recyclable waste are collected in one bin, one type of waste can contaminate another.

However, even in cases where plastic contamination could be dealt with, it is sometimes more economically feasible to divert some loads to landfills. The cost for processing poorly sorted or contaminated plastic waste is more expensive, in some cases outweighing profits from recycled materials (Ritchie, 2018).

Plastic contamination is related to the amount of plastic waste sent to landfills. For example, in the UK, most bottles will usually be managed for reprocessing in the country, but plastic waste that is less valuable which is about two-thirds collected for recycling goes overseas as shown in Figure 7.7. In 2017, the UK exported more than 600 000 tons of plastic waste, and all could be highly contaminated, as reported by The National Audit Office (Rabbit Waste Management, 2020; Stephenson, 2018). This means that with the large number of contaminated plastic waste, the entire load of waste may not be reprocessed and may end up in landfills or contribute to environmental pollution.

7.2.1.1 Contamination of non-recyclable waste or other recycling in plastic waste flow

Waste separation at the source is very important for the effective recycling process. Some recyclable plastic waste becomes non-recyclable waste when contaminated with non-recyclable plastics or food waste, as shown in Figure 7.8. The non-recyclables can lead to contamination of the supply. Although many facilities have automated and/or manual procedures for removing non-recyclables, they are not always 100% effective. If waste streams contain a significant number of non-recyclables, the facilities may not be economically feasible to sort. The same applies to food or liquid waste, uncleaned plastics can contaminate the supply, and the disposal of these loads must be taken directly

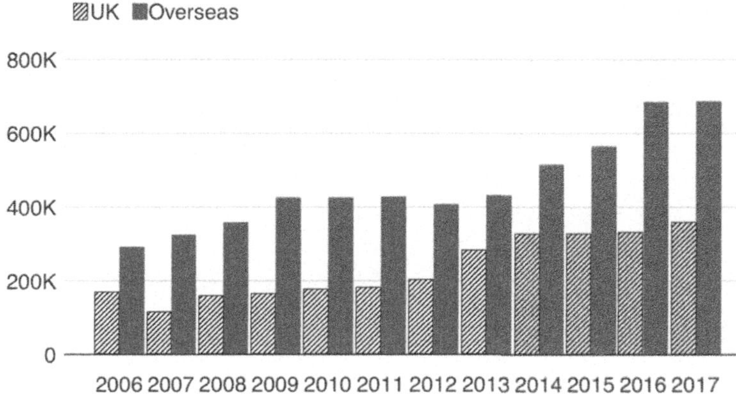

Figure 7.7 Tons of plastic sent for reprocessing (*Source*: https://www.bbc.com/news/science-environment-45496884).

Contamination is a common problem

Figure 7.8 Examples of contaminated plastics (*Source*: https://www.bbc.com/news/science-environment-45496884).

to landfills (Salt Lake City, 2019). The contamination of non-recyclable waste into recyclable plastic streams makes the whole stream unsuitable for recycling. All this plastic waste will become non-recycled plastics which are defined as plastics that are not redirected for recycling and remain in municipal solid waste (MSW) or in materials recovery facility residue. Table 7.3 illustrates the

Table 7.3 Global composition of non-recycled plastic waste consisted in MSW.

Plastic Material	Malaysia Plastic Waste[a] (%)	US plastic Waste[a] (%)	UK Plastic Waste[a] (%)	Thailand Plastic Waste[b] (%)	Global Plastic Waste[a] (%)
PET	16.2	12.4	15.3	5.9	15.4
HDPE	26.2	17.8	13.5	57.4	17.0
PVC	3.9	5.5	3.5	2.2	3.1
LDPE	31.1	19.6	25.0	17.4	34.0
PP	8.2	13.9	22.2	7.3	15.4
PS	13.0	8.7	4.0	4.8	12.4

Source: [a]Sharuddin et al. (2017); [b]Areeprasert et al. (2017).

percentages of plastic materials that are presented in MSW. This table can imply that although there is recyclable waste such as PET, when it is contaminated with other non-recyclable waste, all must be taken to landfills.

The examples of contaminants in the plastic recycling stream are listed below.

(1) Non-recyclable plastics

Types of non-recyclable plastics depend on several factors such as market and city government. If there is a demand in the market, then recyclers and companies will pay for the post-consumer recyclables, and they will become economically valuable materials (Sedaghat, 2018). In addition, the types of recyclable plastics can vary significantly between countries depending on recycling policies, targets and the effectiveness of recycling separation methods. However, mostly plastic bags, shrink wrap, bubble wrap, newspaper bags or trash bags are considered as non-recyclables since they have a high possibility to damage or tangle the machines. In addition, PS is also not usually accepted in recycling program due to its lightness that easily is blown away and creates litter during collection. Moreover, its bulky characteristic is also one of the reasons that PS is not preferred as it requires high transportation cost. Also, PS is hazardous to be in contact with any food as some studies have found that PS can create health problems. Therefore, it is often not profitable to recycle PS (Sharuddin *et al.*, 2017).

(2) Other recyclable items

The combination of different types of recyclable waste can create a complication in waste sorting system. For example, it creates more work to the recycling workers, and the workers can be exposed to hazardous waste through leachate if there is some liquid from bottles and e-waste in the recycling stream. Moreover, plastics can be contaminated with heavy metals from the electronic device. This potentially affects the quality and applicability of reprocessed plastic, as it might be unsuitable for use in certain applications such as food packaging due to elevated metal concentrations (Eriksen *et al.*, 2018).

(3) Other wastes
Food waste and bio-hazardous waste such as tissue paper or diapers are also contaminants found in recycling stream. In addition to complicating the sorting system, they can create unsafe conditions for workers in recycling facilities (Recyclebank, 2014). Most recyclable waste must be switched to landfill when contaminated with organic waste.
Due to the contamination with other wastes as well as thermal breakdown or destruction in processing, the quality of recycled plastics is degraded, resulting in the limitation of repeated recycling. For most recyclable plastics, they are typically only suitable for recycling once. As a result, most recycled plastics eventually reach landfills even if they go through an additional use cycle as another product. The recycling process typically delays rather than prevents plastic disposal to landfills or incineration (Ritchie, 2018).

7.2.1.2 *Contamination of different recyclable plastics*

In addition to the contamination with non-recyclable waste, the contamination of different types of recyclable plastics also needs to be considered in the recycling process as the contaminants can affect the quality of the target recycled plastics. When the waste is contaminated with another type, the quality of the material will drop. For example, a small amount of PVC contaminant presenting in a PET recycling stream will degrade the recycled PET resin due to the evolution of hydrochloric acid gas from the PVC at a higher temperature which is required to melt and reprocess PET. On the other hand, PET plastics in a PVC recycling stream will form solid lumps of undispersed crystalline PET, which significantly reduces the value of the recycled material (Hopewell *et al.*, 2009). For this reason, multi-layer or multi-component articles which consist of several types of polymers, are also not recycled.

Another example is the contamination of biopolymer on the recycling process of petroleum-based plastic waste (Gere & Czigany, 2020). Small amounts of polylactic acid (PLA) have a significant negative effect on the properties of PET plastics when they are mixed together (PET/PLA:95/5% by weight) (La Mantia *et al.*, 2012). At the processing temperature of PET, PLA will be degraded, which leads to the yellowing of the product. In addition, since PET and PLA are thermodynamically immiscible, holes, peaks or clusters can appear in the products. Moreover, the glass transition temperatures of the two polymers are also different, resulting in opaqueness or haziness in PLA-contaminated PET products. These are important problems as optical and surface properties may be even more important than mechanical properties in mass production.

7.2.2 Market value

The selection of recyclable or non-recyclable plastics in each location also depends on the market value of plastics. Although many types of plastics can be recycled, they are not always economical. The price of used plastics depends on both supply and demand of recycling market (Milios *et al.*, 2018). Generally, the plastics that have good quality and sufficient amount and easy processing will be

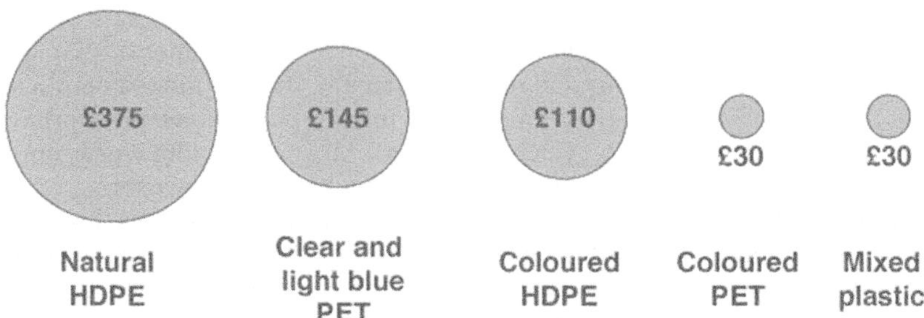

Figure 7.9 Maximum price per ton in November 2018 of used plastic waste (*Source*: https://www.bbc.com/news/science-environment-45496884).

selected by plastic manufacturers. Therefore, recycled PET and high-density polyethylene (HDPE) are more expensive than other plastics. Manufacturers often believe that transparent plastics are often desirable since they can be dyed to transform into new products, resulting in greater flexibility, unlike pigmented or dyed plastics with lower market value. In addition, recyclers also prefer rigid containers made of a single polymer as they are simpler and more economical to recycle over multi-layer and multi-component package (Hopewell *et al.*, 2009).

Normally, high-priced plastics will be collected whereas the plastics that cannot be sold will not be collected and become non-recyclable plastics. Figure 7.9 shows the price of some types of used plastics, in which natural HDPE has the highest price, and the colored PET has the same price as mixed plastic waste.

Based on polymer types, the price of used plastics is presented in Table 7.4. The most recycled plastic worldwide is post-consumer PET, followed by HDPE, low-density polyethylene (LDPE), PP and others. The recycled PET is dominating the market due to its widely used applications such as packaging, electronics, construction and others. In addition, recycled PET and PE accounted for 70% of total post-consumer plastics recycling (Future Market Insights, 2018; Market Research Future, 2019).

Table 7.4 Price of plastic waste based on polymer types.

Type of Plastics	Recycling Rate (%)	Price (Baht/kg)[d]	Recyclability
PET	26.8[a]	1.1 (screened)–9.0 (pure color)	High
HDPE	29.3[a]	1.5–5.0	
PVC	<1.0[b]	0.4 (bottle)–13.0 (pipe)	
LDPE	2.4[c]	1.3	
PP	3.0[b]	1.0–3.0	
PS	–	0.5–1.5	
Other	–	1.0–2.0	Low

Source: [a]USEPA (2020); [b]Seaman (2012); [c]Harper (2003); [d]Wongpanit International Company (2020) (30 Baht = 1US$).

Additionally, the reasons that make some plastics become recyclable or non-recyclable plastics are as follows:

7.2.2.1 PET
PET plastic is recyclable because the polymer chain breaks down at a relatively low temperature, and also there is no degradation of the polymer chain during the recycling process. These properties allow PET plastic to be recycled a large number of times before it becomes unusable. Normally, PET attracts the high prices, especially clear bottle. However, colored plastic is less desirable as the color cannot be removed, thereby complicating the recycling process. For PET plastic, it is suggested for single-use applications. Repeated use can increase the risk of chemical leaching and bacterial growth.

7.2.2.2 HDPE
HDPE is the most commonly recycled plastic and is considered as one of the safest forms of plastic. The recycling process of HDPE is relatively simple and cost effective (Seaman, 2012).

7.2.2.3 PVC
Mostly, PVC products are not acceptable for recycling. Two major problems are the high chlorine content in raw PVC ($\sim$56% of the polymer's weight) and the high levels of hazardous additives added to the polymer to achieve the desired material properties. As a result, PVC requires separation process from other plastics before mechanical recycling (Rubio, 2019).

7.2.2.4 LDPE
LDPE plastics are less toxic than other plastics and are relatively safe for use. However, in the recycling process, plastic wrapping and thin plastic bags can create the risk of clogging the processing machinery if they are collected along with larger, heavier and more rigid recyclable plastics. Also, products made from recycled LDPE are not as hard or rigid as those made from recycled HDPE plastic. Therefore, LDPE is not commonly recycled.

7.2.2.5 PP
PP is considered as a recyclable plastic in some curbside recycling programs. However, it is difficult and expensive to get rid of the smell of the product that the recycled PP housed in its first life. Some scents are particularly offensive, such as gasoline or moldy yogurt. In addition, the recycled material ends up having black or gray color, which makes it difficult to be reused in packaging. Therefore, the recycling rate of PP plastics is still lower than PET or HDPE plastics.

7.2.2.6 PS
PS plastic is not generally recyclable and accounts for about 35% of US landfill material. Because PS is structurally weak and ultra-lightweight, it breaks up easily and is dispersed readily throughout the natural environment. In addition, PS may leach styrene, a possible human carcinogen, into food products. The

chemicals present in PS have been linked with human health and reproductive system dysfunction. Therefore, most curbside collection services do not recycle PS plastics.

7.2.2.7 Other plastics

Because the plastics labelled as no. 7 under RIC system tend to be mixed plastics, combined with other layers of materials such as foil, or contaminated by food waste, these plastics are nearly impossible to be recycled. An effort to recycle this type of low-quality plastic by melting and reprocessing is a costly process that yields lower-value materials that are unlikely to be profitable to recyclers (Ho, 2020).

7.3 ADVANCED SEGREGATING TECHNOLOGY AND PLASTICS WASTE MANAGEMENT

7.3.1 Current situation

The quantity and demand of plastics steadily increase every year since they can be applied to various applications. However, this fact reflects that plastic waste generated will also increase. About 10% of global plastic production is thermoset plastics (Patoski, 2019), and approximately 17% of packaging film production is multilayer materials (Lahtela *et al.*, 2020). These two types of plastics are considered non-recyclable, most of which ends up in landfills and eventually contaminates the environment.

Globally, only 9% of plastic waste is recycled (Parker, 2018). However, the recycling rates of plastics vary significantly across different countries, waste streams and polymer types. Some polymers are more widely recycled than others. For example, recycling rates for PET and HDPE plastics commonly exceed 10%. This is because of the high-volume and relatively clean waste stream of these plastics, making them relatively easy to recycle (OECD, 2018). According to the USEPA report (2019), 8% of 35 370 tons of plastic waste generated in U.S. in 2017 was recycled whereas about 76% was dumped into landfills. Meanwhile, around 6% of plastic waste from India is non-recyclable which accounts for approximately 0.56 million tons per year (Central Pollution Control Board, 2018).

Practically, some of the theoretical recyclable plastics are not accepted by most recycling centers such as plastic bags. This is due to the possibility of machine clogging during recycling and blown-away residuals into the neighborhood near the facility. Several curbside recycling programs do not allow putting plastic bags or films and Styrofoam into the recycling bins, such as in Salk Lake City, San Diego, Bowie, U.S.A. and Ottawa, Canada (Salt Lake City, 2019; The City of San Diego, N.D.; City of Bowie, 2022; City of Ottawa, N.D.). Therefore, in this context, these plastics are considered as non-recyclable waste. Meanwhile, the plastics are recyclable when they are in Los Angeles, U.S (City of Los Angeles, N.D.). The curbside bins in this city allow residents to dispose of any types of cleaned plastics, including plastics labelled as numbers 1–7, including plastic bags and PS products. In the City of Toronto, Canada, PE film plastic bags are accepted in the City's blue bin recycling program where they are separated using hand sorting and a vacuum system from the rest of the recyclable materials such as glass, metals, paper, other plastics (City of Toronto, 2020). Film plastics are then baled and shipped to reprocessors where they are converted into flakes and pellets that can be used to produce new film plastic products. Another example of the city that collects plastic bags is the recycling program in Madison, U.S. (City of Madison, N.D.). The clean and dry plastic bags made from HDPE or LDPE are acceptable, but food wrap, bubble wrap or the shopping bags, which are often thick white or gray bags, are still an exception. For multilayered plastics, mostly they are non-recyclable plastics. However, in Japan, some of them are collected for recycling to make products such as toy blocks, and the rest are sent to incinerators (Thejapantimes, 2019a).

7.3.2 Non-recyclable plastic waste management

The presence of non-recyclable plastics in the plastic recycling stream results in the rejection of the whole waste. All cannot be reprocessed and become garbage which must be diverted directly to landfills or the incineration process. In addition, the contamination of MSW or other non-recyclable waste in the recyclable plastic stream leads to a loss of opportunity to recycle the good plastics. However, with the proper separation of recyclable plastics from general waste, the number of recyclable plastics contained in the total waste will be reduced which means more plastic waste can be reprocessed. Also, the remainder that needs to be disposed of to landfills will be reduced as well.

However, non-recyclable plastics currently are handled in the same manner as general waste, in which they are sent to landfills or incinerators for final disposal. For example, the Japanese government categorizes non-recyclable plastics including dirty plastics as combustible waste and combines them with other combustible items such as food waste and paper waste. Then, all wastes are sent to incinerators for energy recovery (Thejapantimes, 2019b).

Since approximately 6.3 billion metric tons of plastics have become waste (Parker, 2018), and 79% of this (5 billion metric tons) is accumulating in landfills or spreading into the natural environment as litter, this means that at some point, much of it will end up in the oceans, causing significant pollution and threatening aquatic life, especially in coastline-rich nations with less-developed waste management strategies (Benavides *et al.*, 2017).

7.4 CASE STUDIES OF NON-RECYCLABLE PLASTICS MANAGEMENT AND LESSONS LEARNED

Recently, non-recyclable plastics have become a big challenge worldwide. Both developed and developing countries face the problem of how to deal with the residual or non-recyclable plastics waste after segregating processes. To overcome these problems, the government agencies may have to introduce a solid waste policy starting from minimizing plastic waste generation, implementing effective management and conducting proper disposal/recovery methods.

In the case of developing countries, Thailand has become a major plastic polluter of the ocean, with the majority of the plastic debris found in rivers and seas being single-use plastics that are difficult to recycle using conventional methods. Most non-recyclable plastic waste or contaminated plastic waste is combined and managed together with solid waste. The government has implemented the National Solid Waste Management Master Plan (2016–2021) to promote materials recycling and increase the waste collecting and sanitary landfill rate. In addition to that, the government has also been promoting the usage of alternative energy from the residual or non-recyclable plastics waste by 30% within 20 years (2015–2036). This plan to convert residual plastics waste to generate power promotes this alternative energy for the small and medium local administration in Thailand. The first management method has been applied to adjust the ratios of plastic wastes in waste collecting (increasing waste-collecting rate by 40%), recycling (having recycling rate of 21% in industrial waste and 30% in MSW) and disposal (obtaining 40% of sanitary landfill rate), and these are also expected to reduce the ratios of residual or non-recyclable plastics in plastics waste stream. The second management method has been introduced during 2015–2036 to use residual or non-recyclable plastic wastes generating power. The energy recovery rate and alternative fuel used will increase by recovering the electricity generation from industrial waste and MSW, thermal energy from waste and pyrolysis oil. As a result of both managements, in 2020, the plastic material flow analysis (MFA) showed the ratios of plastics in landfill, open environment and recyclable plastics to consumption as follows:

- Baseline figures were 30.55%, 15.78%, and 23.22%, respectively,
- Figures from the implementation of the first management method were 31.88%, 9.47%, and 29.13%, respectively, and
- Figures from the implementation of the second management method were 32.17%, 10.46%, and 22.13%, respectively.

There was an increase in all three ratios when the baseline situation and the implementation of first management method are compared, benefiting the recycling section. However, an increase in the amount of plastic wastes to landfills indicated that some recyclable plastics cannot be included in recycling processes. This would be due to insufficient understanding, lack of cooperation among the relevant stakeholders and some issues in market prices of plastic materials and a degree of contamination. For the second management method, a large amount of plastic waste was disposed of in landfills because a number of plastic wastes cannot be converted to power generating substances or combustible materials. Additionally, the cost of investment for energy recovery facilities and energy price are the main obstacles that prevent the long-term implementation of the energy recovery. To achieve effective management of non-recyclable plastics, sustainable development plans and goals should be promoted by the government sector (Bureecam *et al.*, 2018).

As for developed countries, alternative recycling pathways were conducted in the United States (2017) with effective management approaches, and waste to fuel and waste to energy are frequently applied as a solution to issues related to residual or non-recyclable plastic wastes. However, these processes have specific technical challenges, and it is important to determine the amount of energy recovered from plastic wastes. If all plastic wastes in the US (32.1 Mt) were recovered to energy, a low generation rate of electricity (0.6 EJ or 0.6 quads) would be observed, equivalent to only 4% of total electricity net generated in the US. Also, another concern of CO_2 emission is that combustion of plastics typically results in greater CO_2 emissions than disposing of plastics in the landfill. Alternatively, converting all the landfilled plastics (28.2 Mt) into fuel via pyrolysis could generate more than 26 gigaliters (6.8 billion gallons) of liquid fuels. These equal to 15% of the distillate fuel oil (diesel) consumed annually in the US. The life cycle for the greenhouse gas emissions from plastic-derived diesel fuel is estimated to be 1%–14% lower than conventional diesel, and the plastic to fuel pathway also has lower emissions per ton of plastic than conventional disposal (landfill and waste to energy) (Heller *et al.*, 2020).

Another case study of how the recovery systems can support mechanical recycling is observed in Japan. In 2018, 2080 kt (kt = thousand tons) of plastic waste was used in mechanical recycling (the process for manufacturing new plastic products using recyclable plastic waste as a raw material). Of this amount, 710 kt was domestic plastic waste (17%). Meanwhile, 1370 kt of industrial plastic waste (30%) was found in the mechanical recycling, which was 2 times greater than that of the domestic plastic waste. This is because of its quality and comparative stability in the plastic waste supply in the mechanical recycling as well a large proportion of generated industrial plastic waste. In the mechanical recycling of the plastic waste, used plastic products and the plastics loss in production and processing were observed to be 1470 and 620 kt, respectively. The 1470 kt of used plastic products consisted of 530, 240, 210, 80, 70 kt of PET bottles, wrapping film, home electrical appliance housings, agricultural plastics and electric-wire covering materials, respectively. The success of the mechanical recycling in Japan can be attributed to law enforcements and efficient cooperation among relevant stakeholders in the recycling mechanisms (Plastic Waste Management Institute, 2019).

REFERENCES

American Composites Manufacturers Association (2022). Materials. http://compositeslab. com/composite-materials/ (accessed 23 November 2020).

Areeprasert C., Asingsamanunt J., Srisawat S., Kaharn J., Inseemeesak B., Phasee P., Khaobang C., Siwakosit W. and Chiemchaisri C. (2017). Municipal plastic waste composition study at transfer station of Bangkok and possibility of its energy recovery by pyrolysis. *Energy Procedia*, **107**, 222–226, https://doi.org/10.1016/j. egypro.2016.12.132

Benavides P. T., Sun P., Han J., Dunn J. B. and Wang M. (2017). Life-cycle analysis of fuels from post-use non-recycled plastics. *Fuel*, **203**, 11–22, https://doi.org/10.1016/j. fuel.2017.04.070

Biswas B., Bandyopadhyay N. R. and Sinha A. (2019). Mechanical and dynamic mechanical properties of unsaturated polyester resin-based composites. In: Unsaturated Polyester Resins, S. Thomas, M. Hosur and C. J. Chirayil (eds.), Elsevier, Amsterdam, Netherlands, pp. 407–434.

Bottone G. (2019). Start-up of the day: Turning non-recyclable plastics into synthetic gas. https://innovationorigins.com/start-up-of-the-day-tuning-non-recyclable-plastics-into-syngas/ (accessed 30 October 2020).

Bureecam C., Chaisomphob T. and Sungsomboon P. Y. (2018). Material flows analysis of plastic in Thailand. *Thermal Science*, **22**(6 Part A), 2379–2388, https://doi. org/10.2298/TSCI160525005B

Central Pollution Control Board (2018). Guidelines for the Disposal of Non-Recyclable Fraction (Multi-Layered) Plastic Waste. Ministry of Environment, Forest and Climate Change, Government of India, Shahdar, Delhi.

Chambon F. and Winter H. H. (1987). Linear viscoelasticity at the gel point of a crosslinking PDMS with imbalanced stoichiometry. *Journal of Rheology*, **31**(8), 683–697, https://doi.org/10.1122/1.549955

City of Bowie (2022). Residential Recycling. https://www.cityofbowie.org/1026/ Residential-Recycling (accessed 7 December 2020).

City of Los Angeles (N.D.). Blue bin collection (recycling). https://www.lacitysan.org/ san/faces/home/portal/s-lsh-wwd/s-lsh-wwd-s/s-lsh-wwd-s-r/s-lsh-wwd-s-r-rybb?_ adf.ctrl-state=qo8a8u6q5_5&_afrLoop=19684233644286427#! (accessed 12 November 2020).

City of Madison (N.D.). Plastic bag recycling. https://www.cityofmadison.com/streets/ recycling/plasticbag.cfm (accessed 7 December 2020).

City of Ottawa (N.D.). Recycling. https://ottawa.ca/en/garbage-and-recycling/recycling (accessed 7 December 2020)

City of Toronto (2020). Plastic bags – recycling process. https://www.toronto.ca/311/ knowledgebase/kb/docs/articles/solid-waste-management-services/processing-and-resource-management/processing-recycling/plastic-bags-recycling-process.html (accessed 7 December 2020).

Cleanriver Recycling Solutions (N.D.). What is Recycling Contamination? 4 Ways to Reduce Contamination and Recycle Right. https://cleanriver.com/how-to-reduce-recycling-contamination/ (accessed 27 November 2020).

Eriksen M. K., Pivnenko K., Olsson M. E. and Astrup T. F. (2018). Contamination in plastic recycling: influence of metals on the quality of reprocessed plastic. *Waste Management*, **79**, 595–606, https://doi.org/10.1016/j.wasman.2018.08.007

Forenge Technologies (N.D.). SMC & BMC Composites. http://www.forenge.com/bmc.html (accessed 10 November 2020).

Future Market Insights (2018). Plastic Recycling Market. https://www.futuremarketinsights.com/reports/plastic-recycling-market (accessed 5 December 2020).

Gere D. and Czigany T. (2020). Future trends of plastic bottle recycling: compatibilization of PET and PLA. *Polymer Testing*, **81**, 106160, https://doi.org/10.1016/j.polymertesting.2019.106160

Grand View Research (2019). Europe Composites Market Analysis Report by Product (Carbon Fiber, Glass Fiber), By Resin (Thermosetting, Thermoplastic), By Manufacturing Process, By End Use, And Segment Forecasts, 2019–2025. https://www.grandviewresearch.com/industry-analysis/europe-composites-market (accessed 23 November 2020).

Grand View Research (2020). Composites Market Size, Share & Trends Analysis Report by Product (Carbon, Glass), By Resin (Thermosetting, Thermoplastics), By Manufacturing Process, By End Use, And Segment Forecasts, 2020–2027. https://www.grandviewresearch.com/industry-analysis/composites-market (accessed 23 November 2020).

Harper A. C. (2003). Handbook of Plastics Technologies: The Complete Guide to Properties and Performance. The McGraw-Hill Companies, Inc., New York, USA. https://www.globalspec.com/reference/64215/203279/8-6-recycling-of-low-density-polyethylene-ldpe-and-linear-low-density-polyethylene-lldpe (accessed 5 December 2020).

Heller M. C., Mazor M. H. and Keoleian G. A. (2020). Plastics in the US: toward a material flow characterization of production, markets and end of life. *Environmental Research Letters*, **15**(9), 094034, https://doi.org/10.1088/1748-9326/ab9e1e

Ho S. (2020). Global Waste Crisis: Let's Talk About Low-Value Plastics. https://www.greenqueen.com.hk/global-waste-crisis-lets-talk-about-low-value-plastics (accessed 6 December 2020).

Hopewell J., Dvorak R. and Kosior E. (2009). Plastics recycling: challenges and opportunities. *Philosophical Transactions of the Royal Society B: Biological Sciences*, **364**(1526), 2115–2126, https://doi.org/10.1098/rstb.2008.0311

Ibrahim I. D., Jamiru T., Sadiku R. E., Kupolati W. K., Agwuncha S. C. and Ekundayo G. (2015). The use of polypropylene in bamboo fibre composites and their mechanical properties–A review. *Journal of Reinforced Plastics and Composites*, **34**(16), 1347–1356.

Jacob A. (2011). Composites can be recycled. *Reinforced Plastics*, **55**(3), 45–46, https://doi.org/10.1016/S0034-3617(11)70079-0

Jeffrey G. (2016). UV Curing of Thermosets Part 11: Using UV Rheology to Monitor Curing – 2. https://polymerinnovationblog.com/uv-curing-thermosets-part-11-using-uv-rheology-monitor-curing-2/ (accessed 8 December 2020)

Kaiser K., Schmid M. and Schlummer M. (2018). Recycling of polymer-based multilayer packaging: a review. *Recycling*, **3**(1), 1, https://doi.org/10.3390/recycling3010001

Karuppiah A. V. (2016). Predicting the Influence of Weave Architecture on the Stress Relaxation Behavior of Woven Composite Using Finite Element Based Micromechanics. Doctoral dissertation, Wichita State University.

Lahtela V., Silwal S. and Kärki T. (2020). Re-processing of multilayer plastic materials as a part of the recycling process: The features of processed multilayer materials. *Polymers*, **12**(11), 2517, https://doi.org/10.3390/polym12112517

La Mantia F. P., Botta L., Morreale M. and Scaffaro R. (2012). Effect of small amounts of poly (lactic acid) on the recycling of poly (ethylene terephthalate) bottles. *Polymer Degradation and Stability*, **97**(1), 21–24.

Malhotra M. (2020). Multilayered Plastics: A Persisting Problem in Plastic Pollution. https://medium.com/age-of-awareness/multilayered-plastics-a-persisting-problem-in-plastic-pollution-4015d95230 (accessed 23 November 2020)

Market Research Future (2019). Global Plastic Recycling Market. https://www.marketresearchfuture.com/reports/plastic-recycling-market-2859 (accessed 5 December 2020)

Milios L., Christensen L. H., McKinnon D., Christensen C., Rasch M. K. and Eriksen M. H. (2018). Plastic recycling in the Nordics: A value chain market analysis. *Waste Management*, **76**, 180–189, https://doi.org/10.1016/j.wasman.2018.03.034

OECD (2018). Improving plastics management: trends, policy responses, and the role of international co-operation and trade. *Environmental Policy Papers*, **12**, 20.

Parker L. (2018). A whopping 91% of plastic isn't recycled. https://www.nationalgeographic.com/news/2017/07/plastic-produced-recycling-waste-ocean-trash-debris-environment/ (accessed 11 November 2020).

Patoski A. (2019). Why Is Most Plastic Not Recycled? https://repurpose.global/letstalktrash/why-is-most-plastic-not-recycled/ (accessed 7 December 2020).

Plastic Waste Management Institute (2019). An Introduction to Plastic Recycling. https://www.pwmi.or.jp/ei/plastic_recycling_2019.pdf (accessed 14 December 2020).

Rabbit Waste Management (2020). The problem with contamination and recycling. https://www.rabbitgroup.co.uk/news/the-problem-with-contamination-and-recycling/ (accessed 27 November 2020).

Recyclebank (2014). Why Can't I Recycle Stuff with Food on It? https://livegreen.recyclebank.com/column/because-you-asked/why-can-t-i-recycle-stuff-with-food-on-it (accessed 2 December 2020).

Ritchie H. (2018). FAQs on Plastics. https://ourworldindata.org/faq-on-plastics (accessed 27 November 2020).

Rubio R. M. (2019). Recycling of PVC – Prospects and Challenges. https://www.ecomena. org/recycling-pvc/ (accessed 6 December 2020).

Sabău E. (2018). Recycling of polymeric composite materials. In: Product Lifecycle Management–Terminology and Applications, R. Udroiu and P. Bere (eds.), IntechOpen, London, UK, pp. 103–121.

Salt Lake City (2019). What Can I Put in My Curbside Recycling Container? https:// www.slc.gov/sustainability/waste-management/curbside/recycling-can/ (accessed 12 November 2020).

Seaman G. (2012). Plastics by the Numbers. https://learn.eartheasy.com/articles/ plastics-by-the-numbers/ (accessed 5 December 2020).

Sedaghat L. (2018). 7 Things You Didn't Know About Plastic (and Recycling). https:// blog.nationalgeographic.org/2018/04/04/7-things-you-didnt-know-about-plastic-and-recycling/ (accessed 11 November 2020).

Sharuddin S. D. A., Abnisa F., Daud W. M. A. W. and Aroua M. K. (2017). Energy recovery from pyrolysis of plastic waste: study on non-recycled plastics (NRP) data as the real measure of plastic waste. *Energy Conversion and Management*, **148**, 925–934, https://doi.org/10.1016/j.enconman.2017.06.046

Stephenson W. (2018). Why plastic recycling is so confusing. https://www.bbc.com/ news/science-environment-45496884 (accessed 27 November 2020).

Sustainable Recycling Industries (SRI) India (2016). Co-processing of non-recyclable hazardous plastic waste in cement kiln. https://www.sustainable-recycling.org/ wp-content/uploads/2018/03/Co-Processing-Non-Recyclable-plastics-in-cement-kiln-25-11-2016.pdf (accessed 5 December 2020).

Tartakowski Z. (2010). Recycling of packaging multilayer films: New materials for technical products. *Resources, Conservation and Recycling*, **55**(2), 167–170, https:// doi.org/10.1016/j.resconrec.2010.09.004

The City of San Diego (N.D.). Curbside Recycling Tips (FAQs). https://www.sandiego. gov/sites/default/files/legacy/environmentalservices/pdf/recycling/recyclingfaqs. pdf (accessed 12 November 2020).

Thejapantimes (2019a). Addressing disposal and recycling systems in Japan. https:// www.japantimes.co.jp/2019/04/29/special-supplements/addressing-disposal-recycling-systems-japan (accessed 7 December 2020).

Thejapantimes (2019b). Japan to ask municipalities to dispose of industrial plastic waste as trash piles up due to China ban. https://www.japantimes.co.jp/news/2019/05/17/ national/japan-ask-municipalities-dispose-industrial-plastic-waste-piles-due-china-ban/ (accessed 8 December 2020).

Thomas Publishing Company (2020). Comparison of Thermoset Versus Thermoplastic Materials. https://www.thomasnet.com/articles/plastics-rubber/thermoset-vs-thermo plastics (accessed 8 October 2020).

USEPA (2019). Plastics: Material-Specific Data. https://www.epa.gov/facts-and-figures-about-materials-waste-and-recycling/plastics-material-specific-data (accessed 11 November 2020).

USEPA (2020). Facts and Figures about Materials, Waste and Recycling. https://www. epa.gov/facts-and-figures-about-materials-waste-and-recycling/plastics-material-specific-data (accessed 5 December 2020).

Wongpanit International Company (2020). Price Sunday 6 December 2020 (Retail Price). http://www.wongpanit.com/ (accessed 6 December 2020).

Yang Y., Boom R., Irion B., van Heerden D. J., Kuiper P. and de Wit H. (2012). Recycling of composite materials. *Chemical Engineering and Processing: Process Intensification*, **51**, 53–68, https://doi.org/10.1016/j.cep.2011.09.007

doi: 10.2166/9781789063448_0315

Chapter 8

Recovery of plastics by dumpsite mining

Pawan Kumar Srikanth and Chettiyappan Visvanathan

Table of Contents

8.1 BACKGROUND

8.1.1 Plastics as the magic material

Plastics are considered as the marvel material of the generation. Ever since their production in the 1930s and 1940s, plastic production and consumption have grown exponentially over the last few decades. The global plastic resin production reached 367 million metric tons (MMT) in 2020 from about 1.5 MMT in 1950 (Tiseo, 2021). There are a multitude of factors which contributed to the acceptance of plastics across the globe, key factors being its lightweight, durability, cost and resource efficiency, and versatility in color, touch and shape (Jambeck *et al.*, 2015). Recent research and development in the field of plastic production and design also made plastics as a standalone product with the other consumer counterparts such as metal, glass, and paper. Plastics are widely used in packaging, construction, automobile, electrical and electronics, and other major sectors. Plastics became the 'Skin of Commerce' in food packaging during the 1950s (Hawkins, 2018). The spread of supermarkets, growth of the local plastics industry, and increased use of domestic freezers and consumption of frozen food across the globe were the factors that influenced the growth in food packaging at the time. Figure 8.1 shows the different polymer types, percentage of total quantity produced worldwide in 2015.

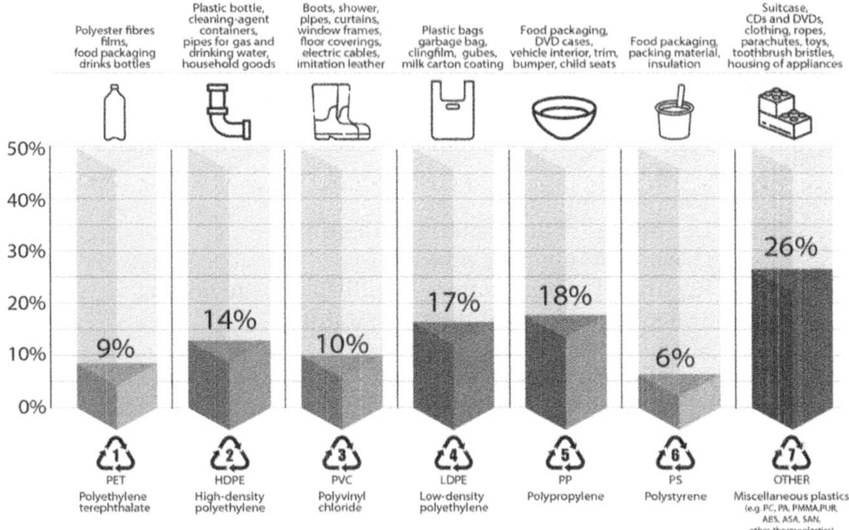

Figure 8.1 Different polymer types, percentage of total quantity produced worldwide in 2015, and their respective use (*Source*: modified from Heinrich Böll foundation; Stiftung, 2019.).

8.1.2 Overview of plastic waste generation and treatment

The unprecedented growth of plastic products and leakage of post-consumer plastic waste have resulted in huge economic and environmental losses threatening various ecosystems. It is estimated that about 2500 MMT of plastics – or 30% of all plastics ever produced – are currently in use. After the consumption of plastic, the post-consumer plastic waste can be classified into managed plastic waste and mismanaged plastic waste, both of which are present in the global ecosystem. Between 1950 and 2015 of 8300 MMT of virgin plastics produced, 6300 MMT of cumulative plastic waste was generated from primary and secondary (recycled) sources. Those 6300 MMT of managed plastic wastes included plastics treated by incineration and recycling. As of 2015, the incineration removed 800 MMT (12%) from the earth, and only 600 MMT (9%) was brought back into the closed loop by recycling. Meanwhile, about 4900 MMT of the remaining plastics produced were discarded and accumulated in the landfill or dumpsites or in the natural environment (Geyer *et al.*, 2017). Packaging-related plastics are short in the use phase, and their composition in municipal plastic waste is higher than other types of plastic waste. Urban litter, open dumps and transportation of plastics via runoff and wind constitute the mismanaged plastic waste. In 2018 alone, out of 454.3 MMT plastics produced, 342.6 MMT of plastic waste was produced, the majority of which was discarded (Figure 8.2).

Studies show that less than 20% of plastic leakage is from ocean-based sources such as fisheries and fishing vessels. This means over 80% of ocean plastic comes from land-based sources. Scientific evidence shows that the top five countries causing land-based plastic waste leakages are China, Indonesia, Philippines, Thailand and Vietnam. These developing Asian countries are in the stage of intensive economic growth and the demand for safe disposal of

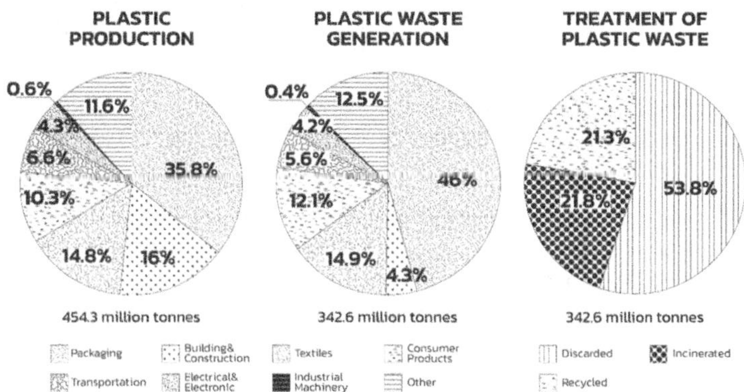

Figure 8.2 Global plastic production, plastic waste generation and plastic waste treatment by sector, 2018 (*Source*: Geyer, 2020).

plastic waste is much higher compared to the previous decades. The problem, coupled with the lack of adequate local waste management infrastructure to manage plastic wastes, results in such magnitudes of plastic waste ending up in dumpsites and landfills.

Information Box – World's 50 Biggest Dumpsites

The first attempt to find the top 50 dumpsites in the World was carried out globally (Waste Atlas, 2014). Based on the report, each of the dumpsites received a higher volume of waste each year; few of them are physically or historically larger than another or have more people working on them, or living within 10 km 18 were in Africa, 17 in Asia, 2 in Europe, 8 in South America and 5 in Central America or the Caribbean. China was not included in the report as getting accurate statistics was impossible. A total of around 60 million people were affected by the top 50 dumpsites, and rivers, lakes and oceans were also polluted.

Difference between a dumpsite and a landfill
Misinterpretation of definitions of dumpsites and landfills
There is a common misconception or misinterpretation in the scientific literature regarding the usage of terms 'dumpsites' and 'landfills. In the developing low- and middle-income countries, dumping of waste in low-lying areas is a common practice.

Dumpsites
Dumpsites are often on-land throw-away areas where solid waste is discarded in an uncontrolled manner, which does not protect the environment. It completely lacks any proper infrastructural elements, operational practices and monitoring. Usually, in many developing countries dumpsites are called 'landfills', although they do not meet the minimum criteria of environmental protection. Dumpsites are often closed and rehabilitated.

Open Dumpsites and Controlled Dumpsites are commonly used approaches globally.

(a) **Open dumpsites:** unplanned and often improperly sited open dumping grounds with no or minimum management.

(b) **Controlled dumpsites:** open dumping of waste with minimum level of planning and management of space and with visible fencing.

Landfills
Sanitary landfills are scientifically planned disposal sites which isolate the solid waste from the environment until the wastes get stabilized and are rendered harmless. In sanitary landfills, the wastes are disposed in a specifically designed infrastructure that involves environmental

protection with engineered equipment and structural elements, and where operational practices and wastes control are properly implemented. It is designed to control and avoid surface water entering the deposited waste by installing a well-designed and constructed surface drainage system. Also, additional lining materials are installed to the site to reduce leakage from the base of the site (leachate) and help reduce contamination of groundwater and surrounding environment.

Among the top five countries polluting the oceans (Jambeck *et al.*, 2015), it is also found that 75% of plastic waste is from uncollected sources while the remaining 25% leaks from waste management systems themselves. Low-quality plastics such as films and composites are rarely picked up by the informal sector. High-quality plastics such as HDPE and PET are picked up for residual value recovery. This segment forms only about 20% of the municipal plastic-waste stream, and the remaining fraction becomes too unattractive for material-to-material recycling. Urban areas in developing countries have high plastic waste generation rates due to intensive economic growth in the last few decades. Higher collection rates in urban locations would make it possible for these countries to meet the high, consistent level of throughput for capital-intensive technologies such as the incineration, pyrolysis and gasification (Ocean Conservancy, 2015).

8.1.2.1 *Journey of plastics to dumpsites and landfills*
A typical waste composition of the raw municipal solid waste (MSW) in developing countries like Thailand, India and Sri Lanka consists of 60% organic fraction (including food wastes, garden and yard trimmings, wood wastes), 10–15% plastics and 10–20% recyclable materials (including paper, glass, metals, rubber/leather products). Waste collection is one of the common services provided at the local levels. Door-to-door and community level collection are the major waste collection types. The collected waste is transported for final disposal. Waste collection rates are higher in urban areas than in rural areas in many parts of many countries. In case of low-income and lower-middle-income countries, nearly 80–90% of waste disposal is by means of open dumping and landfilling (Kaza *et al.*, 2018).

Most of the high-value recyclable fractions such as glasses, metals, PET and HDPE bottles, and paper are captured at a material recovery facility (MRF) or by informal recyclers who recover the valuables and process them for recycling. On the other hand, low-quality plastic types are often disposed of in low-lying areas along with the other invaluable waste fractions. It is highly important to recover the valuable plastics before reaching the dumpsites as their value might be lost. The composition of plastics in dumpsites depends on various factors including the consumption levels and patterns, national policies, technologies for treatment of waste, recycling programs and awareness levels of the citizens. Figure 8.3 shows the typical composition of plastics in MSW.

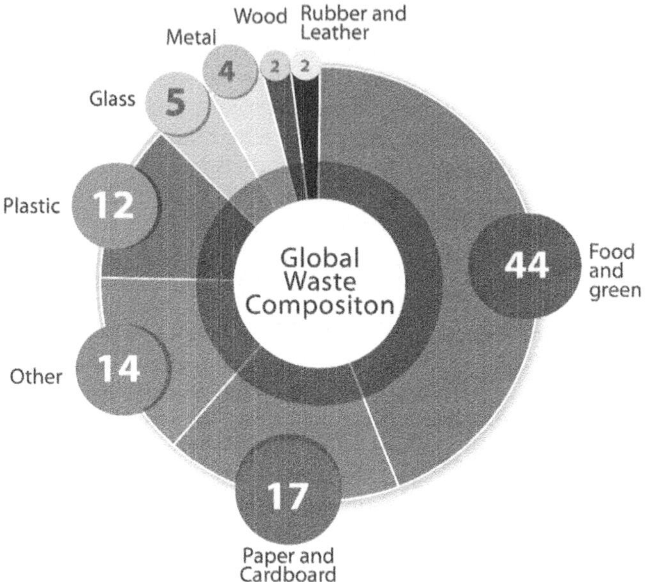

Figure 8.3 Global waste composition and share of plastics in MSW (in %) (*Source*: modified from Kaza *et al.*, 2018).

8.1.3 Issues and opportunities related to dumpsites in developing countries

8.1.3.1 Unavailability of fencing system

Since it is an open land with no fencing, the surrounding community can earn income through waste-picking and contribution to local recycling. Waste pickers practice selective picking (taking valuable items like plastics, metals) and in this way provide waste turning (mixing of wastes) that accelerate the degradation of biodegradable materials. Along with waste pickers, these sites attract animals and rodents (cattle, dogs, rats, crows, etc.). Also, they are breeding grounds for vectors like mosquitos and flies. Particularly, plastic waste holds water and facilitates vector breeding at these places.

Information Box – Plastics Compositional Variations in Dumpsites

Different types of waste degrade at different rates. After years of dumping, the organic biodegradable fraction of the solid waste generally undergoes anaerobic degradation. There are variations in composition of MSW from the place of disposal, at the point of entry in the dumpsites, and in the excavated waste from dumpsites (Sharma *et al.*, 2019).

> (a) Waste composition survey carried out at Nong-Khae and Nakhon Nayok in Thailand showed reduction in composition of organic waste in the excavated waste samples compared to the raw MSW composition. The average compositions of the organic and the plastic fractions in the fresh MSW were about 64% and 18%, respectively. The reduction of organic composition resulted in increases in the composition of plastics in Nong-Khae and Nakhon Nayok dumpsites by 45% and 39%, respectively.
>
> (b) In another study conducted at Klong Sam dumpsite, Pathum Thani, Thailand, the average composition of plastics in fresh MSW increased from 18% to nearly 35–40% in excavated MSW after months of biodegradable waste degradation.
>
> 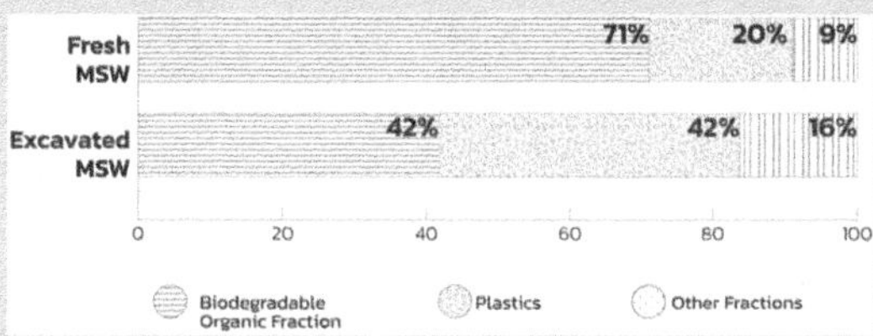
>
>
> Changes in plastic waste composition in fresh and excavated MSW (*Source*: Sharma *et al.*, 2019).

Picking up waste with bare hands and feet may lead to direct exposure to toxic materials in the hot and wet Asian climate. This can impact the health of the waste pickers causing skin corrosion or irritation, respiration hazard, and serious eye damage or eye irritation. Figure 8.4 shows the four main issues related to dumpsites, namely health, economy, safety and environment.

8.1.3.2 Lack of compaction of dumpsite waste

The collected MSW from various locations are spread over a wide area in various parts of the dumpsites and often not compacted using heavy machinery or compactors. There are multiple issues caused by the lack of compaction, leading to multiple hindrances at dumpsites.

(a) **Reduction in dumpsite space**: Lack of compaction leads to reduction of dumpsite space.

(b) **Frequent fires**: The aerobic and anaerobic decomposition of dumpsite waste will result in generation of gases and heat within the void spaces. Especially during the summer periods, the combustible substances in dumpsites catch fire at elevated temperatures and frequent fire may occur. Dumpsite fire also occurs at side slopes, which becomes difficult for firefighters to access and extinguish the fire.

Figure 8.4 Four important aspects of dumpsite issues.

(c) **Lack of liner systems**: Due to lack of liner systems, the level of environment pollution is high. Leachate from the waste can contaminate soil and water, and the greenhouse gases (GHGs) may escape to the atmosphere and contribute to global warming.

8.1.3.3 Resource recovery from the open dumping

Mining of old dumping for refuse derived fuel (or feedstock recovery) is possible, although efficiency is low due to the poor quality of waste, and the process itself may contribute to production of GHGs. The informal waste sector has a major role in plastic recycling and reduces the processing cost in recycling waste. Waste turning or mixing accelerates the fragmentation of plastics contaminating the surrounding environment with microplastics and hazardous chemicals.

8.2 PLASTICS IN DUMPSITES AND LANDFILLS

8.2.1 Phases of degradation of MSW in landfill/dumpsites

The degradation of MSW and plastics in a typical landfill or dumpsite occurs in five phases, namely initial adjustment, transition, acid formation, methane fermentation, and maturation.

(a) *The initial adjustment phase*: Development of sufficient moisture supports the active microbial community resulting in biochemical decomposition of MSW and plastics. In this phase, the degradation of plastics will be slow because the pH is quite low.

(b) *The transition phase*: As the waste is dumped in large quantities, there is an insufficient supply of oxygen to the central layers of waste. Hence, the transformation from an aerobic to anaerobic environment occurs. In this phase, the microplastics generation tends to occur more compared to the initial adjustment phase.

(c) *The acid formation phase*: Continuous hydrolysis and microbial conversion of biodegradable organic content results in the production of carboxylic acids and reduction of pH. The leachate contains high chemical oxygen demand (COD) and biochemical oxygen demand (BOD) during the phase.

(d) *The methane fermentation phase*: Methanogenic bacteria convert intermediate acids to methane and carbon dioxide resulting in an increase in pH.

(e) *The maturation phase*: This is the final stage of waste stabilization which brings back the biological activity to relative dormancy with reduction in gas production while leachate stays steady at lower concentrations.

8.2.2 Degradation of plastics in dumpsites

Due to prolonged exposure to environmental factors such as heat, light, moisture or microbial activity causing degradation of plastics in dumpsites, the plastics may be cleaved into smaller molecules, and lead to the formation of plastic particles less than 5 mm 'microplastics'. Studies show that different plastic types such as HDPE and LDPE have different plastic degradation rates and degrade at a slower rate in the marine environment than in the terrestrial environment. In some situations, plastics under sunlight can experience 'heat buildup', reaching temperatures higher than the surrounding air and resulting in accelerated plastic degradation. Figure 8.5 shows the leakage of microplastics from dumpsites.

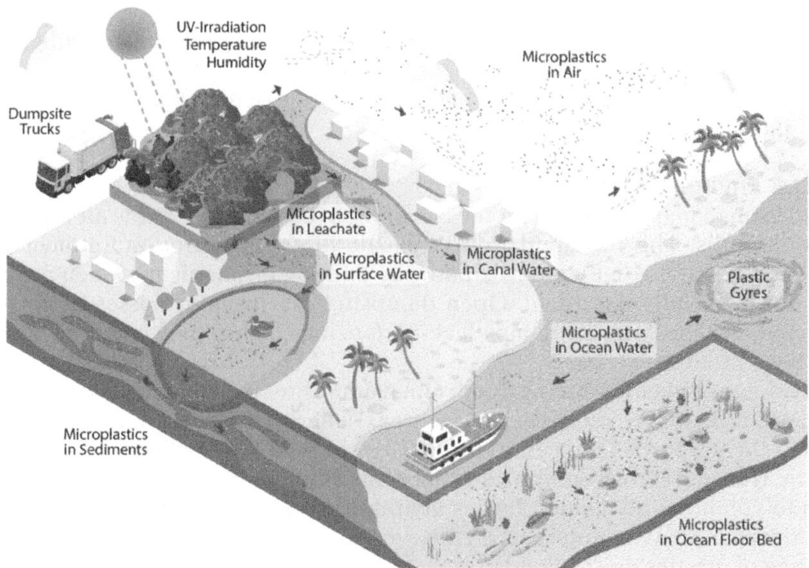

Figure 8.5 Leakage of microplastics from dumpsites.

(a) **Temperatures:** Temperatures in some dumpsites or landfills reach 80–100°C, accelerating degradation, if oxygen and/or moisture are present for the thermal-oxidative degradation and hydrolysis pathways, respectively.

(b) **UV Irradiation:** Solar UV irradiation is necessary to initiate photo-oxidation of most polymers. There are several studies which establish the relationship between photo-oxidation and biodegradation of polyethylene. When polyethylene is subjected to UV radiation for a long period of around 60 hours, photo-oxidation occurs and the polymer increases its carbonyl groups. Once the polymer chain becomes extremely short after a considerable amount of time, microorganisms can start digesting them as their food source.

(c) **Transient thermal treatment:** Moderate heating in air enhances hydrophilicity to incorporate oxygen moieties in the polyolefin (polyethylene or polypropylene). This process subsequently increases oxidative degradation.

(d) **Humidity:** With an increase in humidity or moisture, polyester undergoes hydrolysis. For example, depolymerization of PET in a plastic bottle was observed to be 5 times greater at 60°C and 100%–45% relative humidity. Humidity also causes degradation of PVC, PP and PE by promoting an increase in the concentration of hydroxyl radicals, completely changing the original elemental composition of the polymer.

8.2.3 Leakage of plastics from dumpsites

Plastics potentially leak from the disposal sites. There are no studies accounting for the leak from dumpsites or landfills in a consistent and comprehensive manner. The common pathways are as follows.

(a) Environmental processes including the effects of wind, flooding, and precipitation.

(b) Influence of animals and relevant anthropogenic causes, representing influence of waste pickers.

8.2.3.1 Environmental factors

Wind forms a major pathway for the transport of odors and possible air pollution from landfills. The contamination of plastic in groundwater and atmosphere has been recorded, but wind as a pathway and the mechanism of such process has not been fully described. Open dumpsites are prone to losses of plastics by wind, especially open dumps lacking fencing to stop movement of macro-plastics (Yadav *et al.*, 2020).

Leakage of plastics during flood: Open dumpsites in low-income and middle-income developing countries are in areas where frequent flooding tends to occur. The properties such as mass, density, shape and size of plastics influence plastic transport from dumpsites or landfills through flooding. Higher surface area-to-volume ratio makes plastics float more readily in water compared with other denser MSW counterparts. However, to date, no study has quantified the leakage of plastics due to floods.

Leakage of plastics due to leaching and surface runoff:
There are three main factors which influence the release of macro-plastics and micro-plastics from dumpsites.

(a) Quantity of leachate and surface runoff generated
(b) Concentration of microplastics in the leachate and surface runoff
(c) Presence of physical barriers preventing macro-plastics and micro-plastics from spreading into the environment.

Although substantial evidence is needed, the concentration of micro-plastics in leachate may be minor to negligible compared to urban runoff.

8.2.3.2 Influence of animals and anthropogenic activities

Species with beaks and dexterous hands can pick out food from inedible materials. This may also enable such species to consume plastics by accident as they may not be able to distinguish plastics and food. To date, no study has quantified such impacts on the ecosystem. Further, environmental losses of plastics due to anthropogenic causes can occur as ragpickers visit landfills to scavenge recyclable materials including plastics. They collect only recyclable plastics but in the process expose non-recyclable plastics to other pathways such as flooding or wind.

Information Box – Contribution of Informal Recyclers in Developing Countries

Informal recyclers play a major role in plastic recycling, especially in highly populous developing countries like China and India. The composition of valuable recyclables in dumpsites reduces where the involvement of informal recyclers is higher. In the Philippines, waste pickers recover all the wastes which are of high-value at rates close to 100%. The PET bottles are high-value-residual products which are commonly collected at a rate of 90% and have a resale price. Particularly in the Philippines, waste picking takes place at four distinct points in the waste-management system: pre-collection at the household or street level; during the hauling process; at material recovery facilities; and finally at dumpsites or landfills. The second and third level entry points are considered as the most profitable for the waste pickers. The informal waste recyclers are either prepared to pay for their entry at these points or provide manual MRF function in exchange for the right to operate in these facilities, in which they are not officially allowed to function. The high success of plastics recycling in developing countries like India and China is attributed to informal recyclers in the plastics value chain.

There is a high level of risks associated with recovering valuables in parts of the waste management system. The more the waste gets delayed in getting segregated, the more serious the problem. Often, the informal waste pickers are unequipped with gloves, masks, shoes and other safety equipment to handle products of high levels of toxicity (e.g., household hazardous wastes and sometimes medical waste). A casual interaction with the informal waste pickers might give experiences regarding the garbage slide and injuries due to sharp objects, among others. The waste in the dumpsite can even suddenly catch fire due to methane generation and exposure to sunlight in the presence of combustible substances.

8.3 DUMPSITE MINING

8.3.1 A strategic tool for dumpsite rehabilitation and plastics recovery

Landfill and dumpsite reclamation is a sustainable tool to rehabilitate and close or remediate, operate the landfills or dumpsites upon the scientific assessment involving technical investigations and Environmental Impact Assessment (EIAs) including consultation with the interested and affected parties, specifically adjacent communities. Landfill and Dumpsite Mining is one of the best methods of landfill and dumpsite rehabilitation and helps in perceiving the landfills and dumpsites as a resource to extract valuables from, rather than a risk to be feared.

8.3.2 Dumpsite mining process

The process of landfill and dumpsite mining involves a series of steps ranging from the excavation of waste, stabilizing, screening, sorting and safe disposal of recovered fractions. Figure 8.6 illustrates the various steps involved in the dumpsite mining operations. The eight steps are as follows.

(a) The solid wastes are excavated using heavy machinery such as excavators or bulldozer and stabilized before the sorting and separation of wastes via the mechanical or manual screening process.

(b) The stabilization process is carried out using composting bio culture periodically, after which the wastes are fed into mechanical screeners for segregations.

(c) The mechanical segregation involves the separation of stabilized wastes by trommels, vibrosievers, ballistic separators, or air density separators, in combination with each other depending upon the composition of waste and output of recovered products desired. The screen opening size of the mechanical separators varies with the projects.

(d) At the end of landfill and dumpsite mining, the biodegradable fraction, which dominates the solid waste in landfills or dumpsites, is recovered as bio earth or fine sand. The recovered bio earth or fine sand is rich in nutrients beneficial for plant growth upon microbial action and can be commercially sold as agricultural manure.

(e) The quality of the plastics obtained after mining activities is not good enough to readily be recycled, and pretreatment technologies are required to segregate the plastics from other adhered waste streams which are not commercially viable. The calorific value of recovered

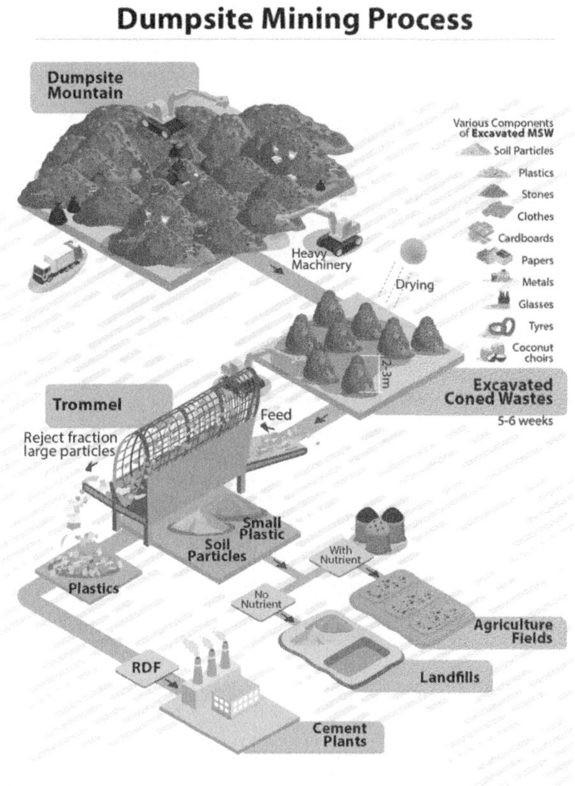

Dumpsite Mining Process

Figure 8.6 Dumpsite mining processes and recovered materials.

plastics is between 16 and 19 GJ/ton (Sharma *et al.*, 2019), and the recovered plastic fractions can be used to generate energy and used as co-fuel in cement kilns.

(f) Upon mining of clusters of dumpsites and unsanitary landfills, a Memorandum of Understanding (MoU) can be signed between the local governments and the cement kilns, and the recovered plastics can be safely disposed of. The environmental issues concerning the waste produced during the re-use of plastics and emissions during cement productions should be considered and addressed sustainably. Hence, through landfill and dumpsite mining, the wastes which are dumped for no specific use have been valorized, and environmental concerns have been addressed sustainably. The reclaimed land should be prohibited from further dumping of wastes as it devalues the purpose of mining activities.

(g) In developing countries, this simple technique can be carried out and incentivized on a larger scale and it is highly recommended for small island developing states (SIDS) such as Sri Lanka, Maldives and

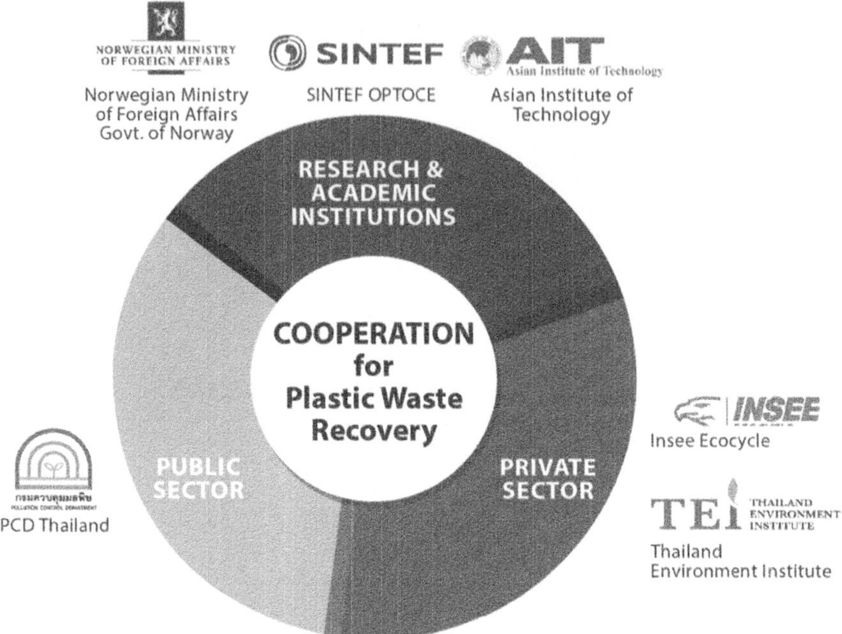

Figure 8.7 Triangular cooperation for plastics recovery in Thailand.

Madagascar, as the risk concerning plastic accumulation in the nearby marine ecosystems are relatively higher. In developing countries, triangular cooperation can be formed between the public sector, the private sector and research & academic organizations to facilitate the recovery of plastics from dumpsites. Figure 8.7 shows the triangular cooperation of Governments of Thailand and Norway collaborating with private and academic organizations in Thailand.

8.3.3 Definitions of dumpsite/landfill mining

Landfill mining is a process of excavating a landfill using conventional surface mining technology to recover metals, glass, plastics, soils and the land resource itself. However, this definition is not complete as multiple aspects of landfill mining are not taken into consideration. Also, landfill mining is defined as 'the excavation and treatment of waste from an active or inactive landfill for one or more of the following purposes: conservation of landfill space, reduction in landfill areas, elimination of a potential contamination source, mitigation of an existing contamination source, energy recovery from excavated waste, reuse of recovered materials, reduction in waste management system costs and site re-development'. In addition to the above-mentioned purposes, creation and development of socio-economic opportunities integrating local knowledge and labor is one amongst them. It helps in strengthening the community involvement in the management of waste.

8.3.4 Drivers for dumpsite mining

Landfill or dumpsite mining was first introduced to obtain fertilizers for orchards in 1953 in Israel. Stricter environmental regulations in U.S. Subtitle D regulations on management of non-hazardous solid waste in the 1990s pushed many landfills to close and manage long-term monitoring of pollutants. The closure of landfills and difficulty in getting permission for the new ones paved the way for solutions to the problem by excavating, processing, treatment and recovery of landfilled materials (Krook *et al.*, 2012).

In European and Asian developing urban cities, the dumpsites were hampering the city-level development and remediation of such sites could potentially reduce the environmental pollution associated with it while reclaiming land for recreational and commercial purposes. Often studies carried out during the 1990s found that it was often difficult to obtain high-quality marketable deposits. The recovered materials sometimes require pre-treatment and pollutant removal treatment techniques to make it suitable for utilization.

8.3.5 Recovered materials from landfill/dumpsite mining

There are various fractions which can be recovered from landfill/dumpsite mining projects. The quantity and quality of recovered products are one of the major drivers of dumpsite mining projects. Investigation of landfill/dumpsites revealed that interesting amount of iron (Fe) plus non-Fe metals, could be recovered from excavated waste if proper segregation equipment and processing techniques are employed. The metals are more observed in European countries' dumpsites than Asian counterparts as the high population of informal recyclers recover the valuables before their disposal in dumpsites.

Apart from metals, soil-like fractions and inert-fractions are commonly observed in dumpsites. The inert fractions consist of stones, glass, construction and demolition waste, and ceramics. It is also observed that the presence of large amounts of fine fractions in excavated waste can be explained by the use of daily or intermediate covers in landfills while a low amount of fine fractions could be related to open dumpsites. The soil-like fractions are often used as ground substitute, such as cover material for operational landfills, soil for non-edible crops and formation of biosoils to be used in environmental remediation activities. This fraction could also be used as fertilizer or composts upon the compliance of regulations.

Particle, moisture content, pH, carbon to nitrogen ratio and electrical conductivity are the main parameters for utilization of compost. Often the dumping of hazardous waste leaches heavy metals to the soil-type particles. Hence, the testing of heavy metals in soil-type particles such as arsenic, cadmium, chromium, copper, lead and mercury should be conducted. Based on the USEPA and EU standards for incorporation as the compost material, fine fractions complied with all the physical and chemical parameters (including the heavy metals) for cultivation of non-edible crops.

The waste to material conversion is a possible option for inert fractions such as stones, glass, ceramics, and construction and demolition wastes if they are

Figure 8.8 Recovered materials from dumpsite mining.

separated properly. When the quality of paper and cardboard, plastics, textiles, and wood recovered from dumpsites is too low or level of contamination is high, waste to energy may be the most suitable option. Figure 8.8 shows the recovered materials from the dumpsite mining.

8.3.6 Trajectory of landfill/dumpsite mining

The practices of landfill/dumpsite mining have evolved over the years. The practices vary in the overall objectives addressed, drivers for landfill mining, materials recovered, technology, environmental pressures and economic feasibility. Modern equipment and machineries aids the dumpsite mining in terms of efficient segregation, transportation and safe handling of recovered materials. The quality control and compliance of recovered materials are

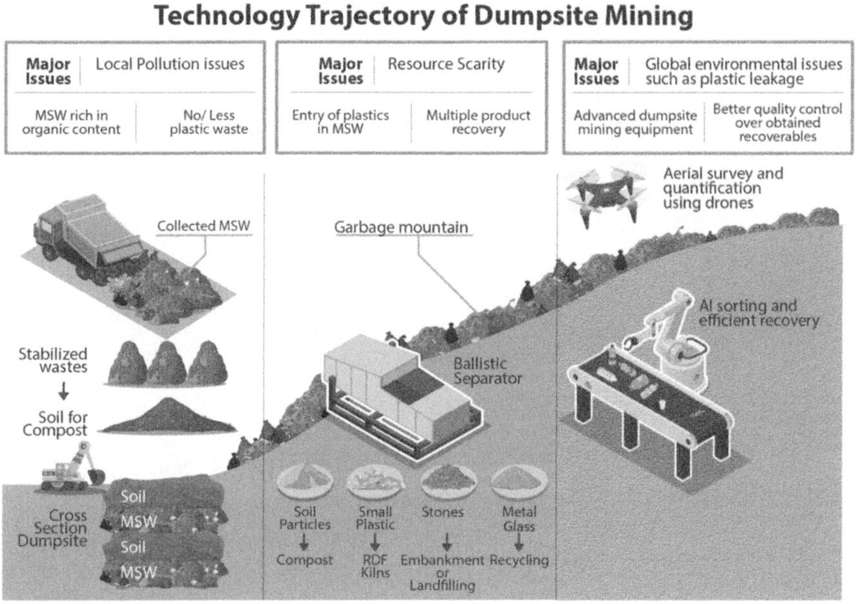

Figure 8.9 Dumpsite mining technology trajectory.

stricter than those of the landfill mining projects done in the past. Figure 8.9 shows the trajectory of dumpsite mining in the past, the present and the future.

8.3.6.1 Landfill mining in the past

The three main objectives of the landfill/dumpsite mining carried out in the past included extension of landfill lifetime, consolidation of landfill areas, and facilitating final closure and remediation. The screening of excavated waste resulted in soil-type material while recovering the marketable recyclables was proven to be ineffective. The composition of plastics was found to be very low or even absent. The common practice of management of MSW is filling the excavated pits or trenches with alternative layers of MSW with soil for stabilization. The quality of recovered soil was agricultural compost-quality. A simple screening equipment such as rotary drum-type screener or vibrosiever were used for the separation of soil-type products. Figure 8.10 shows the typical dumpsite mining carried out in the past.

The local pollution issues were the common drivers of the landfill mining. Manual labors were favored as operation and utilization of advanced dumpsite mining machinery was limited. The resources recovered were used locally and sold without much pre-treatment. During this period, the landfill/dumpsite mining was seen more from an extension of waste management perspective, and the resource estimation and extraction were given a low priority.

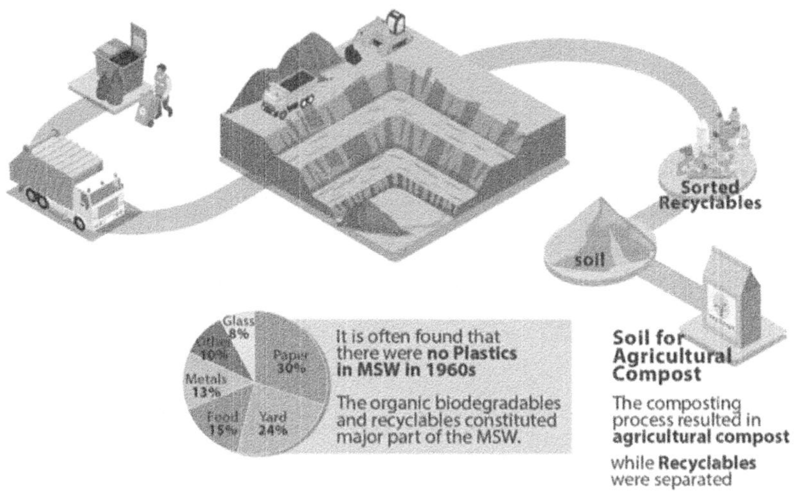

Figure 8.10 Dumpsite mining in the past generation.

8.3.6.2 *Landfill mining in the present*

The increased demand for high-quality alternative materials, global environmental issues, volatile pricing of scarce natural raw materials, availability of heavy machinery for handling volume of excavated wastes, and easy transportation of recovered materials have shifted the way the landfill/dumpsite mining is seen these years. The recovery of raw materials for production of energy, alternative fuel and other valuable products have necessitated the use of state-of-the-art heavy and stationary or mobile machineries to excavate and segregate waste. The recovered materials are then scientifically tested and utilized in compliance with national/local regulations.

The machineries can be remotely controlled with less human intervention. The efficient operation results in larger quantities of recovered fractions with or without pre-treatment depending upon its use. The current dumpsite/landfill mining plants recover a considerable volume of recyclables and secondary raw materials which can be used as waste fuel. The total quantity of plastics in the dumpsites has increased over time, and these plastics are not suitable for material-to-material conversion due to the presence of impurities and poor quality among others. Hence, the focus of dumpsite mining has moved from recovering soil-type fractions to plastic fractions as they can be used as a substitute for coal in cement kilns. Figure 8.11 shows the typical dumpsite mining carried out in the present generation.

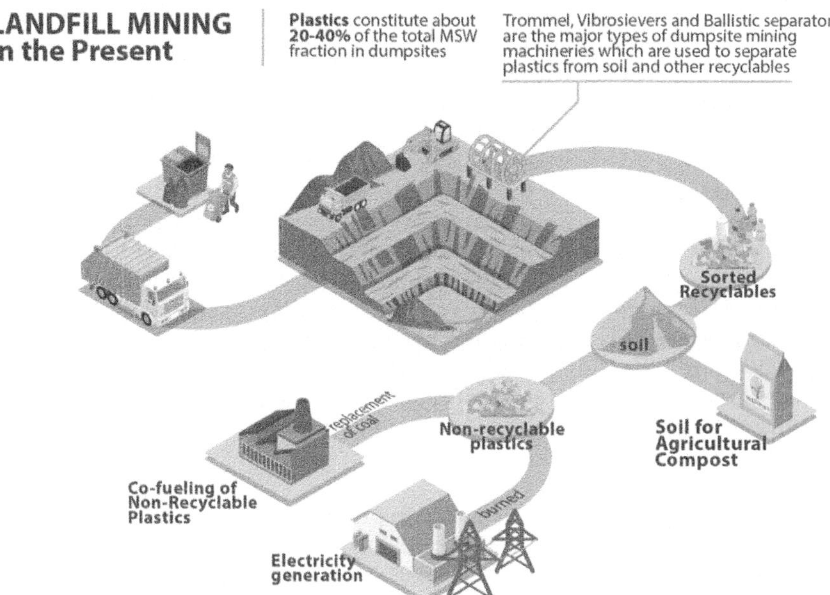

Figure 8.11 Dumpsite mining in the present generation.

8.3.7 Machineries in dumpsite mining

The equipment which is currently used in dumpsite mining plants can be broadly classified into the following categories:

(a) Waste excavation from dumpsites,
(b) Material handling machineries,
(c) Screening equipment,
(d) Transportation of excavated and recovered materials, and
(e) In-plant hauling of excavated fractions.

Among the different machineries, a detailed summary of popular screening machineries is listed below.

8.3.7.1 *Vibrosievers*

Material to be sorted is fed from the top into the machine. Inside the vibrosievers, there are single or multiple inclined plates arranged parallel to each other at a fixed distance in a state of constant vibration. The screen pore size depends on the type of recovered materials needed. In a single plate vibratory screener type, the underflow is usually soil-type fractions while the overflow can be plastics, stones or any other fractions of considerable size. The vibratory screen-type machineries are constantly found to cause operational issues and are frequently clogged with soil particles. This requires frequent scrapping using wire meshes to clean the pores. In addition to that, the vibratory screeners often need loader trucks to feed the recovered materials into the dumped trucks.

8.3.7.2 Trommel systems

Trommel screens are cylindrical drum-type mechanical screening systems attached to the conveyor belts and scrubber meshes. Waste is fed into the trommel systems which are in the state of constant rotation. After sufficient movement of wastes within the trommel, the fractions smaller than the pore size (underflow) of the mesh pass through and are carried out of the system by conveyor belts. The overflow of the trommel screen (fractions greater than the size of the screen) consists of relatively larger fractions such as stones, leather materials, metals, glass products and wooden logs. These are further separated and sold for reutilization. Usually, a series of trommel screener systems is employed to obtain the materials of required sizes and quality. Comparing the vibratory screener–type systems, the trommel systems are more efficient.

8.3.7.3 Magnetic separators

Metals usually constitute a very small proportion compared to the other fractions, and their quality is found to be poor. The overband magnet or electromagnet is typically employed over a running conveyor belt which lifts off the ferrous metals and carries them to a separate container away from the remaining recovered fractions. At times, the quantities of metals obtained do not justify the costs associated with them. It is better to carry out compositional studies in the dumpsites to estimate the metal composition and overall metal recovery prior to the installation of magnetic systems in dumpsite mining plants.

8.3.7.4 Air density separators

Typically, air density separator systems are usually provided at the end of conveyor belts to separate the materials based on the density differences. They include windshifters, separation drums, air classifiers and air knives.

(1) *Windshifters*: Windshifters separate the lighter fraction from the waste exiting the conveyor belts by either sucking or blowing the air.
(2) *Air classifiers*: Air classifiers create vortex flows and centrifugal forces to separate materials based on the cyclone principle. These are rarely used in dumpsite mining plants and are often used in composting operations.
(3) *Air knives*: Air knives create high velocity air operating either vertically or sideways to strip off light materials from the conveyor belts. These are typically used in dumpsite mining plants with fractions such as plastics to separate them from stones, glasses or ceramics.

Inclusion of air technologies requires understanding of the nature of waste to be treated. Understanding waste composition in dumpsites will be key in the installation of suitable air density separator systems. Suitable trial runs can be conducted to optimize the operation, and the volume and pressure of the air knives can be adjusted to separate the required valuables.

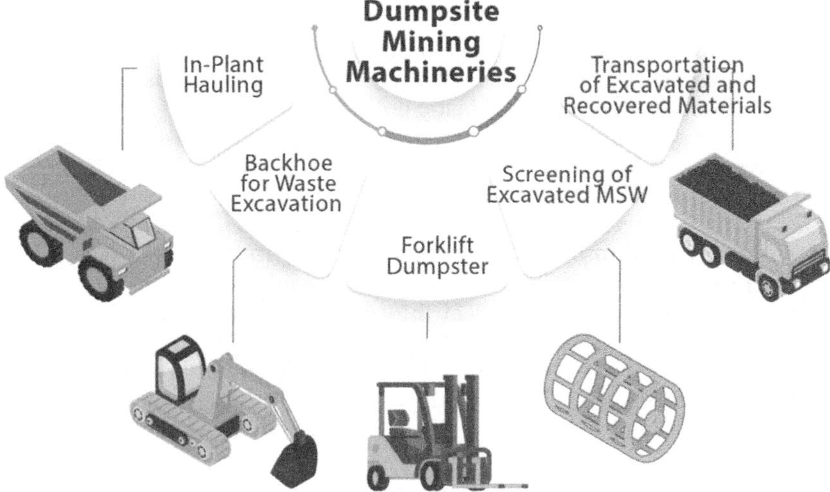

Figure 8.12 Dumpsite mining machineries.

8.3.7.5 Ballistic separators

Ballistic separators are standard MRF processing equipment with oscillating paddles that run the length of the sorting deck. Usually, the ballistic separators take a more compact footprint compared with other screening equipment. They can be highly effective in urban areas where the dumpsite areas take up a lot of space, and less space is available for machinery installation. The material to be sorted is fed from the top into the machine. RDF suitable materials being the lighter fraction are removed from one end while the fraction of higher bulk density such as construction and demolition waste is obtained from the other end. The soil fractions pass through the screen and are separately collected. Figure 8.12 gives the outline for the list of machineries used in dumpsite mining.

8.3.8 Issues due to dumpsite mining

Many issues have been reported due to the current dumpsite mining projects. The aspects of noise, air pollution and workers safety are the least addressed issues for the current dumpsite mining. The policy interventions must be done at the global, regional and national levels in order to quantify amount of plastics from dumpsites, provide technological solutions, and promote state-of-the-art technological tools to be used in mining in an economically sustainable way.

8.4 TECHNOLOGICAL, POLICY, AND FINANCIAL GAPS IN DUMPSITE MINING

8.4.1 Technological gaps in dumpsite mining

Comprehensive data on plastic waste management practices that allow progress towards providing effective plastic waste management for all (including collection systems and appropriate disposal) have yet to be obtained.

There is a lack of key data for dumpsite mining. Some important data are:

(a) Waste composition data including plastic waste in dumpsites which serve as the critical data establishing the need for dumpsite mining and its relationship to marine plastic pollution,
(b) History of dumping and the method of dumping of wastes, and
(c) Presence of heavy metals and other toxic compounds which decrease the quality of plastics over time.

8.4.1.1 *Identifying dumpsite hotspots for intervention*

It is essential to identify key hotspots that require action now (e.g., key dumpsites). It is important to collect information related to cities, rivers and ports to assess hotspots of marine plastic pollution from dumpsites. A simple intervention tool should be developed as it will allow practitioners working at a local level to identify the best intervention point and gain access to best practice and technical guidance.

8.4.1.2 *Advanced technologies to quantify waste*

It is often tricky to measure the volume of plastics mined from dumpsites at a given period of time. The volume of plastics mined and the remaining waste to be mined has to be calculated periodically as dumpsite mining operations take place over a period of time. Drones or other aerial visualizations can be employed to accurately measure the complex site conditions. Figure 8.13 shows the intervention of new technologies for dumpsite mining.

Current machineries for this process are inefficient in separating plastic fractions from the rest of the waste, and subject to wear and tear in a limited period of time. Also, poor quality of recovered materials leads to poor market value of plastic products.

8.4.1.3 *Inadequate data from the source*

Although collecting adequate data on the movement of marine litter from its source and into the wider environment can be a great challenge, this information is critical to shed more light on the effectiveness of interventions and to ensure that the most cost-effective efforts are made to tackle marine litter. One key source of this information is the movement of plastics from the dumpsites to the ocean through nearby waterbodies.

8.4.2 Policy gaps in marine plastic pollution

At the global level, there are various voluntary commitments complying with legally binding international agreements such as MARPOL Convention on Pollution from Ships, the London Protocol and the UN Convention on the Law of the Seas (UNCLOS) and the Regional Seas Programmes such as COBSEA Regional Action Plan on Marine Litter 2019. In these agreements, only maritime sources of pollution are directly addressed. The analysis of international policies highlights key gaps in the response at this level, namely that there are no global, binding, specific and measurable targets agreed to reduce plastic pollution. There are no policies directly establishing dumpsite mining as a measure to reduce marine plastics pollution.

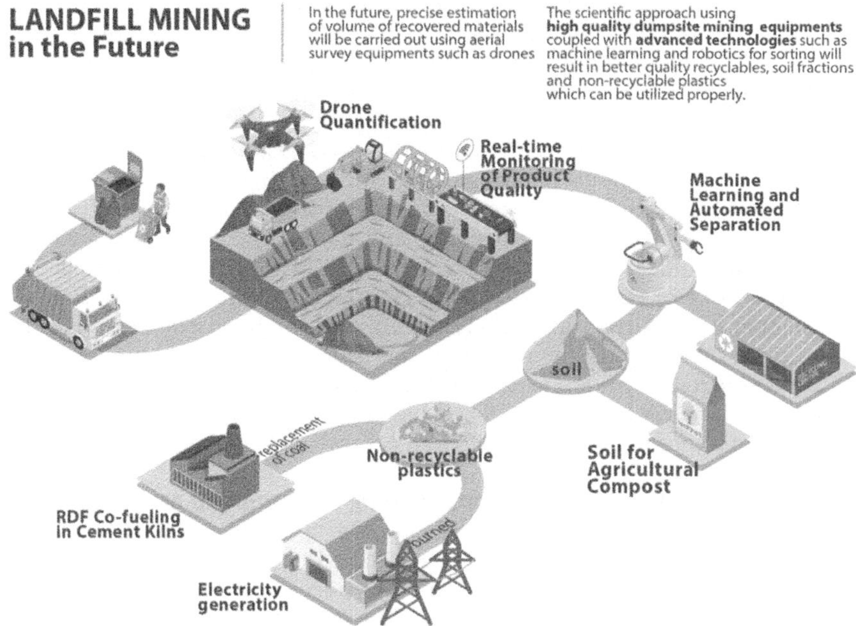

Figure 8.13 Dumpsite mining in the future.

8.4.2.1 Unified guidelines and surveying techniques

There are no detailed operational guidelines for carrying out dumpsite mining processes at the global, regional, or national levels. The operational guidelines are prepared by the environmental consultants hired by the urban local body (ULB) on recommendation of the state or central governments of a country. There are no recommended surveying methods employed to quantify the plastics in the dumpsites.

8.4.2.2 Technology recommendations and prior experience

There are no established technology recommendations from the governmental bodies for efficient recovery of plastics. The dumpsite mining machineries are mostly locally made. The efficiency of the equipment is rarely tested. The quality of plastics is improperly tested. Most of the private players involved in dumpsite mining processes are new to plastic waste management as the dumpsite mining process itself is a new concept for plastic waste recovery and marine plastics abatement.

8.4.2.3 Established markets for recovered plastics

As the dumpsite mining process involves a large volume of recovered plastic products, there are no secondary markets or plastic waste-handling facilities created by the governmental agencies. As it is up to the dumpsite mining operators, it is becoming difficult to safely dispose plastic waste even though they are successfully recovered. Large dumpsite mining firms collaborate with cement factories and easily dispose of their recovered plastic fractions while the small players find it difficult to dispose of plastic waste or convert it into useful products.

8.4.2.4 Institutional framework for policies and research

There is no clear institutional framework for creating a detailed economy-wide action plan to implement new policies, infrastructure (plastic recycling facilities), education, research, funding and communication across various relevant stakeholders in dumpsite mining operations in a country. The dumpsite mining of plastic waste has to be integrated with the holistic waste management of the urban local body (ULB), and the fresh plastics recovered from the households and the industries can be safely disposed along with the plastics recovered from dumpsite mining. There is still a lack of coordinated efforts and planning for marine debris clean-up activities, including clean-up protocols, guidance on methodologies and centralization of marine debris data.

8.4.2.5 Stricter enforcement of 'no dumping' in dumpsite mining zones

One of the primary objectives of dumpsite mining is the reclamation of land with no further dumping of waste. The reclaimed land can be used as a green area, recreational zones or for construction purposes. However, this is not the case in practice. Often, local municipalities in developing countries with limited waste treatment options tend to openly dump waste in the reclaimed land. There is a need for stricter enforcement to stop this practice. More efforts must be taken to develop laws specific to the management of plastic waste and recycling at the household and institutional levels.

8.4.3 Financial gaps for dumpsite mining

The whole value chain of dumpsite mining activities has to be considered for financial planning at national and provincial levels. The major activities that incur major financial costs in a dumpsite mining project are as follows.

(A) Operational costs of dumpsite mining facilities including machinery costs, fuel costs, labor costs, and so on.
(B) Logistics and transportation costs of recovered products.
(C) Technological costs to safely dispose recovered plastic wastes.
(D) Market fluctuations of recovered products.
(E) Research and development, education and public awareness.

Along the whole set of operations, there needs to be multiple financial support, collaborations, innovations, investments are needed to functionally maintain the dumpsite mining operations in a country or a region.

8.4.3.1 Collaboration

New policies and measures on marine debris and dumpsite mining may suffer from a lack of engagement by and constructive feedback from stakeholders during the design and implementation stages. Suitable collaborative involvements must be financially encouraged at the national and local levels to motivate different players into dumpsite mining.

8.4.3.2 Innovation

Despite the gradual growth of dumpsite mining in many economies, there is a lack of innovation among the public and private players. The projects aiming to recover plastics must innovate new technologies and financial business models to share profits and risks associated with dumpsite mining.

8.4.3.3 Recovered plastic fraction and its transportation

As the policies are often unregulated, it is up to the operational owners of the dumpsite mining to enhance their financial income sources. If the dumpsite mining process produces plastics which are poor quality, they cannot be recycled. The transportation of recovered plastic wastes from the dumpsite mining facilities to the cement kilns for co-fueling is costly, and it does not generate any income for the dumpsite mining operators.

8.4.3.4 Assessment of level of investment needed

In contrast to other sectors (e.g., water, sanitation and hygiene), there is no detailed understanding of the levels of investment needed in dumpsite mining projects. Detailed assessment is needed to identify investment needs and to explore the economic benefits of providing plastic waste management for all. This will help make the case for investing in waste and resource management as a means to achieve development goals, which will help galvanize more investors into action.

8.4.3.5 Communication of best practices in preventing marine litter

There is a real need to highlight the role of sound waste and resource management in preventing marine litter by exploring and collating key facts, challenges and opportunities, and identifying, analyzing and communicating examples of best and worst practices. Financial incentives and funding must be provided for establishment of platforms that facilitate the necessary partnerships, links and organizational relationships to facilitate actions and solutions through knowledge transfer and key stakeholder sensitization. Figure 8.14 shows the dumpsite mining gaps which are to be addressed for successful dumpsite mining.

8.5 CONCLUSION

Marine plastic pollution is a global environmental problem affecting the biodiversity, the environment, coastal communities and marine agriculture posing a serious threat to human health and security. In 2016, global plastic waste generation amounted to 242 MMT – 12% of all MSW (Kaza *et al.*, 2018). An estimated 150 MMT of plastics have accumulated in the world's oceans, and the problem has been compounded by overloaded waste management and recycling systems that are unable to cope with rising plastic production. It

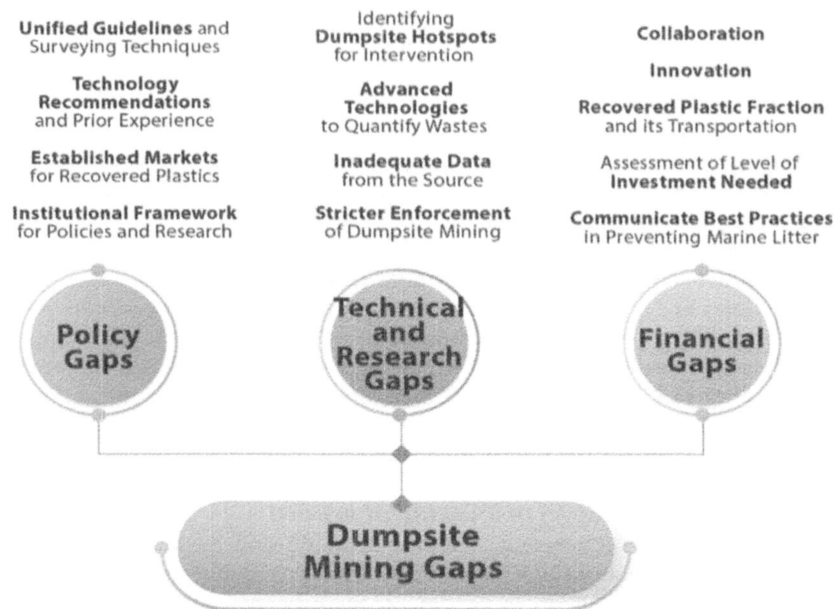

Figure 8.14 Dumpsite mining gaps.

is also significant to note that 80% of marine debris comes from land-based sources, that is, rivers, landfills and industrial sites (Jambeck *et al.*, 2015).

The plastic waste discarded in the dumpsites and other open environments remains in the environment and takes hundreds of years to decompose. According to the report published by Ocean Conservancy and Mckinsey Center for Business and Environment, 55–60% of the global plastic waste leakage is added by these five developing countries, namely China, Indonesia, the Philippines, Thailand, and Vietnam. The report also concluded that open dumps lying near the oceans and in the waterways are a major contributor to the plastic waste leakage. This adds between 1.1 million and 1.3 million metric tons of plastic waste per year. Landfill and dumpsite mining, a dumpsite rehabilitation approach, shifts our perception of viewing landfills and dumpsites as an anthropogenic complication to a resource recovery project. Waste in dumpsites and landfills is one of the biggest untapped sources of energy. The amount of waste accumulated in waste storage facilities has reached such an extensive level that it can be used as a constant source of fuel for waste to energy (WTE) or waste to heat (WTH) plants. As discussed in previous sections, plastic waste has a significant amount of calorific value, and the concentration of plastic waste in dumpsites and landfills is suitably large enough to be used as fuel. Overall, landfill and dumpsite mining helps in

(a) preventing plastics to enter various water bodies and eventually entering oceans thereby ensuring marine plastic abatement,

(b) avoiding or reducing costs of landfill closure and post-closure care and monitoring, and

(c) increasing revenues from recyclable and reusable materials, for example, ferrous metals, aluminum, plastics and glasses.

Hence, the landfill and dumpsite mining can be viewed as a strategic tool in reducing plastic pollution and can help in abating marine plastic pollution from land-based activities. The recovered plastics can also be utilized as a fuel in cement kilns thus resolving a human-made problem in a sustainable way.

Recently, because it has been viewed as a planetary boundary threat, the marine plastics pollution has garnered a lot of scientific attention. The planetary boundaries framework defines precautionary boundaries for several anthropogenic perturbations, set at levels to avoid thresholds or shifts in Earth-system functioning that would generate rising risks for the world societies. There were various control variables defined in terms of emissions, concentrations or effects of chemicals, such as persistent organic pollutants (POPs), heavy metals, or plastics. However, policies on marine plastics have also been emerging, and the need for an international convention on marine plastic debris or pollution is presently being discussed. Key international instruments dealing with sea-based pollution include the London Convention, especially its 1996 London Protocol, and MARPOL 73/78, implemented through national law in signatory nations (Villarrubia-Gómez *et al.*, 2018). Regarding the sources of marine plastic pollution, landfills and dumpsites are the hotspots of origin of the problem. These highly neglected areas, even though far away from the oceans, rivers and other water bodies, are very much linked to the marine environment. Hence, to reduce and control marine plastic pollution, the dumpsites and landfills around the world must be scientifically evaluated and scrutinized.

REFERENCES

Geyer R. (2020). Cited in Baseline report on plastic waste. In: Basel Convention Plastic Waste Partnership working group – First meeting, M. Tsakonam and I. Rucevska (eds.). https://gridarendal-websitelive.s3.amazonaws.com/production/documents/:s_document/554/original/UNEP-CHW-PWPWG.1-INF-4.English.pdf?1594295332

Geyer R., Jambeck J. R. and Law K. L. (2017). Production, use, and fate of all plastics ever made. *Science Advances*, **3**(7), e1700782.

Hawkins G. (2018). The skin of commerce: governing through plastic food packaging. *Journal of Cultural Economy*, **11**(5), 386–403, https://doi.org/10.1080/17530350.2018.1463864

Jambeck J. R., Geyer R., Wilcox C., Siegler T. R., Perryman M., Andrady A. and Law K. L. (2015). Plastic waste inputs from land into the ocean. *Science*, **347**(6223), 768–771, https://doi.org/10.1126/science.1260352

Kaza S., Yao L., Bhada-Tata P. and Van Woerden F. (2018). What a Waste 2.0: A Global Snapshot of Solid Waste Management to 2050. World Bank, Washington, DC, USA. https://pubdocs.worldbank.org/en/429851552939596362/What-a-Waste-2-Revised-version.pdf

Krook J., Svensson N. and Eklund M. (2012). Landfill mining: A critical review of two decades of research. *Waste Management*, **32**(3), 513–520, https://doi.org/10.1016/j.wasman.2011.10.015

Ocean Conservancy (2015). Stemming the Tide: Land-based strategies for a plastic-free ocean. https://oceanconservancy.org/wp-content/uploads/2017/04/full-report-stemming-the.pdf

Sharma A., Aloysius V. and Visvanathan C. (2019). Recovery of plastics from dumpsites and landfills to prevent marine plastic pollution in Thailand. *Waste Disposal & Sustainable Energy*, **1**(4), 237–249.

Tiseo I. (2021). Global plastic production 1950–2020|Statista. https://www.statista.com/statistics/282732/global-production-of-plastics-since-1950/ (accessed 23 September 2021)

Villarrubia-Gómez P., Cornell S. and Fabres J. (2018). Marine plastic pollution as a planetary boundary threat – The drifting piece in the sustainability puzzle. *Marine Policy*, **96**, 213–220, https://doi.org/10.1016/j.marpol.2017.11.035

Waste Atlas (2014). The World's 50 Biggest Dumpsites. https://nswai.com/docs/World's%20Fifty%20biggest%20dumpsites,Waste%20Atlas%202014.pdf

Yadav V., Sherly M., Ranjan P., Tinoco R. O., Boldrin A., Damgaard A. and Laurent A. (2020). Framework for quantifying environmental losses of plastics from landfills. *Resources, Conservation and Recycling*, **161**, 104914, https://doi.org/10.1016/j.resconrec.2020.104914

doi: 10.2166/9781789063448_0343

Chapter 9

Business development from plastic wastes toward circular economy

Atitaya Panuvatvanich and Nantamol Limphitakphong

Table of Contents

9.1 FUNDAMENTALS OF CIRCULAR ECONOMY

9.1.1 Why circular economy

Since the early days of industrialization, a linear model of resource consumption that follows a 'take-make-dispose' pattern has taken place as a fundamental of things. Raw materials are extracted from the environment, energy and labor are applied for manufacturing, products and services are launched in the market and are transferred to be consumed by the end consumer who later discards them as waste and disposes them back to the environment either properly or improperly. Simply put, raw materials that are extracted from the environment are produced and thrown away after use. Coupling with an over-consumption, the humanity's demand on nature has exceeded the capacity of the Earth's annual biological. Consequently, this planet is now facing the impacts of ecological overspending everywhere, for instance, soil erosion, desertification, reduced cropland productivity, overgrazing, deforestation, rapid species extinction, fisheries collapse and increased carbon concentration in the atmosphere.

In 2021, the Earth Overshoot Day (Figure 9.1), which represents the day of the year when the humanity's demand on nature exceeds the Earth's annual biological capacity to regenerate, fell on July 29th. It is implied that between January 1st and July 29th, humanity's demand for biological regeneration is equivalent to the planet's entire annual regeneration. In other words, 1.7 earths a year is needed to support the entire world demands on natural capital. Consequently, carbon dioxide emissions increased 6.6% whilst the global forest biocapacity was estimated to decrease around 0.5% (Lin *et al.*, 2021).

To cope with the unsustainable linear paradigm, recently, a concept of circular economy has been gaining interest rapidly in all sectors. 'Circular Economy' as a theoretical concept is not new. In 1988, Kneese used this

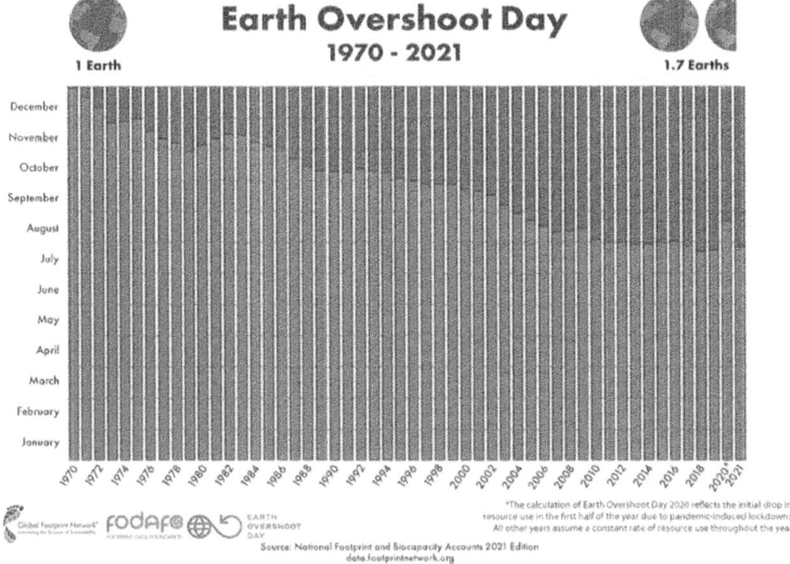

Figure 9.1 Earth overshoot day 1970–2021. Note: Earth's biocapacity/humanity's ecological footprint) × 365 = earth overshoot day (*Source*: https://www.overshootday.org/newsroom/visuals/).

Table 9.1 An overview of CE principles identified in the literature.

R-Principles	Description
3Rs	Reduce, reuse, recycle
4Rs	3Rs + recover
5Rs	3Rs + rethink + repair
6Rs	4Rs + remanufacture + redesign
7Rs	4Rs + rethink + resilient + regulate
8Rs	5Rs + redesign + return + refuse
9Rs	6Rs + refuse + repair + refurbish
10Rs	5Rs + refuse + refurbish + remanufacture + repurpose + recover

Source: Reike *et al.* (2018).

word for the first time in '*The Economics of Natural Resources*' to describe an economic system where waste at extraction, production and consumption stages is turned into inputs. However, its definition and concept are still vague and amorphous. More than a hundred academic articles were found in peer-reviewed publications attempting to define the term. Below are some examples.

- It is a model developed for promoting the responsible and cyclical use of resources to maintain its value in the economy and to minimize the environmental impacts (Geissdoerfer *et al.*, 2017; Ghisellini *et al.*, 2016; Murray *et al.*, 2017).
- It is a booster for achieving sustainable development goals (SDGs) (Kristensen & Mosgaard, 2020); Schroeder *et al.*, 2019; Suárez-Eiroa *et al.*, 2019).

In addition, the principles of the circular economy can also be found in a wide range of R-principles in the literature. Although there are several terms and principles defining the concept, it is undeniable that its root originates in the 3Rs philosophy (Table 9.1).

A circular economy is a global economic model that decouples economic growth and development from the consumption of finite resources. By design product, components and materials at their highest utility and value, at all times.

the Ellen MacArthur Foundation (2013)

The circular economy is a model of production and consumption, which involves sharing, leasing, reusing, repairing, refurbishing and recycling existing materials and products as long as possible.

European Parliament (2015)

The circular economy is an economy that is restorative and regenerative by design, and which aims to keep products, components and materials at their highest utility and value at all times, distinguishing between technical and biological cycles.

BS 8001:2017

In a circular economy, materials circulate in two separate cycles: *the biocycle and the techno-cycle*. The distinction between these cycles helps to understand how materials can be used in a long-lasting and high-quality way (Figure 9.2).

The butterfly diagram inspired by the Cradle to Cradle (C2C) theory is a tool that helps to understand the components of the model in practice. The three principles that must be considered in order to transform towards the circular economy are as follows.

- ***Preserving natural capital***: promoting the effective use of finite resources and balance the use of renewable resources.
- ***Enhance the usefulness of products, components and materials***: keep circulating the products, components and materials in the cycle up to their capacity limit.
- ***Develop effective systems***: – minimizing the volume of waste that ends in landfills and negative externalities (Figure 9.3).

In brief, the difference between the circular economy and the current linear economy is that the cycles of all materials are closing. Not only recycling, the

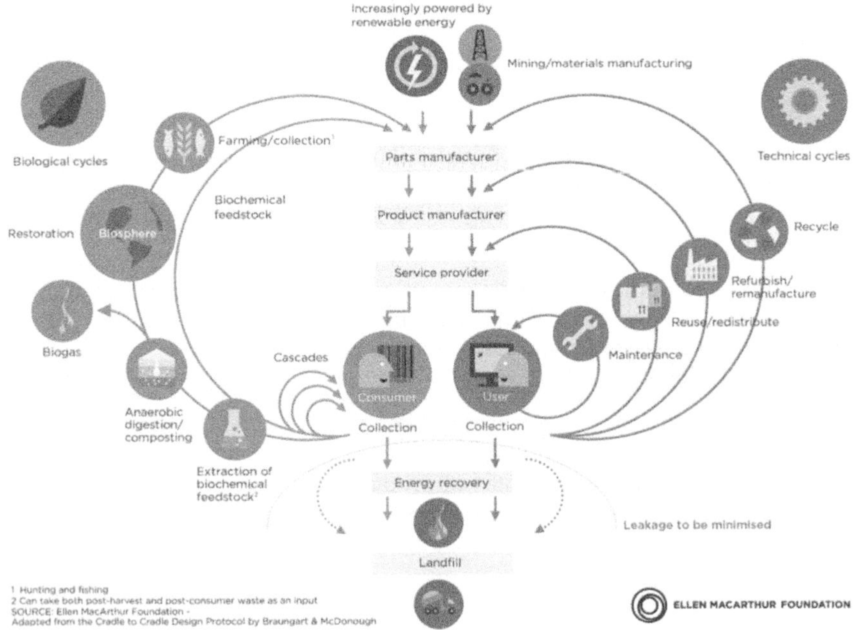

Figure 9.2 Butterfly diagram (*Source*: Ellen MacArthur Foundation, 2016).

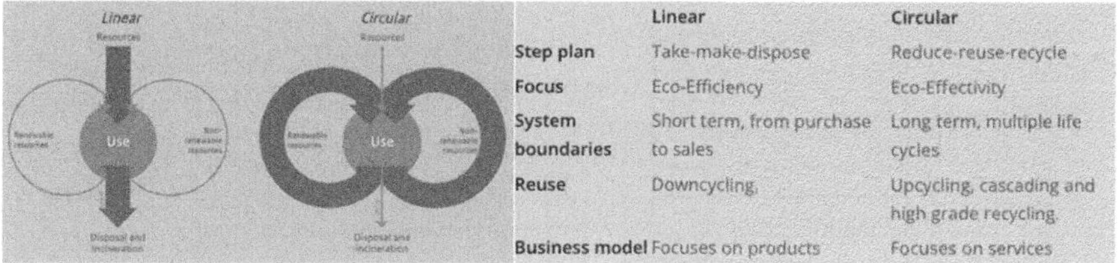

Figure 9.3 Linear vs circular economy (*Source*: Ellen MacArthur Foundation, 2016).

way that the products and services are created and maintained is also changing towards sustainability. Value of products and services is created by concentrating on resources conservation and preservation. Lesser virgin materials are put into the system whilst more products and services are produced and sold using the recirculated materials in the system as many as possible.

9.1.2 Circular economy and business model

At the individual company level, implementing circular solutions involves rethinking products and services using principles based on durability, renewability, reuse, repair, replacement, upgrades, refurbishment, and reduced use of materials. Given this strong positive sentiment, moving towards the circular economy can help business gain significant benefits such as

- increased growth from new customers and deeper relationship,
- innovation and competitive advantage cost reduction,
- lower material and production costs,
- reduced energy consumption and CO_2 emissions, and
- risk mitigation by improving supply chain and resource security.

Following the journey of a circular economy, three types of circular innovations can be observed with progressing degrees of complexity that should be tackled in stages (the easiest to the hardest).

- **Process innovation**: the development and implementation of new or significantly improved production, logistic or recycling methods.
- **Product innovation**: the development and introduction of new or significantly improved goods or services.
- **Business model innovation**: significant changes and/or the creation of a new logic around how a company generates value (Figure 9.4).

Table 9.2 demonstrates five business models that shed light on the implementation of the circular economy for the business.

9.1.3 Circular metrics

As the circular economy is a broad concept, in order to drive the organization performance and to justify the implementation achievement, the impact of circular activities implemented should be measured and evaluated to determine how such initiative can reduce costs, enhance customer and employee

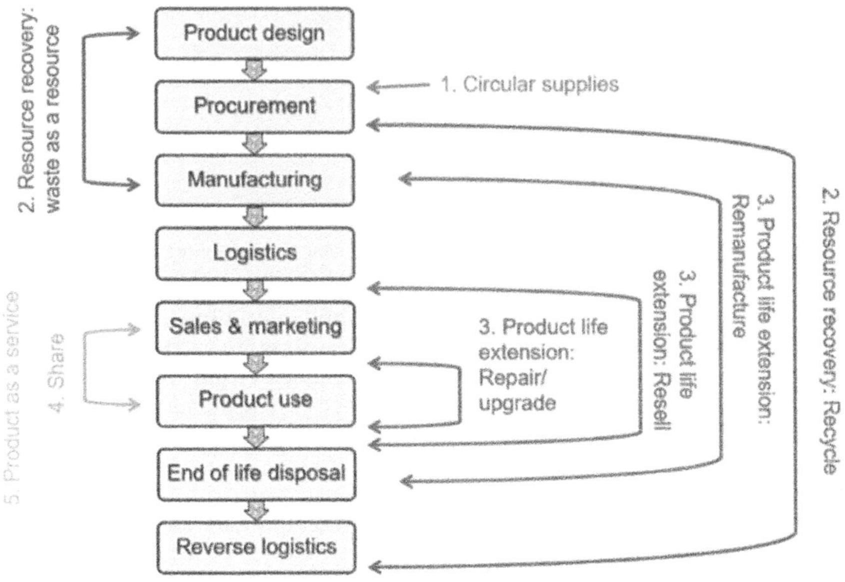

Figure 9.4 Types of circular innovations (*Source*: Lahti University of Applied Sciences Ltd, 2016).

relationships, differentiate from competition and spur innovation. To do so, the metrics and its scope for implementation should be identified clearly.

- **Scope**: What are elements that should be involved in a circular measurement framework?
- **Level**: On which level that the implementation of circularity from the organization has had impact?
- **Value chain and lifecycle factors**: In which stages of products and services life cycle that the implementation has token on?
- **Category**: How circular-related metrics can change as a company moves from operational efficiency aspirations to value creation ones?

Within the scope of the circular economy, elements that could be measured are materials, energy, water, emissions, land, mineral elements and governance. However, selecting what should be measured is all relative. Not all organizations can choose elements that indeed reflect the organization performance to measure how circular activities affect such elements significantly (Figure 9.5).

To implement the circular economy concept in consistency with the related activities according to the core business of organizations, it is important to distinguish roles of different stakeholders and understand how relevant stakeholders are in the improvement.

- Macro: the highest level where cities, countries and international agencies reside

Table 9.2 Business models for the circular economy.

Business Model	Business Case
CIRCULAR SUPPLIES Using renewable energy, bio-based or recyclable materials to replace single lifecycle inputs	Biomass balance innovative has been developed by **BASF** – the world's largest chemical producer – to replace fossil resources with renewable resources such as bio-naphtha or biogas derived from organic waste or vegetable oils. In this process, renewable raw materials are used as feedstock at the very beginning of production and allocated to the respective sales products using a novel certification method. The formulation and quality of the end products remain the same.
RESOURCE RECOVERY Recovering materials, resources and energy from discarded products or by-products	Niaga® Technology has been developed by **DSM** – one of the world's leading producers of essential nutrients – to produce full recyclable carpets using only one or two materials. It has been found that the new process consumes 90% less energy and zero water during manufacturing, and allows full material recovery after use, without losing material quality.
PRODUCT LIFE-EXTENSION Extending working lifecycle of products and components by reselling, repairing, remanufacturing and upgrading	**Renault** – a leader in the European electric vehicle industry – extends the life cycle of EV batteries by selling them as a service through battery leasing campaign that successfully gains interest from more than 90% of company customers. Later, the company also extends the lifetime of EV batteries by using them for stationary energy storage.
SHARING PLATFORM Enabling increased utilization rate of products by making possible shared use/access/ownership	**TATA** – one of the biggest steel companies in India – has created an e-market platform for selling by-products and idle business assets. Buyers and sellers have price transparency, and what would be waste is now feedstock for companies, saving money while ensuring environmental benefit.
PRODUCTS AS A SERVICE Offering product access and retaining ownership to internalize benefits of circular resource productivity	**Philips Healthcare** has developed a value-based healthcare model by providing long-term solutions to hospitals and other care providers instead of selling equipment alone that helps reduce cost for their customers as well as extend the product lifetime. It has been found that about 50–90% of materials depending on product types have been reused through refurbishing activities.

Source: WBCSD (2017).

- Meso: representing all inter-industries and inter-firm networks
- Micro: the level where companies and consumers stand
- Nano: the lowest level of analysis possible at which stand products and components (Figure 9.6)

For the private sector, the metrics may involve only three levels: nano, micro and meso. Each one is comprised of a series of levels beneath it. Organizations

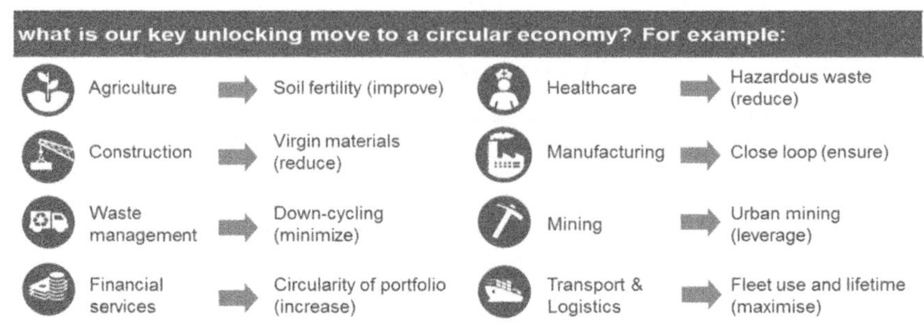

Figure 9.5 Sector priorities (*Source*: adapted from WBCSD, 2018).

can measure circularity at only one or at more than one level. For instance, an organization may measure the circularity of each material and then aggregate the types at the organization level to understand how its portfolio performs overall whilst an investor may want to understand the circularity of its investment portfolio by aggregating each organization's individual circular performance within it.

In addition to resolving the scope and level differences in circular metrics, a common framework for measuring circularity at the organization level will need to resolve the nuances caused by the position in the value chain or life cycle. Figure 9.7 emphasizes the phases in the life cycle and provides example indicators for measuring circularity in each phase.

Organizations are at different stages of maturity with respect to their ambitions in the circular economy. Many are beginning their circular journey by weaving a circular economy narrative into their current operations. A smaller

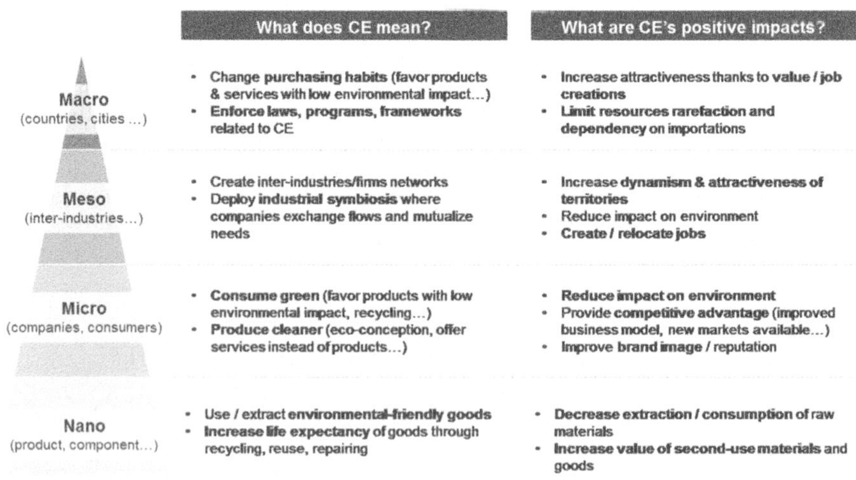

Figure 9.6 Levels of circular metrics (*Source*: adapted from WBCSD, 2018).

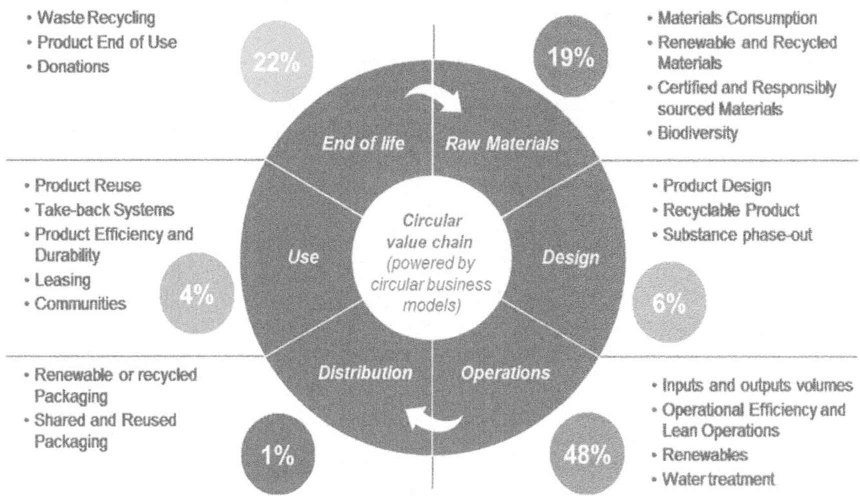

Figure 9.7 Circular metrics along the life cycle (*Source*: adapted from WBCSD, 2018).

portion is integrating circular thinking into their sustainability strategy, taking a more ambitious step in their circular journey.

- Operational efficiency metrics are often standard performance metrics that may be tracked even before a corporate sustainability program is adopted. Examples include resource efficiency, energy consumption, water and waste.
- *Sustainability performance metrics* take it a step further by looking at some of the environmental endpoint and social impacts of company activities and products. Example metrics include greenhouse gas emissions, local stakeholders engaged or biodiversity impact.
- Sustainability performance metrics goes yet another step further by looking at some of the environmental endpoint and social impacts of company activities and products. Example metrics include greenhouse gas emissions, local stakeholders engaged or biodiversity impact (Table 9.3).

9.1.4 Circular economy and SDGs

The SDGs encourage businesses to minimize their negative impact whilst maximizing their positive impact to sustainable development. Companies can use the SDGs as an overarching framework for their existing corporate sustainability or responsibility strategies. The SDGs also enable companies to communicate their efforts on sustainability and responsibility in a consistent manner with their stakeholders. This is due to the fact that the SDGs provide a consensus on the priorities across all dimensions of sustainable development. The SDGs may also aid in establishing more effective partnerships with governments, civil society organizations and other companies (Figure 9.8).

Table 9.3 Sample circular metrics by category.

	Operational Efficiency	Sustainability Performance	Circular Value Creation
Environmental	• Energy efficiency • Water efficiency • Material efficiency	• Recycled content • Circular projects • Waste diverted from landfill	• Valorization of residues • Preserved value • EP&L intensity
Social	• Labor hours per unit • Productivity level • Supply chain transparency	• Local stakeholders engaged • Customers reached • No. of accidents or incidents	• Jobs created (direct and indirect) • Social enterprises started • Total economic contribution
Financial	• Energy cost per unit • Price per resource unit • Landfill tipping fees	• Carbon credits • Circular procurement • Resource cost saving	• Circular revenue • Circular percentage of portfolio • Remanufactured goods sold

Source: adapted from WBCSD (2018).

Companies are expected to address all 17 SDGs in a holistic and integral manner; however, they can identify which goals are most materialistic or relevant to their value chains. The following guidelines on implementing the SDGs are loosely based on the guidance offered by the SDG Compass in combination with the authors' own knowledge and experience. It is worth

Figure 9.8 Sustainable development goals (*Source*: https://sdgs.un.org/goals).

mentioning that the SDG Compass consists of key actors related to corporate sustainability and responsibility including Global Reporting Initiative (GRI), the UN Global Compact and the WBCSD, who together developed their guidance on adopting the SDGs from three consultations with companies, government representatives, academics, and civil society organizations worldwide. The following guidance is broken down into seven key steps, which do not need to be adhered to in a chronological order.

- **Senior leadership buy-in**: Without senior leadership's vision and enthusiasm, little progress can be made in the long term by a company in adopting the SDGs as part of the corporate strategy. Senior leadership must be on board with addressing the SDGs from an organizational and value chain perspective.
- **Defining priorities**: The SDG Compass recommends companies start their impact assessment by doing a high-level mapping of their value chain to identify areas with high likelihood of either negative or positive impacts on the issues that the SDGs represent. It is important to note that at this stage, companies are not required to conduct a detailed assessment of each SDG at each stage of the value chain.
- **Taking the local context of their own operations, as well as that of their suppliers into consideration**: Companies should always seek to obtain the input from external stakeholders such as local communities, relevant expert civil society organizations and government officials at this stage to help ascertain the priority areas for tackling the SDGs. It is also important to select several key performance indicators (KPIs) for each indicator that can help with monitoring and providing specific, measurable and time-bound targets.
- **Further priority definition**: Once impact areas and indicators for key SDGs have been identified, companies should consult with external (expert) stakeholders to discuss the magnitude, severity, and likelihood of current and potential negative impacts. Opportunities to strengthen competitiveness or gain advantage from its current or potential positive impacts across the SDGs should be assessed.
- **Embedding the SDGs within the business**: It is highly recommended for companies not to centralize their SDG efforts with their sustainability departments. Instead, companies should create cross-functional sustainability working groups or committees with representatives from business areas crucial to the SDGs. This may include human resources, supply chain and operations, among others.
- **Engage in cross-sector partnerships**: Cross-sector partnerships are now commonplace in the quest to attain sustainable development. Companies can aim for three different types of partnerships: value chain partnerships, sector initiatives, and multistakeholder initiatives.
- **Reporting and communicating**: Companies are strongly encouraged to report using international standard and communicate the message via a policy commitment. This will help steer the company to live up to its public commitment.

9.2 VALUE CHAIN OF PLASTIC WASTES MANAGEMENT

9.2.1 Overview of the plastic waste value chain

Plastics are often made from fossil fuels and do not usually decay in nature. This makes it even more important to close the plastic cycle. The plastics industry is highly reliant on finite stocks of oil and gas, which make up more than 90% of its feedstock.

Today, 95% of plastic packaging material value or USD 80–120 billion annually is lost to the economy after a short first use. More than 40 years after the launch of the well-known recycling symbol, only 14% of plastic packaging is collected for recycling. When additional value losses in sorting and reprocessing are factored in, only 5% of material value is retained for a subsequent use while another 14% is sent to an incineration and/or energy recovery process, mostly through incineration in mixed solid waste incinerators, but also through the combustion of refuse-derived fuel in industrial processes. Furthermore, an overwhelming 72% of plastic packaging is not recovered at all, 40% is landfilled, and 32% leaks out of the collection system – that is, either it is not collected at all, or it is collected but then illegally dumped or mismanaged (Figure 9.9).

The plastic value chain is complex, touching most (if not all) business sectors globally. Therefore, investor portfolios are exposed to an array of risks and opportunities associated with plastic.

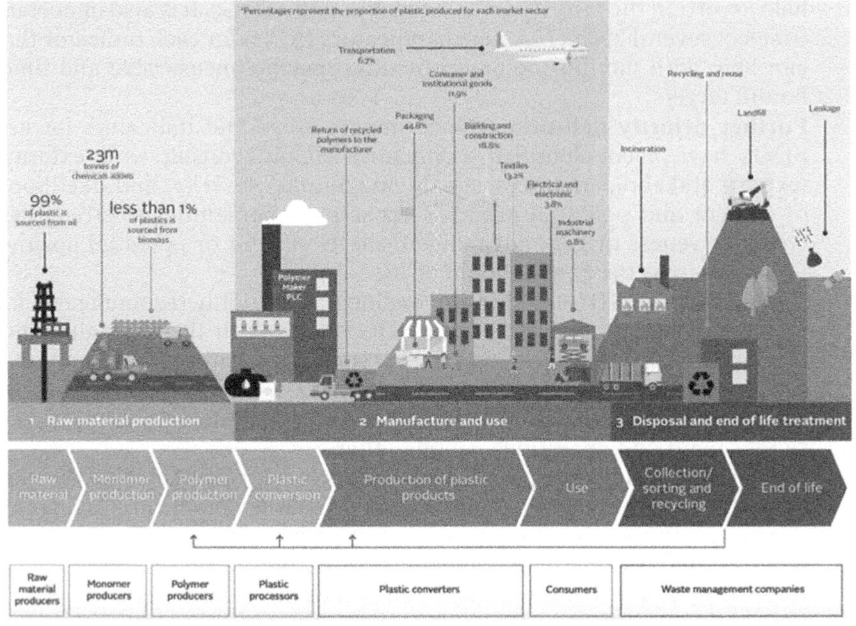

Figure 9.9 Plastic value chain (*Source*: adapted from Principles for Responsible Investment, 2019).

- **Raw material production**: Nearly all (97–99%) plastics come from petrochemicals sourced from fossil fuels with the remaining produced from bio-based plastic. Renewable feedstock for bio-based plastics typically comes from corn and sugarcane.
- **Primary plastics production**: Over 30 types of primary plastics are commonly used globally. However, five primary plastic types account for almost three-quarters of plastic used: polyethylene terephthalate (PET), high-density polyethylene (HDPE), polyvinyl chloride (PVC), low-density polyethylene (LDPE), and polypropylene (PP).
- **Manufacturing and use of plastic products**: Four sectors account for three-quarters of plastic used in the world today: containers and packaging, infrastructure (construction), automobiles, and electrical and electronic equipment. Plastic plays a major role in packaging because (1) it helps to prevent food waste, (2) it keeps food fresher for longer, enabling a wider variety of food availability, and (3) it protects goods during transport and distribution.
- **Disposal and end of life**: Poor management of global plastic waste at end of life adds to the challenges facing the plastics landscape. Plastic packaging accounts for over 141 million tons of annual global plastic waste. There are three main routes for plastic disposal: recycling, incineration or landfills. To achieve a circular economy will require eliminating incineration and landfill streams.

9.2.2 Plastic value chain in circular economy

Figure 9.10 illustrates different types of recycling in plastic packaging. As a key principle of the circular economy is that products and materials are circulated at their highest value at all times, in the technical cycle, this implies that plastic packaging is re-used, when possible (circulating the packaging product), and then recycled (circulating the packaging materials). Within recycling, this principle results in the following general order of preference:

- Mechanical recycling in closed loops,
- Mechanical recycling in open loops, and
- Chemical recycling.

Mechanical recycling in closed loops: This is the most value-preserving loop since it keeps polymers intact and hence preserves its value. The quality of the materials is still at a similar level by cycling materials into the same application (e.g. from PET bottle to PET bottle) or into applications requiring materials of similar quality. As such, mechanical closed-loop recycling not only preserves the value of the material, it also maintains the range of possible applications in future, additional loops.

Mechanical recycling in open loops ('cascading'): Polymers are also kept intact, but the degraded quality and/or material properties require applications with lower demands. Cascading to the highest-value applications each cycle can help maximize value preservation and the number of possible loops.

Chemical recycling: This breaks down polymers into individual monomers or other hydrocarbon products that can then serve as building blocks or feedstock

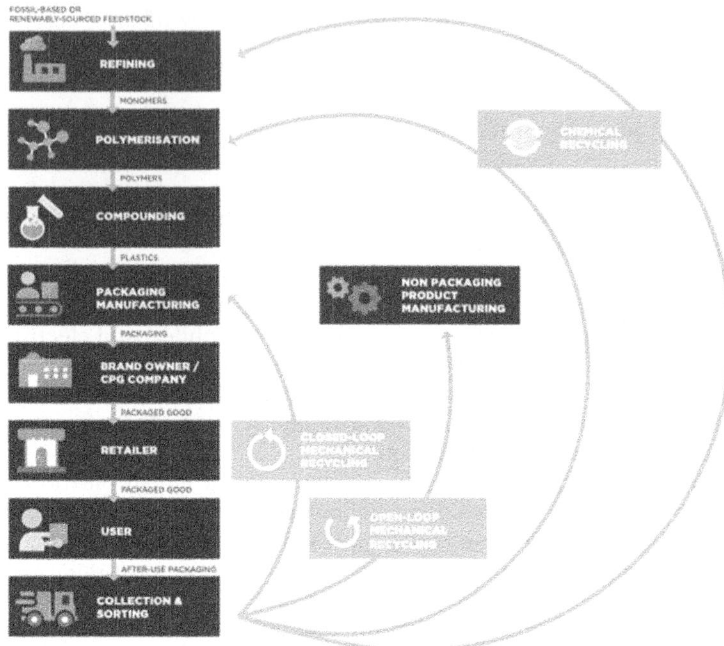

Figure 9.10 Plastic packaging recycling by type (*Source*: Ellen MacArthur Foundation, 2016).

to produce polymers again. As such, it is less value preserving than mechanical recycling. Chemical recycling technologies are not yet widespread and/or not yet economically viable for most common packaging plastics. However, as they can enable after-use plastics to be upcycled into virgin-quality polymers again, they can become an option for materials for which mechanical recycling is not possible (e.g. most multi-material packaging or plastics that cannot be cascaded any further).

9.2.2.1 Circular economy practices in product design

- **Design without packaging**: TerraCycle has launched Loop, an e-commerce platform, in partnership with several brands in EU. Loop brings the old 'Milkman' model by delivering cleaning, personal care and food products bought online in reusable packaging. Packaging is returned in tote bags provided by Loop.
- **Design with bio-degradable packaging and single-use products**: Evoware in Indonesia uses seaweed to make single-use food sachets and wrappings that, at the end, can be dissolved or eaten.
- **Design for less material use**: The use of lightweight plastics in place of other heavier materials means less plastic is introduced into the environment.

- **Design for easier recycling**: The University of Pittsburgh has used nanotechnology to create a multilayer food packaging from a single polymer, namely, polyethylene (PE), combining it in layers with different properties. The material replaces a multi-layer packaging containing PET, PE and aluminum. The new technology enables easy re-processing of the materials without separation steps.

9.2.2.2 Circular economy practices in manufacturing

Marine litter can be caused by the manufacturing sector through inappropriate or poor handling of packaging waste. The application of resource efficient and cleaner production methods and demonstration of practices show convincing results in multiple sectors such as food processing, building materials, tourism, textiles and chemicals across the developed and developing world.

9.2.2.3 Circular economy practices in service delivery

Marine litter can be caused by the service sector through laundries, retail establishments and tourism businesses, and during transportation. Service providers can reduce the release of microplastics by observing a number of good practices in the selection of materials and operations (good housekeeping), and elimination or substitution of plastics with other materials

9.2.2.4 Circular economy practices for product and service use

An example of retailers' role in the area of plastic is the role different stores play in response to introduction of bans on plastic bags in various countries. Such responses include provision of alternatives, installation of plastics sorting facilities and consumer information. For example, Starbucks, since 2020, has not been providing plastic straws. Other food and drink companies that have joined the move toward elimination of single-use plastic are KFC, Danone and Nestle.

- Circular economy practices at end-of-first-life: In the case of durable, longer-living plastic packaging and longer-lived plastic products, end-of-first-life may, in fact, be extended by strategies for reuse and repurposing and by extraction of materials through recycling. However, for single-use plastic packaging and short-lived plastic products, this is usually not the case due to lack of required policy frameworks, systems and infrastructure for their collection, sorting, recycling and/or repurposing. For example, DEMETO, a European consortium is working on chemical recycling of any waste plastic, including waste recovered from the oceans or from other production processes to make food-grade plastic (PET). Rays Enserv in India has developed an 'Advanced Supercritical Thermal treatment technology' to convert PE, PP and polystyrene plastic waste into usable low-sulfur synthetic fuel (Figure 9.11).

9.2.3 Successful circular economy in organizations

Better Future Factory is a sustainable product design and engineering studio based in Rotterdam, the Netherlands (Figure 9.12). The company helps clients

Figure 9.11 Circular economy practices in plastic recycling (*Source*: Ellen MacArthur Foundation, 2016).

finding new ways to transform waste streams into valuable and scalable products, in particular plastic waste streams. Nowadays, Better Future Factory has three startups, and has many on-going projects related to plastic recycling for clients.

Figure 9.12 Sample successful circular economy organization (*Source*: EU-LAC Foundation, 2018).

9.2.3.1 Perpetual plastic

A mobile, interactive recycling installation: it is the world's first mobile interactive recycling machine (Figure 9.13), where people can convert their own plastic waste into a 3D printed object like a ring, going from trash to jewellery. Recently, a lesson program for kids between the age of 10 and 14 has been developed. This aims to motivate kids to be enthusiastic about the possibilities of plastic recycling, stimulate their creativity and involve them in future challenges.

9.2.3.2 Refil

Selling recycled 3D printing filament around the world: it is the industrialized version of Perpetual plastic, making high-quality 3D printing filament (input materials for 3D printers) from plastic waste. This enables people with a 3D printer to use a sustainable alternative. The portfolio offers different plastics from different waste sources like ABS from car dashboards, HIPS from

Figure 9.13 Perpetual plastic (*Source*: EU-LAC Foundation, 2018).

Figure 9.14 Recycled 3D printing product (*Source*: EU-LAC Foundation, 2018).

refrigerators, PLA from packaging and PET from plastic bottles. The aim of the company is to become the leading brand in recycled filament, offering a broad variety of materials, colors and advice for using the material in an optimal way. Figure 9.14 shows an example of a recycled 3D printing product.

9.2.3.3 *New marble*
Selling marble looking wall tiles made from plastic bottles: the technique was initially developed as a low-tech solution for developing countries. As a result of positive feedback in the Dutch market, it is now also being further developed for large-scale production in the Netherlands. The product (Figure 9.15) is now an officially certified building material in the Netherlands, and the technique is further refined in production. After an investment round, the building of a pilot factory was planned for the beginning of 2019, which could produce over 3000 m² per month (30 tons of PET). After a successful trial period, more production locations have been planned.

9.3 BUSINESS MODEL FOR THE 'NEW ECONOMY' OF PLASTIC WASTE MANAGEMENT
9.3.1 What is a business model?
In short simple explanation, a business model is 'a plan that describes all the involve resources, partners and operational methods of the organization to create, deliver and capture value' (Osterwalder & Pigneur, 2010). The most common way to express these components of business is via **Business Model Canvas** (see Figure 9.16) as developed by Alexander Osterwalder. 'While the

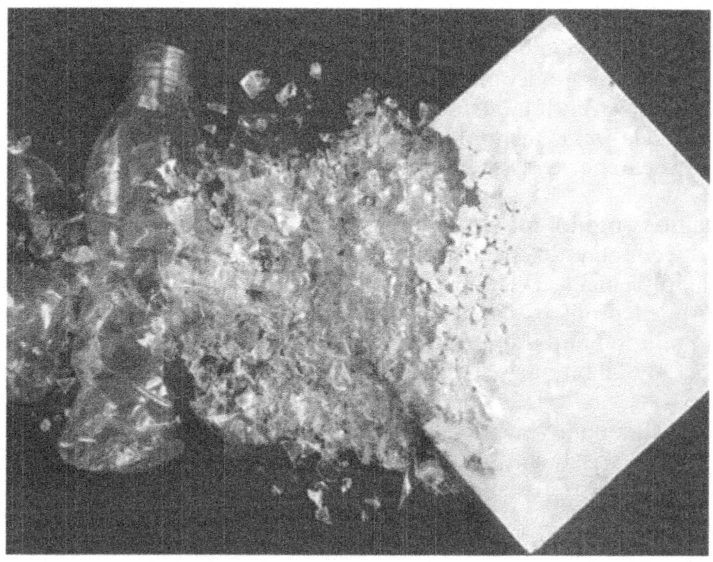

Figure 9.15 New marble by better future factory (*Source*: EU-LAC Foundation, 2018).

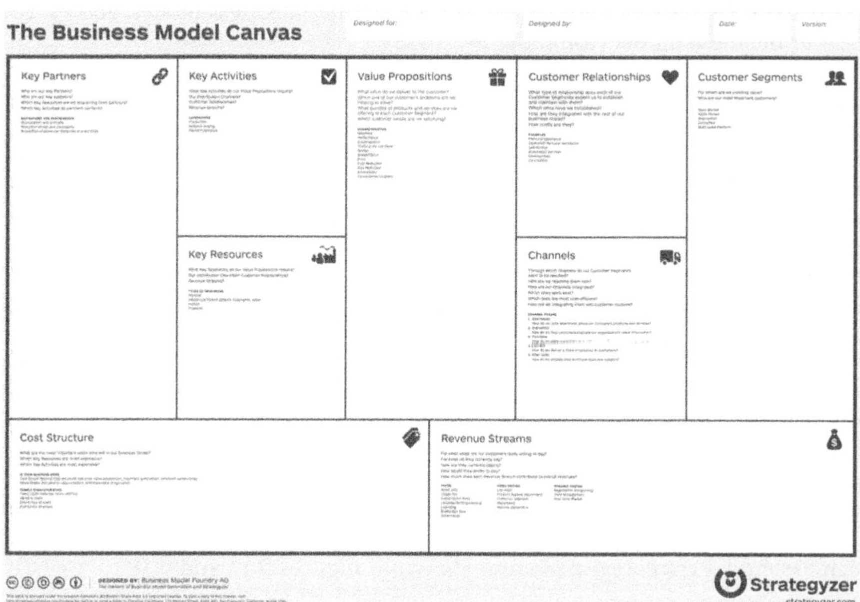

Figure 9.16 Business model canvas (*Source*: http://www.businessmodelalchemist.com/tools).

Business Model Canvas universally use in business world, it is only a summary and require some understanding of business terms.'

A good business model must logically connect with each building block of the canvas. However, if a business model targets the elderly, yet their channel of customer relation is technologically heavy, it will not be considered logical. These things can be easily seen on the Business Model Canvas and addressed.

9.3.2 Business model for the 'New economy'

The word 'economy' itself means wealth and resources management. In the past, humanity focused heavily on creating more wealth in an organization in the form of profit or assets. However, in the new era, we become more aware of pollution and the resources that are scarcer by an alarming rate. The New Economy ambitiously addresses both wealth and resources in the word 'Economy'.

The New Economy business model entails plans to make businesses use resources more wisely and to the point of disintegration if possible. By doing this, the resources are used to its fullest value, allowing more time for nature to replenish its resources. However, New Economy businesses have been in existence since the rise of machinery. Expensive machineries are utilized to its fullest to breakeven its cost and reap more profit for the surplus of their machine output capabilities and utilization rate. There are many business models that share the same goal of reducing wasted resources (materials, times, machine capability, etc.). These business models are categorized mainly into three types: (1) the platform business model, (2) the community business model, and (3) the circular business model (Jonker *et al.*, 2018).

9.3.2.1 *The platform business model*

The platform business model is a business model that creates value by facilitating exchanges between two or more interdependent groups, usually consumers and producers (Moazed & Johnson, 2018). The word platform itself is used often in business in this era. Internet-based businesses such as ride-sharing applications or e-commerce platforms are common these days, but they all share the same concept in which resources are shared among many people (customers) using one single asset. Some simpler and more direct examples of this business model are laundromats, equipment rental businesses or car rental businesses.

For example, if a person buys a hand drill to hang a picture frame and each house in the neighborhood each bought one hand drill for their own use, this will result in a lot of hand drills collecting dust after one or two uses. However, if a person buys a hand drill and tells the neighbors that anyone can rent this hand drill, the equipment owner can recover some cost, and the neighbors or customers also get to use the machine that they possibly need once a year without paying a full price for it. Examples of business using the platform business model are categorized and shown in Figures 9.17 and 9.18.

9.3.2.2 *The community business model*

Most of these businesses are usually (but not always) concerned with the congregation of finance toward something the community believes in. It might

Figure 9.17 Example of platform business models.

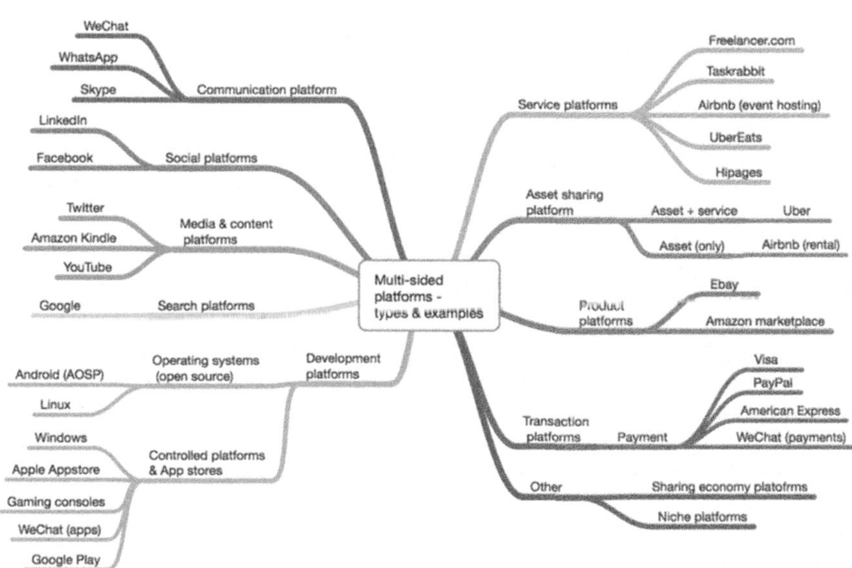

Figure 9.18 Types and examples of platform businesses (*Source*: https://innovationtactics. com/platform-business-model-complete-guide/).

not even be a business; for example, the neighborhood plans to open a small energy farm, and those who have invested can enjoy a certain reduction in their energy bill based on their contribution. All profit from a community business is reinvested in the local area. Some people find that raising their level of community benefits them in the long run such as higher security or even higher overall property values. This is why the community business model has its own charm, although the main concept is still the same: using an asset to the fullest.

Kickstarter is an example of a business that uses the community business model. It is where creators share new visions for creative work with the communities that will come together to fund them. Another example is GoFundMe (GoFundme, 2010), a website that allows users to raise money by sharing their project with people, and people can then donate to the project.

9.3.2.3 *The circular business model*
The most ambitious business model aiming to reduce waste of resources is the circular business model. Not only that its implementation directly results in the complete use of resources, but its complexity can also determine whether the whole circle stays or collapses.

> A circular business model articulates the logic of how an organization creates, delivers, and captures value to its broader range of stakeholders while minimizing ecological and social costs. (Ellen MacArthur Foundation, 2013)

The circular business model involves many other organizations or partners, making it complex as any plan will have to leave room for cooperation, speed and accuracy. Most companies are not open to sharing their knowledge and strategies with outsiders as they view such knowledge and strategies as competitive advantage for their business. The more secretive and rigid the business is, the more likelihood that the circular model will fail, as cooperation is key. Organizations which are involved in the circular business model must cooperate fully for a chance to thrive.

9.4 INNOVATIVE BUSINESS MODELS*

Based on many circular economy business models (CBM) approached, Jonker *et al.* (2018) have integrated CBM, the central principals of the circular economy, and related building blocks as presented in Figure 9.19, in approach provide a clear guideline for the reader to develop the circular economy business.

9.4.1 Principle I: organizing cycles
Organizing cycles are the first step of transition from the linear economy approach to the circular economy approach. The organization of materials, products and processes in cycles is the main activity. This enables value creation and retention because the parties involved cooperate in such a way that (raw) materials and products can be used as optimally as possible (Jonker *et al.*, 2018).

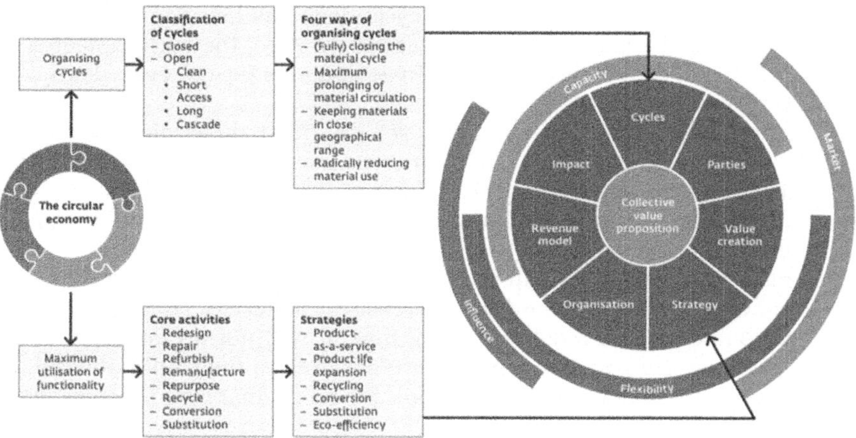

Figure 9.19 Integrated circular economy business (*Source*: Jonker *et al.* 2018).

9.4.1.1 Classification of cycle

A cycle is the collection of entities; we refer to these entities in a cycle as parties. Parties involved cooperate with each other for the goal of using (raw) material as optimally as possible. A common classification is to divide these cycles into two types: closed and open cycles.

- Closed: the materials in a closed cycle can be recycled, usually for use in the same type of product. (Raw) materials mostly return to their original state to be used again.
- Open: the materials in an open cycle are converted to a new (raw) material and can be used for other purposes. The materials will likely leave the cycle. The open cycle is also referred to as downcycling or reprocessing.

This classification can also overlap with another cycle classification which includes five further cycle types (Jonker *et al.*, 2018; Kraaijenhagen *et al.*, 2016).

(1) Clean cycle: the purpose of a clean cycle is to design a cycle specifically to retain the purity of the material. One example is glass or synthetic materials that contain no 'junk'. It is suitable for products that require purity of the materials, and it is its point of value that is reflected in its price.
(2) Short cycle: the purpose of a short cycle is to focus on the (raw) material that is used in the cycle in a very short period. This cycle concerns mostly packaging materials. The materials used for packaging are barely degraded and slightly polluted. The cycle normally concerns the frequency of reuse and the cost. The main problem of the short cycle is logistics.
(3) Access cycle: the focus of this cycle is obtaining materials and their quality. It is all about acquiring access to the stock and availability of

materials in a circulation. This cycle concerns law and regulation of acquiring the materials, and its feasibility. Also, the cycle concerns its worth and availability after use. In genera, this type of cycle deals with rare materials.

(4) Long cycle: the long cycle focuses on (raw) materials which have longlife cycle time. This cycle concerns mostly construction materials. The materials used for construction are often up in buildings, roads, or art works long time. Unlike short cycles where it is wise to just invest in big machinery for high capacity and constantly churn out refurbish materials, long cycle will render that action a big faulty one. The core concept is the same, and that is to reuse the materials. However, because the materials are used for a long time, it will not do any good if the machines are collecting dust or are not used to their full potential. Of course, the party will have to look for other financing schemes while waiting for the materials.

(5) Cascade cycle: This cycle is likely the hardest to implement as it requires the parties to work closely with each other in order to break away from just value retention (materials do not lose value) to 'value creation'. For example, the cycle implies turning prawn shells which is otherwise useless, into sources of calcium for the chemical industry, or turning woodchipper, otherwise low in value, into pellets. However, the core of this cycle type is all parties working together to the point where they even share a revenue model, which can be risky for all parties.

9.4.1.2 *Four ways of organizing cycles*

The nature of each cycle is different, and the cycles have different requirements to complete the cycle.

(1) Completing or closing a material cycle (as) fully (as possible); focusing on bringing the material in the same quality back to the cycle so that can be reused or refurbished without a loss of material.

(2) Utilizing the value of materials for as long as possible; keeping materials in circulation for a prolonged amount of time.

(3) Acquiring materials as close as possible (geographically speaking); maintaining a high level of accuracy; keeping them in a close geographical range.

(4) Radically reducing material use (with and without degradation) in the various phases of the cycle.

9.4.2 Principle II: maximizing utilization of functionality

The second principle is to maximize utilization of material functionality by selecting appropriate core activities and identifying supporting strategies.

9.4.2.1 *Core activities*

- Redesign: designing products that the components can be easily disassembled and modified for example houses, cars, polymer.

- Repair: slight modifications or upgrades, aimed at product life extension. Many electronics companies have a policy to offer repair with warranties instead of a provision of a new item.
- Refurbish: upgrading and updating of products and product parts for the purpose of a 'second-life' and the ability to market them 'as new'. Refurbishing is usually a situation specific method; it is not often used as companies often have different product designs when they roll out new items or versions.
- Remanufacture: this includes the remanufacturing of the entire product, reusing parts of products or on the basis of 'second hand' materials and parts. Companies usually undergo a remanufacturing process when they face expansion or downsizing, or introduction of new technology.
- Repurpose: using products, parts or even raw materials for other purposes than originally intended. The word 'Makeshift' best represents this activity for example fixing a wobbly table with a bottle cap glued to the table leg, or breaking glass bottles and sprinkling the glass pieces on top of the walls to ward off thieves or animals.
- Recycle: recovering (raw) materials for the purpose of reusing them, ideally in high-value application by which they become 'as new'.
- Conversion: converting raw materials and waste to new materials and resources such as CO_2 to methane, the conversion of surplus electricity into hydrogen or sewage sludge as the basis for energy.
- Substitution: replacing some materials with sustainable, usually bio-based materials as it is less harmful to the environment if it leaves the cycle.

9.4.2.2 Strategies
There are six basic strategies.

(1) Product-as-a-service: instead of people owning a product (washing machines, lawn mowers or cars), the company becomes a renter and rent out the products instead.
(2) Product life expansion: products should remain in use for as long as possible. This strategy aims to extend the life of a product by making the product more durable and easier to maintain, or by developing innovative alternatives that allow the product to adapt to the users' changing needs or to allow the product to have more than one owner during its lifetime.
(3) Recycling: partially or fully recovering the (raw) materials. This means in the end the material itself doesn't change form and can go through the cycle again and again.
(4) Conversion: unlike recycling, conversion changes material form. Some examples are converting old tires to carbon black or sewage sludge to energy. Companies will most likely have to obtain the (undesired, waste) raw material from somewhere else rather than it being in the cycle again and again.
(5) Substitution: companies focus on looking for cheaper, better and environmentally friendlier material to replace the original material.

(6) Eco-efficiency: most often, many companies practice this because being efficient means lower cost. This strategy is very common and suitable for almost every type of company. It is just a matter of doing things more efficiently. Even though the company publicly and overtly employ other strategies, this strategy is always part of the practice.

9.4.3 Collective value proposition

Value proposition of a company concerns 'the value or what does the company's products address the customer's pain point.' The circular economy is still difficult to achieve by one particular company. Thus, one can look at the 'collective value proposition' which refers to the value proposition of the whole band of companies. Some companies are created solely for the purpose of the circular economy such as a company that focuses on recovering raw materials from used products. Such companies can also be considered to have a collective value proposition. However, the same could not be said about those long-established companies or organizations would have their value proposition emanate such as 'bringing the best (product/service) to market' or 'Innovative faster'. Such values can be the double-edged sword to the circular economy as these values can also focus more on creating new things while the circular goals are secondary. The processes of fast innovations or the best products can create a lot of waste.

9.4.4 Contextual factors

Four important contextual factors must be considered to complete a circular business plan.

9.4.4.1 Market

Buyers are the first factor to consider. Businesses need to understand their own target customers. Not all end users are buyers. Setting marketing plans to appeal to the spenders is key. Identification of customer segmentation is the first step in marketing planning. The customer segments can be categorized into five groups.

(1) Mass market: there is no specific segmentation for a company that follows the mass market element as the organization displays a wide view of potential clients.
(2) Niche market: customer segmentation is based on specialized needs and characteristics of its clients.
(3) Segmented: a business distinguishes its clients based on selected segments such as gender, age, and/or income.
(4) Diversify: a business serves multiple customer segments with different needs and characteristics. An example is a company that serves both business-to-consumer (B_2C) and business-to-business (B2B) markets.
(5) Multi-sided platform/market: for a smooth day-to-day business operation, some companies will serve mutually dependent customer segments. For example, a credit card company will provide services to credit card holders while simultaneously assisting merchants who accept those credit cards.

Forging a new market is possible. However, those are mostly considered luxurious. Luxury goods easily create waste as luxury products tend to include packaging and scarce materials. However, the word 'new market' does not always mean luxurious as being eco-friendly or environmentally concerned is also considered a new market, however niche it may be. The question is then how to sustain that niche market in terms of both demand and reasonable revenue streams.

9.4.4.2 Capacity

Most of the time capital and human talents are the two aspects on which businesses focus. Capacity is the limit to what the company can extend to at that moment. Resources, technologies, knowledge, human talents and capital that company have determine how much the company can do and how much they can produce, which in turn creates limit of sales. These limits are more pronounced when the company products are services that are performed by humans.

Businesses need to be aware of what they have on hands and understand that these are their capacity at that time. With the capacity identified, the company would be less likely to overexpand or spend too much of the limited capital on aspects such as inventory or machines that have high capacity that cannot be achieved with the amount of manpower the company can currently hire.

9.4.4.3 Flexibility

What room for maneuver exists in terms of available financing, rules and regulations, and technological developments? The company ideas and activities could be halted or limited by laws or politics. What companies cannot control is referred to as external factors. Possible external factors are listed below.

- **Politics**: how and what degree a government intervenes in the economy. This can include – government policy, political stability or instability in overseas markets, foreign trade policy, tax policy, labor law, environmental law and trade restrictions among others.
- **Economic conditions**: a significant impact on how an organization does business and also how profitable they are. These factors include economic growth, interest rates, exchange rates, inflation, disposable income of consumers and businesses.
- **Socio-cultural factors**: the areas that involve the shared belief and attitudes of the population. These factors include population growth, age distribution, health consciousness and career attitudes among others. These factors are of particular interest as they have a direct effect on how marketers understand customers and what drives them.
- **Technological advances**: technological factors affect marketing and the management thereof in three distinct ways: (1) producing goods and services, (2) distributing goods and services, and (3) communicating with target markets.
- **Environments**: seasonal or climatic. This is important due to the increasing scarcity of raw materials and pollution target.

- **Legal factors**: such factors include health and safety, equal opportunities, advertising standards, consumer rights and laws, product labelling, and product safety.

9.4.4.4 Influence

It is an unspoken truth that connections make things easier. Sometimes an employee is hired based on who they are connected to. Their aunts or uncles may be working as a government official who can influence the company's industry and have power to make change in the policy that they need. It could be for the benefit for all, and it could be the benefit solely for the company. All in all, it is about using connection to change the rule in their favor. Moreover, if certain companies can generate a lot of GDPs (gross domestic products) to the country, policies from the government are likely to serve them if they ask for it to happen.

9.4.5 The building blocks

The building blocks refer to the elements that complete the circular economy. Each building block needs to be discussed by all the companies involved in the circle. These building blocks are the extension of the collective value proposition. They explain the details of each company's contribution to the circular economy (Figure 20).

9.4.5.1 Cycles

The heart of a circular business is how well the product, parts or (raw) materials can be used multiple times. The amount of times the material can be reused

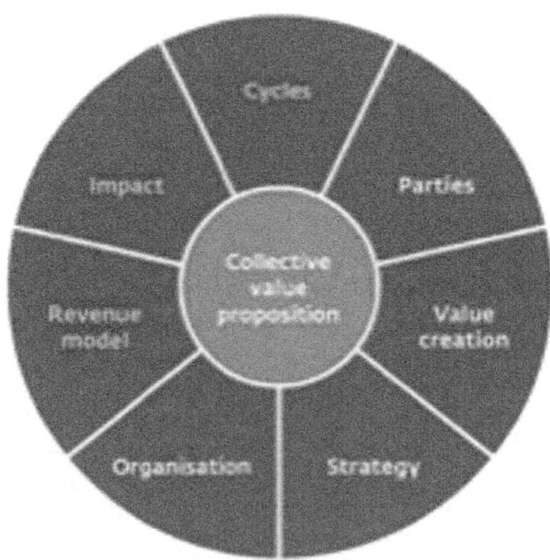

Figure 9.20 Building blocks in circular economy (*Source*: Jonker *et al.* 2018).

can impact cost. As used materials can be partially lost or damaged when they circulate back, it is difficult to get 100% return. For example, clothing items lose their fiber due to wear and tear. Plastic packages can also be torn, disintegrated and broken down into pieces. To recover such materials, the materials need to be clean, and return to their original form, the process of which can generate cost.

9.4.5.2 *Parties*

Companies or organizations that work together to close a cycle are referred to as parties. Companies each have their own working method and culture, and these differences can hinder the chance to close the cycle (organization model is one of the parties). Adaptation must be made between the parties to close the cycle, whatever the differences. Otherwise, the cycle will not work. It is preferable if closed cycles address core activities of the circular economy objectives rather than those on an organization's periphery.

9.4.5.3 *Value creation*

Companies in the circular economy must accept the environmental approach and be comfortable in it. Companies should ask themselves the following questions.

- Does using the recycle materials create more value?
- Do consumers agree with you reusing recycle materials?
- Is it beneficial to the environment or the ecology?
- Does the company benefit from it financially?

Companies that can consistently answer yes to these questions over time are circular businesses. Value creation means transforming these activities and viewpoints toward the company into monetary value. Both the good attitudes of the people on product recycling or the good cost-effective benefits of the environmentally friendly technologies can promote people's willingness to purchase the products.

9.4.5.4 *Strategy*

Strategies are fluid in the real business world. It is ever changing, and no one concrete strategy can fit all companies even though these companies have the same goals or problems. However, most strategies are just derived from a main solid plan, much like a guideline, and companies can pick the strategy that addresses their problems the most. They can also adjust existing strategies so that such strategies fit their own situation. However, it should be noted here that tailoring strategies requires in-depth knowledge and understanding.

9.4.5.5 *Organization*

As no company can close the cycle all by itself, it requires collaboration among those involved in the process to jointly close the cycle. Thus, the organization mission and vision must align both internally and externally. However, conglomerates or giant corporations that normally build their empire in their industry normally control all the aspects of their supply chain and have a party of one, themselves. This allows them to close the cycle alone, but doing so also requires a great amount of capital and investment.

(a) **Internal organization**: How the internal organization (in terms of current strategy, processes, culture, competences, and systems) fits with the demands of organizing in cycles

(b) **External organization**
   - Organizational ecology: without legal binding, companies in the cycle work together as equal to achieve the same goal. Since they are not legally bound, the cycle success and failure all depend on verbal commitments.
   - Coordinating body: a task force which includes representatives from the parties involved can be formed to manage and organize the cycle. This task force are not in the building blocks nor will it participate within the economy. Rather, the task force simply manages the cycle ensuring progress.
   - Chain management: there is a need for a clear leader who leads and makes decisions on who can join the cycle and how the circular economy progresses. Often, it is the manufacturer who extends their influence or power, expands horizontally and closes the cycle, or a manufacturer's product designed for the materials to easily return to them with reverse logistics. Whether the ecosystem of this circular economy lives or breaks all depends on this leader.

9.4.5.6 *Revenue model*

In CBM, it is no longer a simple selling. The revenue model usually shifts away from just selling the inventory away and replenishing the stocks. There are many revenue models in the business world, but five of them are likely to fit CBM much readily than others.

(1) The sales model: this is the most basic of commerce. However, this can and should be combined with a take-back or service agreement (or both). We often see the strategy of eco-efficiency being combined with sales, but a strategy of product-as-a-service in the form of sales with an extensive service agreement is also a good example.

(2) The servitization model: selling the function of the item. This is suitable for the product-as-a-service strategy.

(3) The cascade model: this model is quite complex as the model itself, and it does not have concrete definitions or methods. The core of this model stems from adding efficiency and recycling strategies into the mix. For example, WesterZwam receives coffee ground from cafés, uses it to grow mushrooms and sells the mushrooms back to the cafés. In this case, the cost of soil used to grow mushrooms is eliminated, waste is reduced, and more revenue is gained from selling the mushrooms.

(4) The revival model: the same materials, or parts or even the finished products are reused multiple times. Molds cast from iron are melted down to make new molds for new projects or products. Used or broken glasses are turned into sand dust which is then used to produce new glass.

(5) The hybrid model: this model encompasses revenue methods that do not employ the conventional purchasing practice (money as payment for products or services). If a company sells their product and receives another product (other materials, crypto currency), instead of money in return, and then uses these products for other revenue models, then it is considered hybrid, commonly referred to as the barter system. This model has become more and more common due to the rise in access to the Internet and globalization.

9.4.5.7 Impact

Creating multiple value creation can yield more benefit than just financial gain. To create impact is to go above and beyond. For example, the overall volume of garbage sent to landfill decreases dramatically. Sweden uses trash to generate electricity which has helped encourage the rest of the world to follow. The circular economy measurement of success is the impact it generates. Measuring these impacts is the purpose of this building block.

REFERENCES

Ellen MacArthur Foundation (2013). Towards the circular economy – Business rationale for an accelerated transition, https://ellenmacarthurfoundation.org/towards-a-circular-economy-business-rationale-for-an-accelerated-transition (accessed 20 May 2021).

Ellen MacArthur Foundation (2016). The New Plastics Economy – Rethinking the future of plastics, http://www.ellenmacarthurfoundation.org/publications (accessed 20 May 2021).

EU-LAC Foundation (2018). Case studies on Circular Economy models and integration of Sustainable Development Goals in business strategies in the EU and LAC. Available at https://eulacfoundation.org/en/system/files/case_studies_circular_economy_eu_lac.pdf (accessed 15 December 2021).

European Parliament (2015). Circular economy: definition, importance and benefits, https://www.europarl.europa.eu/news/en/headlines/economy/20151201STO05603/circular-economy-definition-importance-and-benefits (accessed 20 May 2021).

Geissdoerfer M., Savaget P., Bocken N. M. and Hultink E. J. (2017). The circular economy – a new sustainability paradigm? *Journal of Cleaner Production*, **143**, 757–768, https://doi.org/10.1016/j.jclepro.2016.12.048https://doi.org/10.1016/j.jclepro.2016.12.048 (accessed 14 July 2021).

Ghisellini P., Cialani C. and Ulgiati S. (2016). A review on circular economy: the expected transition to a balanced interplay of environmental and economic systems. *Journal of Cleaner Production*, **114**, 11–32, https://doi.org/10.1016/j.jclepro.2015.09.007

GoFundme (2010). About GoFundMe, https://www.gofundme.com/c/about-us (accessed 18 May 2021).

Jonker J., Stegeman H. and Faber N. (2018). Circular Economies: Denbeelden, ontwikkelingen en business modellen, https://repository.ubn.ru.nl/bitstream/handle/2066/198996/198996.pdf (accessed 20 September 2021).

Kraaijenhagen C., Oppen C. and Bocken N. M. P. (2016). Circular Business – Collaborate and Circulate. Circular Collaboration, Amersfoort.

Kristensen H. S. and Mosgaard M. A. (2020). A review of micro level indicators for a circular economy–moving away from the three dimensions of sustainability? *Journal of Cleaner Production*, **243**, 118531.

Lahti University of Applied Sciences (2016). Circular design-design for circular economy. Lahti Cleantech Annual Review, 32.

Lin D., Wambersie L. and Wackernagel M. (2021). Estimating the Date of Earth Overshoot Day 202: Nowcasting the World's Footprint & Biocapacity for 2021, Global Footprint Network, https://www.overshootday.org/content/uploads/2021/07/Earth-Overshoot-Day-2021-Nowcast-Report.pdf (accessed 15 February 2021).

Moazed A. and Johnson N. L. (2018). Modern Monopolies: What It Takes to Dominate the 21st Century Economy. St. Martin's Press, New York, USA.

Murray A., Skene K. and Haynes K. (2017). The circular economy: an interdisciplinary exploration of the concept and application in a global context. *Journal of Business Ethics*, **140**(3), 369–380, https://doi.org/10.1007/s10551-015-2693-2

Osterwalder A. and Pigneur Y. (2010). Business Model Generation: A Handbook for Visionaries, Game Changers, and Challengers. John Wiley & Sons, Hoboken, New Jersey, USA, Vol. **1**.

Principles for Responsible Investment (2019). The plastics landscape: Risks and opportunities along the value chain, https://www.unpri.org/plastics/risks-and-opportunities-along-the-plastics-value-chain/4774.article (accessed 14 March 2022).

Reike D., Vermeulen W. J. V. and Witjes S. (2018). The circular economy: new or refurbished as CE 3.0? – exploring controversies in the conceptualization of the circular economy through a focus on history and resource value retention options. *Resources, Conservation and Recycling*, **135**, 246–264, https://doi.org/10.1016/j.resconrec.2017.08.027

Schroeder P., Anggraeni K. and Weber U. (2019). The relevance of circular economy practices to the sustainable development goals. *Journal of Industrial Ecology*, **23**(1), 77–95.

Suárez-Eiroa B., Fernández E., Méndez-Martínez G. and Soto-Oñate D. (2019). Operational principles of circular economy for sustainable development: linking theory and practice. *Journal of Cleaner Production*, **214**, 952–961.

WBCSD (2017). CEO guide to the circular economy. Available at https://docs.wbcsd.org/2017/06/CEO_Guide_to_CE.pdf (accessed 10 February 2021).

WBCSD (2018). Circular Metrics – Landscape Analysis. A joint report on the current landscape of circular metrics use and recommendations for a common measurement framework, https://docs.wbcsd.org/2018/06/Circular_Metrics-Landscape_analysis.pdf (accessed 10 February 2021).

doi: 10.2166/9781789063448_0375

Chapter 10

Bioplastics: potential substitution to fossil-based plastics

Wenchao Xue

Table of Contents

10.1 INTRODUCTION TO BIOPLASTICS

Extensive production and universal use of conventional plastics have posed a persistent threat to both the limited fossil fuel resource and the environment. To address the emerging environmental nuisance, sustainable solutions have been on the rise. One of such long-term solutions is seeking substitute materials for fossil-based plastics such as bioplastics and other eco-friendly materials. The production of such substitute plastics takes into account consumer's needs for the products, addresses sustainability demand, reduces reliance on fossil fuels, and relieves the environmental burden due to over-consumption and mismanagement of plastic litter. According to an industry survey by the International Solid Waste Association (ISWA), development of new materials is expected to have the most important impact on the waste management industry in the next decade, and gains the highest investment interest by the industrial sector (ISWA, 2017).

10.1.1 What is bioplastic?

Plastic materials that are wholly or partially bio-based, biodegradable, or both are known as bioplastics (European Bioplastics, 2020a). Broadly, bioplastics can be classified into bio-based plastics and/or biodegradable plastics. According to the European Commission (2021), bio-based plastics are fully or partially made from biological research, rather than fossil raw material. They are not necessarily compostable or biodegradable. The American Society for Testing Materials (ASTM) also defines a bio-based material as 'an organic material in which carbon is derived from a renewable resource through biological processes' (BPI, 2015). In a more layman's term, alternate to fossil-based plastics, bio-based plastics are made from renewable biomass resources such as agricultural by-products, ligno-cellulosic feedstock or microbial fermentation products. Bio-based plastics can be a key to sustainability as it helps to shift the burden from the limited fossil fuels to the renewable resources or readily available by-products.

On the other hand, biodegradable plastic refers to the organic polymers that can be broken down by the microorganism in the presence of oxygen to simple products such as carbon dioxide, water, mineral salts and new biomass, or in the absence of oxygen to carbon dioxide, methane, mineral salts and biomass. When microbial fragmentation and degradation occur within a designated duration (e.g., 180 days) in a composting environment and the end products meet the relevant compost quality criteria, it is considered as compostable plastic (ASTM International, 2019). It is worth noting that the terms 'bio-plastics' and 'biodegradable plastic' are usually used interchangeably, which might be misleading in the terminology. It is important to clarify that bio-based plastics are not necessarily to be biodegradable, and similarly, not all biodegradable plastics are bio-based.

In comparison to the conventional fossil-based plastics which are typically produced from petrochemical products and non-biodegradable in the environment, bioplastics are classified into three categories based on their

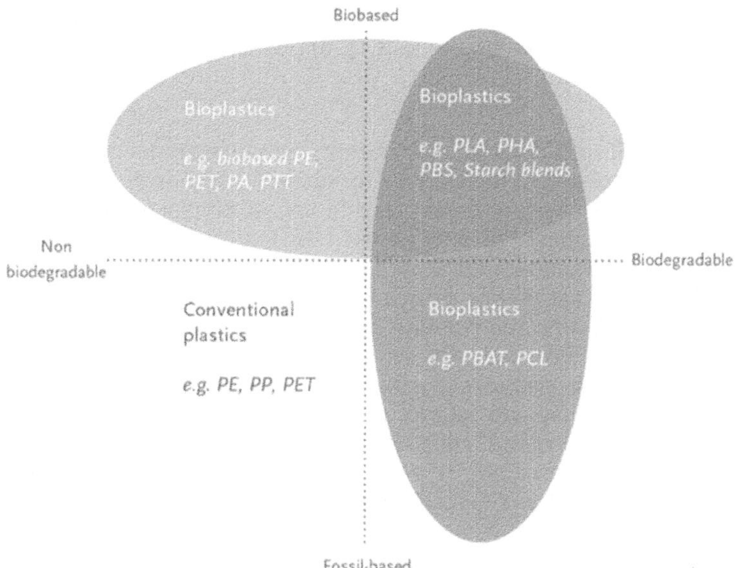

Figure 10.1 Materials coordination system of plastics based on their feedstock sources and biodegradability (*Source*: European Bioplastics, 2020a).

source of feedstock and biodegradability in the material coordinate system (see Figure 10.1).

(1) Bio-based or partly bio-based, but non-biodegradable plastics, which are also called 'drop-in' bioplastics, sharing the same chemical compositions and polymer structures as conventional fossil-based plastics, whereas the feedstock comes from renewable resources like biomass and agricultural products. Examples of this category include bio-based PE, bio-based PET, bio-based PA, and bio-based PTT among others.

(2) Bio-based biodegradable plastics, typically with the natural-based polymers that are degradable in the environment, meanwhile, produced from renewable resources. Examples of this category are starch blends, polylactic acid (PLA), polyhydroxyalkanoates (PHAs), and polybutylene succinate (PBS) among others.

(3) Fossil-based biodegradable plastics, with the polymers degradable in the environment though produced from petrochemical products. Examples of this category are polybutylene adipate terephthalate (PBAT) and PBS (Spierling *et al.*, 2018).

Packaging Digest enunciates the classification of bioplastics in more detail, as plastics that are partially bio-based, fully bio-based, non-bio-based, biodegradable, compostable or non-biodegradable, as long as it is not both non-bio-based and non-biodegradable.

10.1.2 Benefits of bioplastics

Potential benefits that can be brought by applying bioplastics to address the sustainability issues caused by plastic over consumption and mismanagement are as follows.

- Saving fossil resources by substituting them using renewable sources.
- Increasing resource efficiency by means of (1) resources being cultivated on an annual basis, and (2) cascade use of bioplastic products, which means that biomass can first be used for materials production and then for energy generation. Figure 10.2 provides a conceptual diagram of using bioplastics in a cascade principle in order to maximize the resource efficiency and close the material loop.
- Reducing carbon footprint and greenhouse gas (GHG) emissions associated with plastic materials and products. Spierling *et al.* (2018) compared the impacts related to the global warming potential of the global plastic demand caused by fossil-based plastics and bio-based plastics (see Figure 10.3). It was estimated that bio-based plastics would emit 241–316 Mt less CO_2-eq. per year (substituting 65.8% of all plastics). This calculation reveals the potential of approaching global climate goals through applying bio-based plastics.
- Shortening the composting process of plastic contained waste.
- Relieving the accumulation of plastic litter in the environment.

10.1.3 Global market trends of bioplastics

According to European Bioplastics (2020b), bioplastics represent about 1% of more than 368 million tons of plastic produced annually. However, since

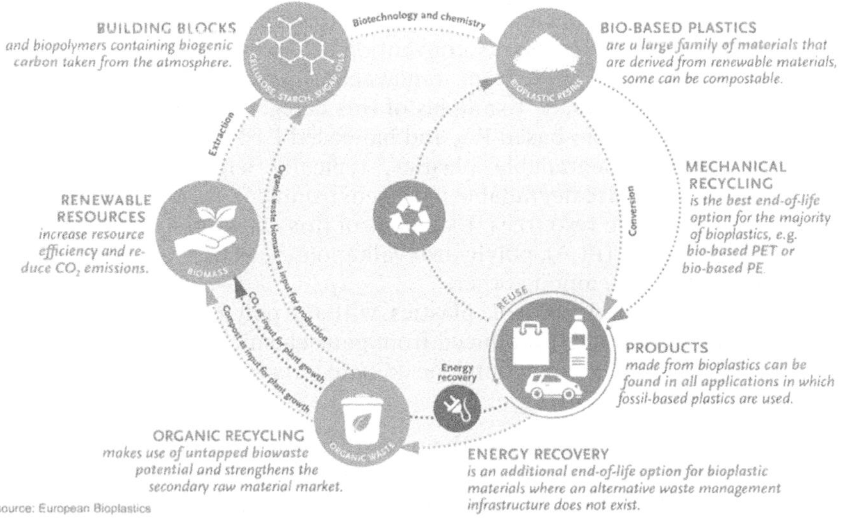

Figure 10.2 Closing the material loop of bioplastic via maximizing the resource efficiency using the cascading principle (*Source*: European Bioplastics, 2020b).

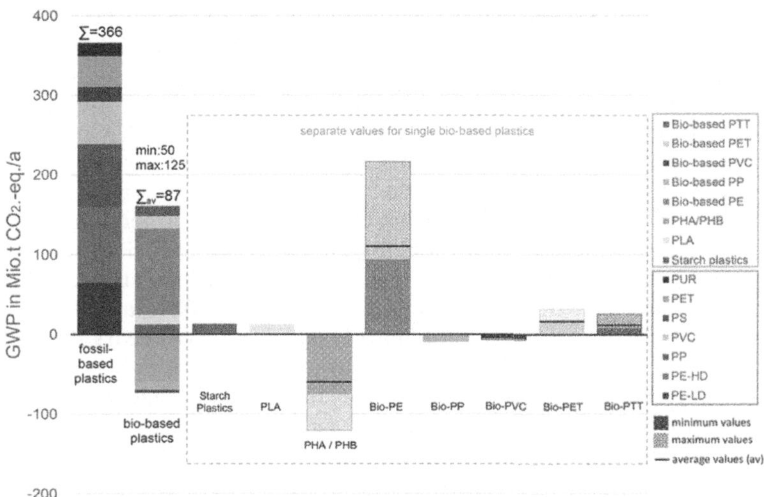

Figure 10.3 Global warming potential of fossil-based and bio-based plastics for 65.8% of the global plastic demand (*Source*: Spierling *et al.*, 2018).

green consumerism and eco-awareness is on the rise, and more sophisticated biopolymers, applications, and products emerging, the market for bioplastic has been continuously growing and diversifying. The global bioplastic production capacities are projected to increase from around 2.11 million tons in 2020 to approximately 2.87 million tons in 2025 (Figure 10.4). Innovative polymers like PLA, bio-based PP and PHAs are continuing to grow at high rates. Bio-based PP entered the market on a commercial scale in 2019 but has been forecast to quadruple by 2025 due to widespread applications of PP in numerous sectors.

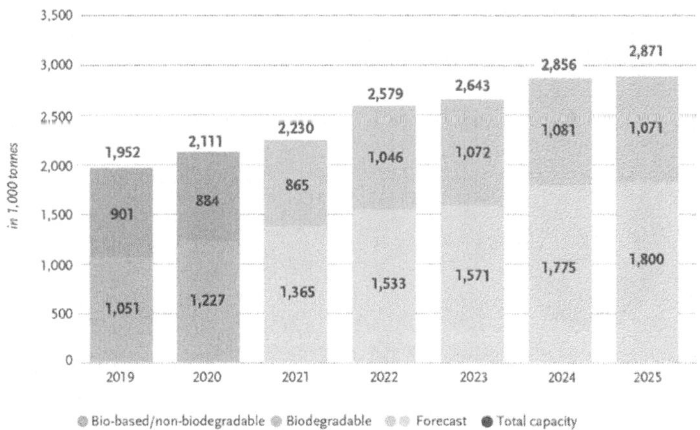

Figure 10.4 Projection of global production capacity of bioplastics by 2025 (*Source*: European Bioplastics, 2020c).

By 2020, biodegradable plastics, including PLA, PHA, starch blends and others accounted for 58% (over 1.2 million tons) of the global bioplastic production capacities, slightly surpassing the production capacity of bio-based non-degradable plastics. This is expected to continue in the next five years. Due to PHA's significant growth rates and new investments for PLA productions, the production of biodegradable plastics is expected to increase to 1.8 million in 2025 (European Bioplastics, 2020b).

According to an earlier global market observation and projection, the global production capacity of bioplastics was forecast to reach over 6 million tons by 2018 (Ashter, 2016a). Bio-based non-degradable bioplastics such as Bio-PET were expected to share the highest popularity in the market, far exceeding the biodegradable plastics. However, the market trend in recent years showed that the growing rate of the overall bioplastic market was not as high as expected in early 2010s. By 2020, the total global capacity of bioplastics reached 2.11 million tons, lower than the earlier forecast. This may be mainly attributable to the relatively low cost-efficiency of bioplastics, slowly progressing policy supports and hesitating market attitude. On the other hand, bio-based biodegradable plastics become the most popular bioplastic in the market, owning to the rapid development in material science and technology. The production capacity of several novel bioplastics such as PHA and PLA has been largely advanced, which drives the robust growth and market confidence in increasing the production and consumption of these substitute plastic products (Figure 10.5).

The latest market data compiled by European Bioplastics in cooperation with the nova-Institute (European Bioplastics, 2020b) pointed out that 46% of the global production capacity is located in Asia by 2020, due to its huge plastic market demand and ease of access to the feedstock supply. On the other hand, Europe produces approximately one fourth the bioplastics, ranks second and surpasses South America. The main reason could be that Europe strengthened its position as a major hub for the entire bioplastics industry. It has ranked highest in the field of research and development and become the largest market of the bioplastic industry worldwide. The bioplastic production capacities in North America and South America account for 17% and 10% of

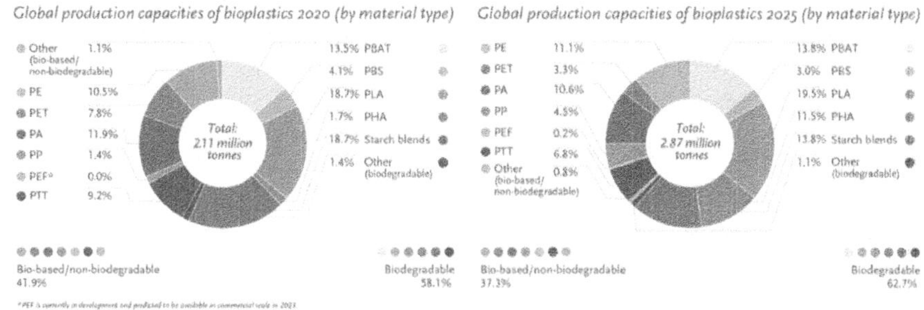

Figure 10.5 Global production capacity projection of bioplastics in 2020 (left) and 2025 (right) (*Source*: European Bioplastics, 2020c).

the global share, respectively. It is noted that the interests in investment and production of bioplastic in South America significantly decreased compared with the data shown in 2012 in which South America shared over a quarter of the global production of bioplastics and was estimated to be one of the most robust markets for the industry.

Nowadays, the application of bioplastics can be found in many market segments such as packaging, food-services, agriculture and horticulture, textiles, and some electronics and automotive parts (see Figure 10.6). According to the market data, bioplastic production for packaging including both flexible and rigid packages amounted to approximately 0.99 million tons, sharing almost half of the global bioplastic market. In addition to several bio-based conventional polymers such as bio-PE and bio-PET, novel biodegradable polymers such as PLA are attracting increasing interests and gaining pace in the packaging market, owing to their high capacity for both industrial composition (for

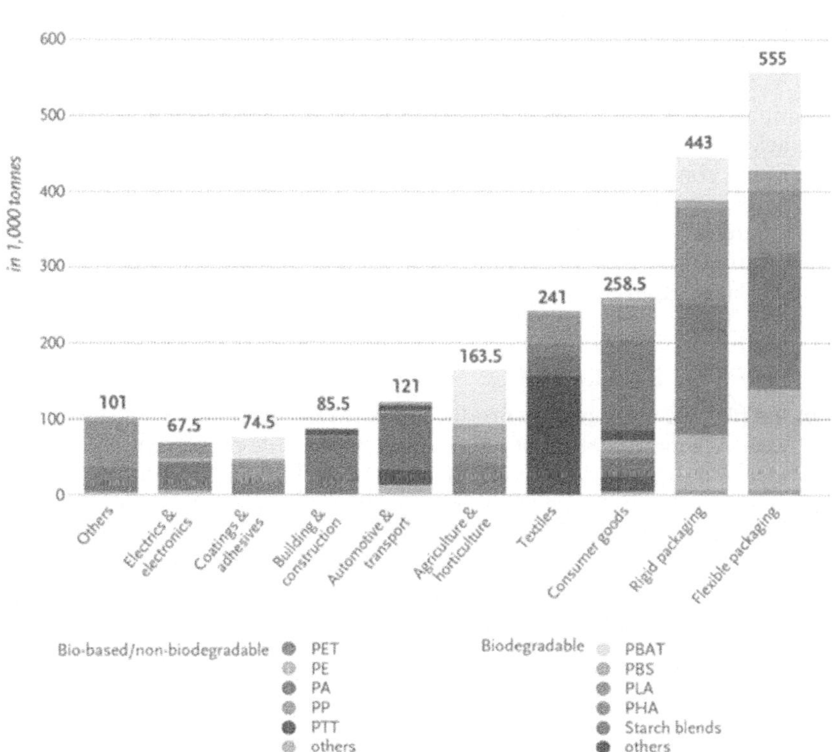

Figure 10.6 Global bioplastic production capacity by market segment, 2020 (*Source*: European Bioplastics, 2020c).

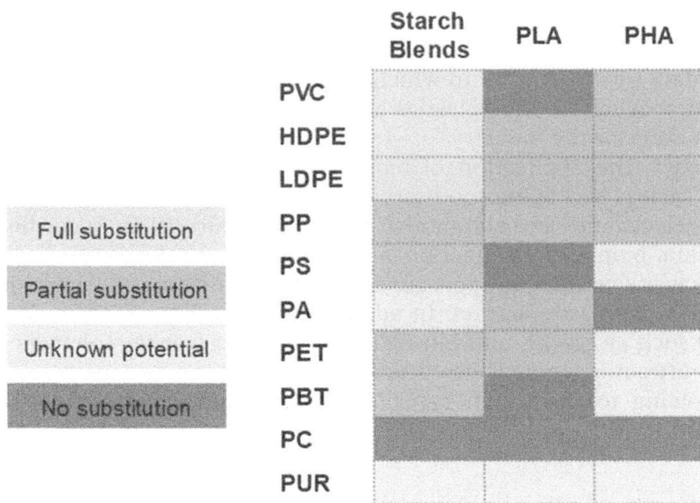

Figure 10.7 Potential substitution of fossil-based polymers by biopolymers (*Source:* modified from Spierling *et al.,* 2018).

contaminated packaging wastes) and potential for mechanical recycling (for rigid packaging purpose). Consumer goods, textiles and agricultural products are enjoying comparable market shares of around 8–12% in today's bioplastic industry. It is possible to conclude that today, for almost every conventional plastic material and application, there is a bioplastic alternative available on the market that has the same properties and potentially offers additional advantages (see Figure 10.7).

10.2 TYPES OF BIOPLASTICS

Bio-based plastics are typically produced via three routes (see Figure 10.8). The first route is via modification of natural polymers while preserving the polymer backbone (mainly) intact. This is the most important method today, used for the production of starch and cellulose-based plastics and for various other bio-based polymers. The second route comprises a two-step biomass conversion process starting with the production of bio-based monomers by means of biochemical and/or chemical transformation followed by the polymerization of the monomers in the final step. If the monomers obtained are bio-based versions of conventional monomers, they are called drop-in (replacements). They can readily enter existing processing and recycling systems. Bio-based monomers that have new structures or have not been applied to the markets in the past are also used to produce plastics. They often show an improved functionality, and thus additional markets and applications. However, novel bio-based plastics require the development and implementation of new recycling systems. The third route comprises the production of a polymeric material, which can be

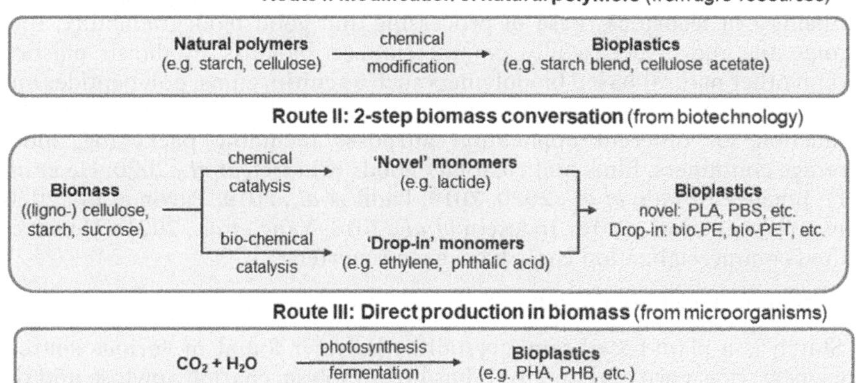

Figure 10.8 Classification of bio-based polymers based on their production routes (*Source*: modified from Brodin *et al.*, 2017; Storz & Vorlop, 2013).

used as a plastic without further modification, directly in microorganisms or plants. The modern genetic technology has allowed transfer of the function from microorganisms to crops, so this kind of bioplastics can be generated through photosynthesis. Based on their different production routes, bio-based plastics are categorized into three categories: (1) polymers from natural biomass (including agricultural resources); (2) polymers from biotechnology conversion; and (3) polymers from microbial fermentation.

10.2.1 Biopolymers from natural biomass

Numerous carbon-rich precursors from the natural biomass can be used in the production of bio-based polymers through polymer blending and/or chemical modification methods (Figure 10.9). Polysaccharides including starch and cellulose are commonly present in agricultural products and/or by-products

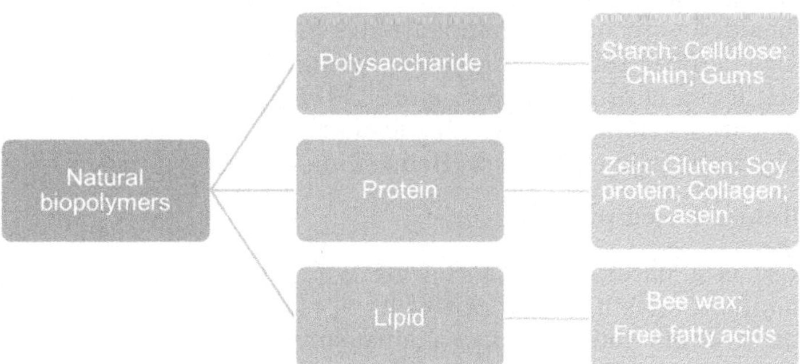

Figure 10.9 Natural biopolymers applied for bioplastic production.

such as corn, potatoes, maize, cassava and straws. Due to their low cost, high availability of feedstock, ease of processing and good biodegradability, they become the most successfully commercialized biomass produced plastics. Several other natural-based biopolymers such as chitin, gums, polypeptides and lipids have also been extensively studied and proven to be useful in bioplastic production for different application purposes including packaging, food/beverage containers, films and customer goods (Chaireh *et al.*, 2020; He *et al.*, 2017; Jiménez-Rosado *et al.*, 2020, 2019; Padil *et al.*, 2019; Pavon *et al.*, 2020; Ramakrishnan *et al.*, 2018; Tedeschi *et al.*, 2018; Yang *et al.*, 2021). However, limited commercialization cases have been reported.

- **Starch-based bioplastics**

Starch is a plant-based semi-crystalline polymer found in various sources like wheat, rice, corn and potato. It has linear polysaccharide amylose and the highly branched polysaccharide amylopectin. However, the complex and partly non-linear structure of the starch results in issues of ductility, which makes the starch-based plastics difficult to commercialize. Starch-based plastics can suffer from the phenomenon of retrogradation-increased brittleness due to increase in natural crystallinity over time. Therefore, plasticizers such as glycerol, glycol and sorbitol need to be added to thermoplastic processing of starch. The resulting plastics are also called thermoplastic starch (TPS) and can be tailored to specific needs by adjusting the amount of the additives. Addition of plasticizers can create biodegradable polymers with sufficient mechanical strength, flexibility and water barrier properties suitable for commercial packaging and consumer production (Avella *et al.*, 2005; Storz & Vorlop, 2013).

Starch-based plastics are used in biodegradable films for shopping, bread, fishing bait bags, flushable sanitary products, packaging material and special mulch films (Xie *et al.*, 2013). They can be processed via blow molding, extrusion and injection molding. The global market of start-based bioplastics was valued at $424 million in 2016, and is projected to reach the value of $561 million by 2023, registering a 3.7% increase in CAGR (Roy, 2017). As one of packaging solutions with the highest potential, starch-based bioplastics have been improved and applied in different ways including (1) blended with other compostable polymers such as PLA, PBS and PHAs to reduce the production cost; (2) blended with natural nanoparticles for reinforced nanocomposite to enhance mechanical, thermal and water resistance properties (Anugrahwidya *et al.*, 2021; Hong *et al.*, 2021; Sun *et al.*, 2014; Zhang *et al.*, 2014). Potential future applications could be loose fill packaging and injection molded products such as take-away food containers (Babu *et al.*, 2013).

- **Cellulose-based plastics**

Cellulose is a natural polysaccharide with crystalline morphology. It is the predominant constituent in cell walls of plants. Cellulose possesses good mechanical strength and resistance against solvents due to its strong hydrogen bonding in the crystalline regions of the polymeric structure. Cotton fibers and woods are the important raw material sources used to produce cellulosic plastics. It is difficult to be used directly for packaging because of its poor solubility, hydrophilic nature and highly crystalline structure. These features make the polymer film made of cellulose have poor moisture barrier property and relatively brittle. Hence, extensive research has been conducted to develop different cellulose derivatives for bioplastic packaging. Several main derivatives of cellulose for industrial purpose are cellulose acetate, cellulose esters (molding, extrusion and films) and regenerated cellulose for fibers (Babu *et al.*, 2013). The three main chemically modified cellulosic polymers are cellulose esters, cellulose ethers, and regenerated cellulose.

- Cellulose esters refer to a big family of cellulose derivatives that are found in many pharmaceutical applications. They can be produced through esterification of cellulose under acid catalysis. Three types of cellulose esters, namely cellulose acetate (CA), cellulose acetate propionate (CAP) and cellulose acetate butyrate (CAB), are commonly applied in plastic feedstock. As cellulose esters are inexpensive in cost, ease to produce transparent or colored products, and possessing good water resistance and toughness, they are commercialized for various bioplastic products such as films and fibers, wearables, automobile parts as well as other customer goods.
- Cellulose ethers are a group of organic polymers derived through etherification process of cellulose with different reagents such as methyl/ethyl chloride and ethylene/propylene oxide. They have particular value in pharmaceutical applications due to several characteristics including high glass transition temperatures, high chemical and photochemical stability, solubility, limited crystallinity, hydrogen bonding capability, and low toxicity (Arca *et al.*, 2018). Different cellulose esters including methyl cellulose (MC), ethyl cellulose (EC), hydroxyethyl cellulose (HPC), and carboxymethyl cellulose (CMC) have shown their values in film and coating, solar cell and battery, food, cosmetics and personal care, and especially in pharmaceutical applications such as tablet and drug emulsion (Dai *et al.*, 2019).
- Regenerated cellulose is a class of cellulose materials that are manufactured by the conversion of natural cellulose to a soluble cellulosic derivative and subsequent regeneration. It is typically produced in the form of fibers (also known as 'Rayon') or films (known as 'cellophane'), and widely applied in textiles, hygienic disposables, and home furnishing fabrics because of its good thermal stability and modulus (Edgar *et al.*, 2001). Nonetheless, due to their diverse applications in the textile industry, they are viewed as a major contributor of a significant proportion of microplastics in the marine environment due to the discharge of domestic sewage (Niaounakis, 2017).

Recent research and development on cellulose-based bioplastics have further enabled this inexpensive material towards a wider range of potential. For instance, intensive research has been conducted to develop cellulose-based biopolymers as a promising 'ink' for 3D printing. Considering the abundance of cellulose in the nature, and the disruptive innovation brought by 3D printing technology, it is believed that to combine these two emerging technologies is of critical interest for further promoting the sustainable plastic industry (Dai *et al.*, 2019; Giri *et al.*, 2021). Wang *et al.* (as cited in Edgar *et al.*, 2001) reported a sustainable hydro-setting method for the processing of a hydro-plastic polymer – cellulose cinnamate. Except for excellent tensile strength and Young's modulus, the most attractive characteristic of this bioplastic is that it can be processed into either 2D or 3D shapes with only contact with water. This innovative technology substantially reduces environmental burden during plastic production and extends the lifetime of plastic for practical applications.

- **Chitin and Chitosan**

Chitin is a nitrogen containing polysaccharide and ranks the second most abundant biopolymers in nature. Its molecular structure appears as well ordered crystalline microfibrils (Figure 10.10). Chitin can be extracted from shrimp, prawn and crab shells in which its content could reach 15–40% (Bi *et al.*, 2021). Considering the huge annual generation of shrimp and crab shell waste (approximately 6–8 million tons globally) which are discarded into the sea or ending in landfills, applying chitin-based biopolymer to produce bioplastics provides an attractive option to practice the 'waste-to-resource' principle. Chitins cannot be used directly as an initial feedstock for bioplastics because of its low solubility. However, they can be modified to chitosan through the alkaline deacetylation process. Chitosan has several unique characteristics including biodegradability, biocompatibility, chemical inertness, high mechanical strength, good film-forming properties, antibiotic

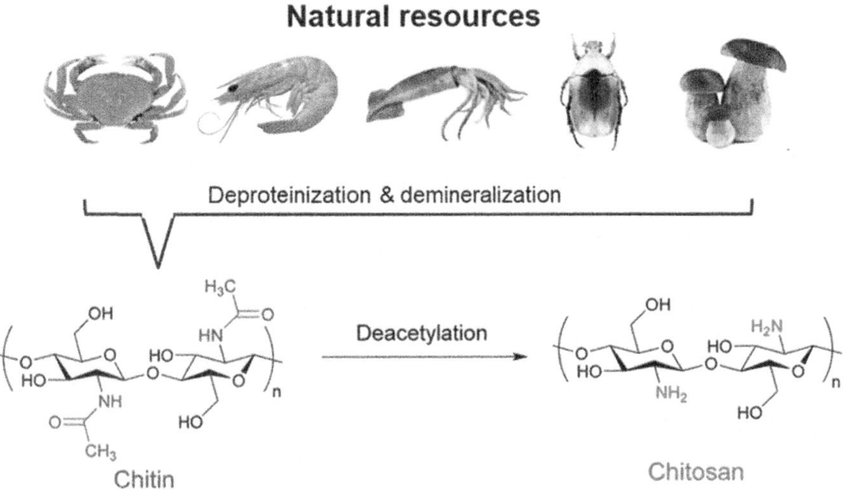

Figure 10.10 Schematic of conversion from chitin to chitosan and its natural sources (*Source*: modified from Jin *et al.*, 2021).

activity, low toxicity and low cost (Shukla *et al.*, 2013; Wang *et al.*, 2018; Zargar *et al.*, 2015). Due to its preferable features, chitosan has been used to develop a number of functional materials such as bioplastic films, medical devices and pharmaceutical ingredient, and pollutant adsorbents are utilized in various fields including food preservation, biomedicine, and environmental protection (Bi *et al.*, 2021; Duan *et al.*, 2018; Mujtaba *et al.*, 2019).

The method to fabricate large 3D plastic objects from chitosan has been developed by Fernandez and Ingber (2014). This method can preserve the strong mechanical properties of chitosan, and produce tough, transparent, and large plastic products with complex shapes using traditional, large-scale manufacturing techniques such as casting or injection molding. According to the report, 'chitosan-based plastic is cheap and easy to make on larger scales and would not take up land like plant-based bioplastics. The material can be dyed by changing the acidity of the chitosan solution; the dyes can be collected and reused when the material is recycled' (Fang, 2014). Recent research on chitin nanomaterials (i.e., nanocrystals and nanofibers) offers new opportunities to explore the novel applications of chitin and chitosan biopolymers in biomaterial, electrochemistry, energy storage, separation/absorption, catalyst carrier and textile materials (Duan *et al.*, 2018).

- **Gum-based polymers**

Algae gum is a natural polysaccharide rich in seaweeds ($\sim$50% of dry weight) such as blown algae, red algae, and green algae. Applying seaweeds as the raw feedstock to produce bio-based plastics offers more benefits compared with other plant-based polysaccharides. For instance, seaweeds have fast growth rate and extensive environmental tolerance; they require neither large land use nor fertilizers/pesticides; and the harvest of seaweeds is usually easy and cheap (Lim *et al.*, 2021; Zhang *et al.*, 2019). Furthermore, seaweeds can uptake CO_2 through photosynthesis which helps to address the global warming issue. Seaweeds can be used to form bioplastic films either directly or via their derivatives such as alginate and carrageenan. They can also be composed with other polymers to improve the characteristics and properties of bioplastic products. Though seaweed films formed directly without chemical treatment show great potential as a green technology, they are still new and require more research (Lim *et al.*, 2021). Most typically, seaweed-based biopolymers are produced from their derivatives.

Alginate is a polysaccharide that accounts for 22–44% of the dry cells weight of marine brown seaweed (Zhang *et al.*, 2019). It is a linear copolymer containing mannuronic acid (M) and guluronic acid (G) linked by 1,4-glycosidic linkages. Alginate has been widely used in various biomedical applications, such as drug delivery systems and cartilage (Sun *et al.*, 2012). When cross-linked with multivalent cations such as Ca^{2+}, alginate can form hydrogels or packaging films with good mechanical strength and water barrier. Cations-associated alginate has been proven as a potential feedstock in the development of various environmental-friendly packaging materials (Benselfelt *et al.*, 2018; Cazón *et al.*, 2017; Xue *et al.*, 2019).

Carrageenan is a common vegetal polysaccharide that is composed of D-galactose and 3,6-anhydo-D-gactose, mainly extracted from marine red algae (30–50%). Similar to alginate, the double helix arrangement in carrageenan provides anions-hosting position for facilitating a better gel structure. Many carrageenan bioplastics are aimed at developing edible packaging films (Cazón *et al.*, 2017). The global carrageenan market is the largest share of the global hydrocolloids market, which is predicted to reach approximately US$ 1 billion by 2024 (Zhang *et al.*, 2019). Recently, some viable products from seaweed have been provided a readily accessible margin for markets. For instance, the Loliware company based in New York have commercialized edible cups and straws which are made of seaweed gum (Loliware, 2020). The Mojito straw comes from the UK along with the Ooho water capsule, delivers a serving of water in a thin pouch made from edible brown algae (Villaluz, 2017). However, over 80% of the products are only concerning application sectors such as processed meats, dairy, desserts and jellies (Campbell & Hotchkiss, 2017). In terms of the number of new bioplastic products appearing, it will emerge as a more lucrative market in the future.

- **Protein and polypeptide-based bioplastics**

Various crop proteins (corn zein, wheat gluten, soy proteins, etc.) and animal proteins (collagen, gelatin, keratin, casein, etc.) have been used to manufacture bioplastics (see Figure 10.11). A protein-based polymer is composed of 3D macromolecular structure which is strengthened and stabilized by hydrogen bonds, hydrophobic interactions and disulfide bonds. Since protein itself does not have sufficient plasticity and malleability, a plasticizer is typically added during the manufacturing to modify the flexibility and mobility of polymeric chains (Jerez *et al.*, 2007).

Corn zein has been used to form fibers and thin films for various applications such as clothing production, biomedicine and food grade plastics (Zhang *et al.*, 2015). It is a class of alcohol-soluble prolamine proteins present in maize

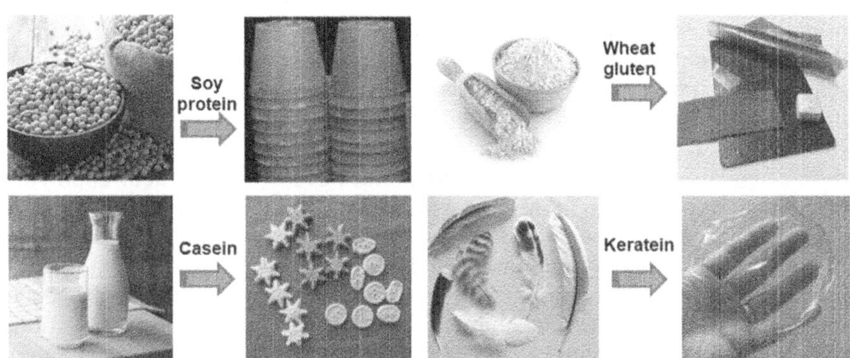

Figure 10.11 Various protein and polypeptide-based bioplastics.

endosperm, and has been approved as a safe material for food and pharmaceutical applications by the United States Food and Drug Administration since 1985.

Collagen is a major insoluble fibrous protein in the extracellular matrix and in connective tissue (Babu *et al.*, 2013). Abundant collagen can be found in pig skin, bovine hide, and pork and cattle bones. It is one of the important biomaterials due to its biocompatibility, biodegradability and weak antigenicity. Applications of collagen films in drug delivery systems and tissue engineering have been also explored since 1970s.

Keratin extracted from chicken feathers has been tested to produce bioplastic films (Ramakrishnan *et al.*, 2018). Keratin is a non-burning, hydrophilic, biocompatible and biodegradable protein that accounts for 90% of the feather protein. Instead of being disposed of in landfills, keratin can be extracted from chicken feather waste through chemical processing and recycled for bioplastic production.

Despite extensive research explorations on various protein-based bioplastics, its industrial scale production and application have yet to emerge. One concern is the impact that using protein-rich crops for bioplastic manufacturing may cause risk on food security. To address this concern, researchers proposed to utilize genetically modified crop products such as corn and potatoes for bioplastic production (Thiruchelvi *et al.*, 2021). In addition, proteins recycled from agricultural by-products or organic waste can be an option to avoid competition of protein-based bioplastic feedstock with valuable food crops.

10.2.2 Biopolymers from biotechnology via conventional synthesis

This second way of producing bioplastic consists of a two-step biomass conversion approach that starts with the production of bio-based precursors (monomers) via biochemical and/or chemical transformation, followed by monomer polymerization in the second step. However, both steps can be complex and can be further divided into several sub-steps. The bio-based version of conventional monomers is called drop-in (replacements). Drop-in monomers are advantageous as they share the same chemical and physical properties with conventional plastic and hence can readily enter existing processing and recycling systems. One example of this type of bioplastics is bio-polyethylene. On the other hand, novel bio-based monomers which have new chemical structures or have not yet been applied to the markets can be formulated through monomer polymerization process. These new polymers often show improved functionality and are able to gain the favor of the markets. One classic example of these novel monomer-derived bioplastics is polylactic acid (PLA).

- **Bio-polyethylene (Bio-PE)**

Traditionally produced from fossil resources, polyethylene (PE) is an important engineering polymer manufactured via polymerization of ethylene under pressure, temperature, and in the presence of catalysts. Ethylene can be formulated through steam cracking of naphtha or heavy oils, or through ethanol dehydration. The former process results in ethylene from petrochemical feedstock, while the latter process is used to synthesize ethylene from bio-ethanol. Bio-ethanol is the fermentation products of agricultural feedstock such as sugarcane and corn. Extracted sugarcane juice with high sucrose content is anaerobically fermented to produce ethanol in a typical industrial fermentation process. The ethanol is distilled in order to remove water and to yield a zeotropic mixture of hydrous ethanol at the end of the process. Then, it is dehydrated at high temperature over a solid catalyst to produce ethylene and subsequently, polyethylene (Chen *et al.*, 2007; Yakovleva *et al.*, 2016). Bio-PE has exactly the same chemical, physical and mechanical properties as petrochemical produced polyethylene. It is currently the most widely used bio-based nondegradable bioplastics. Bio-PE and their blends are used in engineering, agriculture, packaging, and many day-to-day commodity applications because of its low price and good performance (Babu *et al.*, 2013).

- **Polylactic acid (PLA)**

PLAs are aliphatic polyesters with the basic constitutional unit of lactic acid. The monomeric lactic acid is a hydroxyl carboxylic acid that can be obtained by bacterial fermentation of corn, tapioca and sugarcane. Corn outperforms other renewable feedstocks owing to its high productivity of qualified lactic acid. PLA can be synthesized from lactic acid by direct polycondensation reaction or ring-opening polymerization of lactide monomer. However, it is difficult to obtain high molecular weight PLA through polycondensation reaction due to water formation of the reaction (Babu *et al.*, 2013). Figure 10.12 illustrates the fabrication process of PLA.

PLA shares some similarities with hydrocarbon polymers like PET. Its unique characteristics such as transparency, glossy appearance, high rigidity and ability to tolerate various types of processing conditions equips it with a high potential to replace several traditional polymers such as PET, PS and PC for packaging. Although the mechanical properties of PLA and PET are similar, the thermal properties of PLA are not attractive due to its low glass

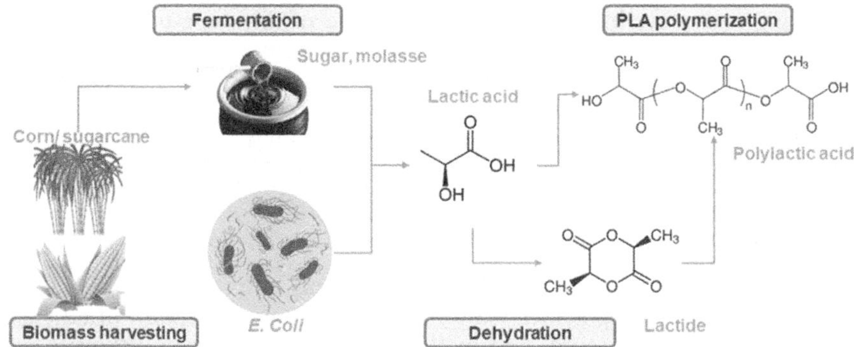

Figure 10.12 Schematic of PLA fabrication process.

transmission temperature (T_g) at approximately 60°C. However, this can be overcome by changing the stereochemistry of the polymer and blending it with processing aids and other polymers. For instance, NEC corporation of Japan has developed a reinforced PLA with carbon and kenaf fibers which improves thermal and flame retardancy properties (Serizawa *et al.*, 2006). PLA has also been grafted with chitosan to enhance the degradability (Elsawy *et al.*, 2017). PLA is widely used in food industry to provide disposable packaging, for example, food trays, tableware, water bottles, candy wrappers and cups. It is also applied for soil retention sheathings, agriculture films, waste shopping bags, disposable garments, feminine hygiene products and diapers (Atiwesh *et al.*, 2021; Elsawy *et al.*, 2017). It is worth noting that PLA is currently known as the most easily biodegradable thermoplastics. It can be degraded by hydrolysis or thermally.

According to a report by Mordor Intelligence (2021), the global market for bio-PLA is expected to reach |USD 1638.76 million by 2026, registering a CAGR of around 14.96% during the 2021–2026. Currently, the bio-PLA market is highly concentrated, and the four major players have a share of over 80% market in terms of production capacity. Several leading PLA manufacturers in the market include NatureWorks LLC, Zhejiang Hisun Biomaterials Co. Ltd, Sulzer Ltd, Futerro, and Total Corbion PLA. Table 10.1 summarizes several leading industries for PLA production. Similar to the cellulose-based bioplastics, the growing production of genetically modified corn and the growing usage of bio-PLA in 3D printing are likely to act as opportunities in the future.

10.2.3 Bioplastics from microorganism fermentation

This includes the production of polymeric materials, in microorganisms and plants that can directly be used as plastic with slight modifications. Advances in genetic engineering and biotechnology have made it possible to manipulate the functional genes responsible for production of biopolymers such as polyhydroxyalkanoates (PHAs) in microorganisms. Several bacterial-derived PHAs including polyhydroxybutyrate (PHB), poly(hydroxy-butyrate-hydroxyvalerate) (PHB/HV), and poly(ε-caprolactone) (PCL) have been investigated as novel renewable feedstock for bioplastic production. However, even though direct production of bioplastic through microbial fermentation seems feasible, it has not been produced extensively due to the relatively low microbial productivity and the related complex environmental and regulatory issues.

• **Polyhydroxyalkanoates (PHAs)**

PHAs refer to a class of bio-based polyhydroxyesters which can be synthesized by certain bacteria, microalgae, and plants from renewable sources. PHAs are biocompatible, biodegradable and non-toxic, hence, presenting the potential to replace conventional hydrocarbon-based polymers. A number of bacteria including heterotrophics and photoautotrophics have

Table 10.1 Global leading industries for PLA production.

PLA Manufacture	Description	Products and Applications
Luminy by Total Corbion PLA (Total Corbion, 2021)	Total Corbion PLA is a global leader in marketing, sales and production of polylactic acid (PLA). Its Luminy® PLA portfolio, which includes products in four categories. • High heat PLA for demanding applications • Standard PLA for general purpose applications • Low heat PLA typically used as seal layer • PDLA used either as a nucleating agent or to create full stereocomplex compounds It is an innovative material that is used in a wide range of markets including fresh food packaging, consumer goods, fibers, food service ware and 3D printing.	Rigid food packaging, flexible packaging, food service ware, durable goods, non-wovens, 3D printing
Ingeo by NatureWorks LLC (NatureWorks, 2021)	NatureWorks is now a world-leading biopolymer supplier and innovator with its Ingeo portfolio of naturally advanced materials made from renewable, abundant feedstocks with performance and economics that compete with oil-based intermediates, plastics, and fibers. The company is jointly owned by Thailand's largest chemical producer, PTT Global Chemical, and Cargill, which provides food, agriculture, financial and industrial products and services to the world. Its manufacturing facility, located in Blair, Nebraska, USA, has a name plate capacity of 330 million pounds (150 000 metric tons) of Ingeo biopolymer.	3D printing, beauty & household, building & construction, food service ware, electronics & appliances, food & beverage packaging, agriculture, medical & hygiene, and so on.
Futerro PLAs by Futerro (Futerro, 2021)	Futerro was established in September 2007 to bring to fruition the project to develop technology for PLA production from renewable vegetable resources. Futerro PLA grades have been successfully used in various transformation techniques, such as cast film, blown film, injection, thermoforming, blown moulding, ISBM and those employed to produce fibers. Due to its high purity, the Futerro biopolymer is especially well suited for the fibers production or any other application that requires a semicrystalline grade.	Films, packaging, fibers, and so on.
REVODE by Zhejiang Hisun Biomaterials Co., Ltd. (HISUN, 2021)	Zhejiang Hisun Biomaterials Co., Ltd (hereafter referred to as Hisun Biomaterials) is located in Taizhou, southeastern China. It is a high-tech enterprise engaged in production of PLA. HISUN is the first enterprise engaged in the industrialized production of polylactide in China and high-temperature-resistant polylactide in the world.	Tableware, stationery, household, textile, 3D printing, films, baby care

been found to produce PHAs as an energy and carbon storage materials in cells typically under nutrient-limited conditions in the presence of excess carbon source (Anjum *et al.*, 2016; Laycock *et al.*, 2013). The organic carbon sources that can be used to synthesize PHAs include various crop biomass, lingo biomass, and organic-rich waste and wastewater. The biosynthetic process of PHAs involve directly or indirectly with many central metabolic pathways including glycolytic and pentose phosphate pathways, Krebs cycle, and pathways for the biosynthesis and degradation of amino acids and fatty acids (Lu *et al.*, 2009). Figure 10.13 shows the metabolic pathways of PHAs in microbial cells.

A general process to produce PHAs through microbial fermentation process is composed of anaerobic fermentation, isolation, and purification from the fermentation broth. Microbial seed culture is inoculated in the fermentation vessel together with necessary mineral medium. Excess carbon source is provided for the microorganisms for cell growth and PHAs accumulation. Extraction and purification of biosynthesized PHAs is the critical step to determine the quality and efficiency of biopolymer production at the industrial scale. Typical extraction approaches such as solvent extraction, flotation, digestion, supercritical fluid extraction, mechanical extraction (e.g., ultrasonic or microwave), and enzyme-assisted extraction have to be developed and optimized in order to ensure high quality and purity of the products (Chong *et al.*, 2021; Muneer *et al.*, 2020). In addition to microbial fermentation, PHAs

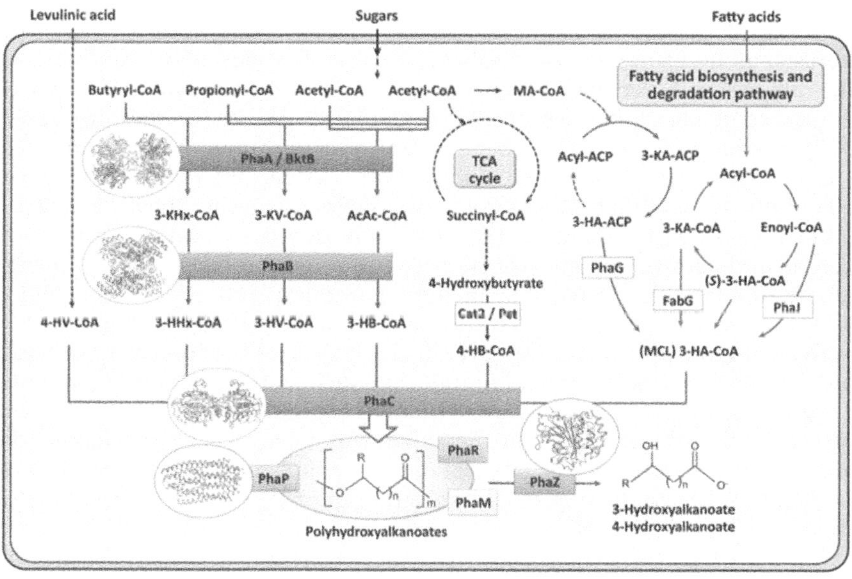

Figure 10.13 An overview of the metabolic process of PHAs by microorganisms (*Source*: Sagong *et al.*, 2018).

may also be produced through chemical synthesis or genetically modified plants (Mannina *et al.*, 2020). These technologies, however, are less attractive than biological synthesis process.

The first bacterial PHA identified was poly-3-hydroxybutyrate which has received the greatest attention for pathway characterization and industrial scale production. It has thermal and mechanical properties similar to polystyrene and polypropylene (Savenkova *et al.*, 2000). However, it may have shortcomings like slow crystallization, narrow processing temperature range and tendency to 'creep', and hence not suitable for many items (Gigante *et al.*, 2020). Depending on their monomer compositions, PHAs can be thermoplastic or elastomeric, exhibiting various physical and chemical characteristics, such as tensile strength, elongation to break, biocompatibility and biodegradability. It has been reported that poly (3HB-co-3HV) can be completely degraded in anaerobic sewage, soil and sea water within 6, 75 and 350 weeks, respectively (Castilho *et al.*, 2009). The potential in natural degradation by marine microorganisms makes PHAs more attractive as a way to address the marine plastic debris issue (Atiwesh *et al.*, 2021). PHAs can be processed by existing polymer-processing equipment and can be converted into injection molded components such as fibers, laminates, non-woven fabrics, synthetic paper production, feminine hygiene products, adhesives, waxes, binders and foam. Items such as forks, spoon, knives, tubs, trays, hot cup lids, and products such as housewares, cosmetics, and medical packaging can be made from it (Philp *et al.*, 2013).

Notably, utilizing agro-food industrial organic waste and wastewater as the feeds for PHAs fermentation and production is attracting increasing interests, as it can reduce the feedstock cost to produce PHAs and possibly integrate resource recovery from organic waste and wastewater into plastic circular economy (Bhatia *et al.*, 2021; Chong *et al.*, 2021; Mannina *et al.*, 2020). A wide range of waste organic substrates including food waste, sugar cane molasses, waste activated sludge, mill wastewater, milk/cheese whey, dairy manure, and different types of food process wastewater have been tested and demonstrated for the potential of PHA biosynthesis. Both heterotrophic bacteria and autotrophic microalgae have been found to be able to synthesize and accumulate the PHAs for further extraction and processing. Figure 10.14 illustrates a concept of PHB production process based on a wastewater microalgae treatment plant. However, biological PHA production from waste and wastewater is facing several bottlenecks such as the complex constitution of organic substrates, relatively low productivity, and the lack of efficient and low-cost PHAs extraction and purification technologies. These drawbacks limit technical and economic feasibility and pose major challenges to production scale-up of PHAs to the industrial level (Rodriguez-Perez *et al.*, 2018). Further research efforts are still needed.

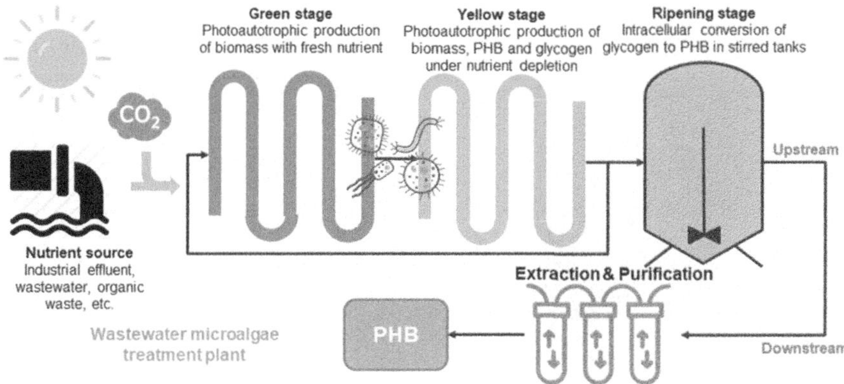

Figure 10.14 A conceptual process of utilizing wastewater to produce PHB.

10.3 INNOVATIVE APPLICATIONS OF BIOPLASTICS

Biodegradable and compostable polymers are predominantly used in sectors where the property of biodegradability is an added advantage during their use and their end-of-life, such as agriculture, horticulture, cosmetics, industrial and flexible packaging. In contrast, they are almost absent from transportation and building materials. However, modern bioplastics with largely enhanced physicochemical properties and innovative applications are now suitable for an impressive range of applications (Figure 10.15). Some of the innovative applications of bioplastics have been summarized in Table 10.2.

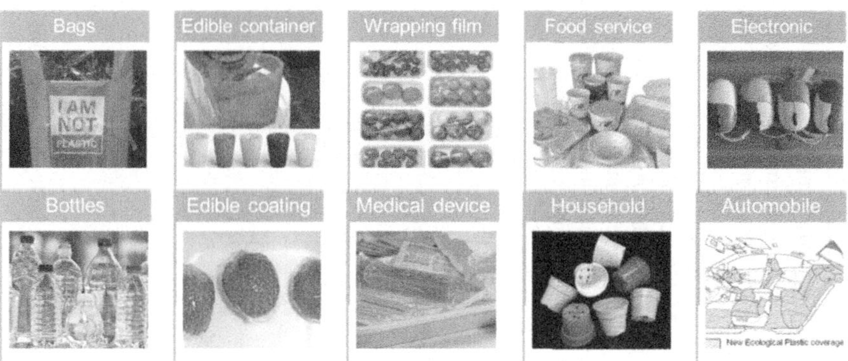

Figure 10.15 Various applications of bioplastics.

Table 10.2 Examples of recent innovative application of bioplastic.

Case I: Plastic car made from soybeans, U.S. (Benson Ford Research Center, n.d.)

In the early 1940s, Henry Ford experimented using bioplastics made from soy beans for some of the car parts. *The New York Times* in 1941 states that 'the body of the car and fenders were made from a strong material derived from soy beans, hemp, wheat, and corn' (The New York Times, 1941). One of the reasons that Henry Ford decided to build this car was to look for a project that would combine the fruits of industry with agriculture. Popular in the history as the 'Soyabean Car', it was much lighter and more fuel efficient than the cars made from steel. However, the experiment of plastic car was suspended due to the breakout of World War II. Although this automobile never made it to the market, it demonstrates a good example of innovative attempt using bioplastics.

Henry Ford's soybean car. Credit: The Henry Ford

Case II: Vegan, biodegradable and compostable fashion accessories, Australia (Barrett, 2018a)

Conventional glitter is made up of plastic and aluminum bonded with polyethylene terephthalate (PET). However, a 10-year-old Australian girl, Sofia Rizzo launched a vegan, biodegradable and compostable glitter brand named 'Glitter Girl'. The glitters are primarily made from trees, primarily eucalyptus, sourced from responsible managed resource.

Eco glitters made from biodegradable plastic. Credit: glittergirl.com.au

(Continued)

Case III: 3D glasses made from Cereplast's bioplastic, U.S. (nova-Institut GmbH, 2010)

With the popularity of 3D cinema today, it is estimated that the consumption of 3D glasses for each movie reached approximately 10 million pairs across the globe. These 3D glasses, typically made of fossil-based plastic, resulted in the CO_2 emissions equivalent to burning 50 000 gallons of gasoline, and eventually ended up in the landfills. In 2010, Cereplast Inc. and the Oculus 3D joined hands to manufacture the biodegradable/compostable 3D glasses. These 3D glasses featured Cereplast's compostable resin, made with Ingeo® poly-lactic acid. These resins allowed for the manufacturing of glasses made of renewable material and create a truly compostable product. If discarded at a compost site, the 3D glasses will return to nature in less than 180 days with no chemical residues or toxicity left in the soil.

Case IV: The ecolactifilm packaging by Lactips company, France (Barrett, 2018b)

Lactips' milk-based biodegradable plastic. Credit: Bioplastic magazine.com

The French start-up company 'Lactips' developed a milk-based biodegradable and water-soluble packaging material which composed of 100% natural ingredients without chemical treatment. This packaging film breaks down harmlessly in water or home compost, and takes about three weeks to biodegrade. It can be printed with labels or usage instructions, and is fully water soluble at low temperature. With slight adjustments, it can be produced using existing plastic processing machinery.

(Continued)

Table 10.2 Examples of recent innovative application of bioplastic (*Continued*).

Case V: Recyclable bioplastic membrane to clear oil spills from water, the Netherlands (Ye *et al.*, 2021)

Polymer scientists from the University of Groningen and NHL Stenden University of Applied Sciences, the Netherlands, have developed a superamphiphilic vitrimer epoxy resin (SAVER) membrane to separate oil from water. The polymer membrane was fabricated from bio-based malic acid, hence fully degradable and recyclable. It is expected to be a robust membrane material for efficient oil-spill remediation. When the pores are blocked by foulants, it can be depolymerized, cleaned and subsequently pressed into a new membrane.

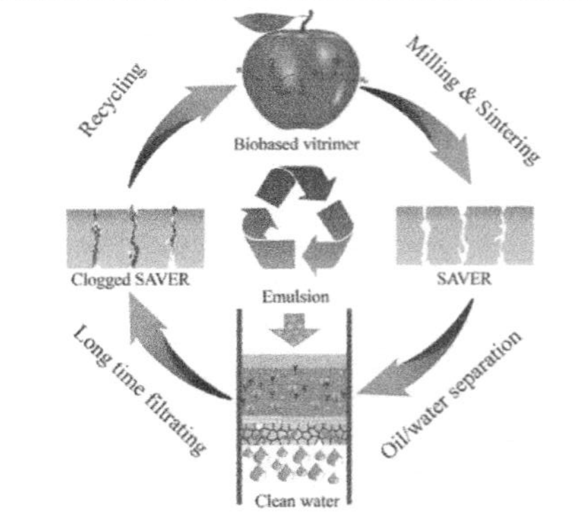

The closed-loop life of SAVER membrane (*Source*: Ye *et al.*, 2021).

Case VI: Edible bioplastic packaging (Patel, 2020)

Global concern on plastic waste is growing rapidly, especially for the food packaging sector. According to Patel (2020), the most common plastic items that end up as litter and accumulate in the ocean are linked to food: wrappers, straws, cutlery and bottles among others. Edible packaging offers an innovative and sustainable alternate to address such issue.

A UK start-up company Notpla produced waterproof beverage capsules made from seaweed. The capsules can be gulped and swallowed together with the beverage. In the case it is spitted out and disposed of as a waste, complete biodegradation can be accomplished within 4–6 weeks. Another US-based company Loliware turned alginate and agar extracted from seaweed and red algae into flavored straws. The products have attracted great interests from business partners such as Marriott Hotels that have started using their straws since 2019. An Indonesian company Evoware founded in 2016 has developed its edible seaweed-based films for food packaging and tested them as burger wrappers, instant noodle sachets and coffee pouches.

(Continued)

Despite several technical issues and barriers, increasing efforts are carried out by companies and academic researchers across the globe to convert extensive natural-based biopolymers such as potato starch and milk proteins into edible plastic packaging.

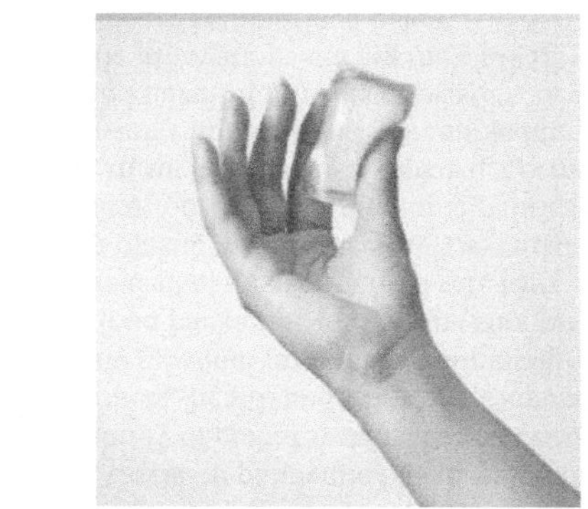

Edible plastic package by Notpla (*Source*: Credit: Notpla).

Case VII: Bioplastic silverware from avocado pits, Mexico (UNEP, 2019)

A Mexican entrepreneur, Scott Munguía Biofase, successfully recycled waste avocado seeds into bioplastic products. Unlike other types of bioplastics, this alternative does not consume crops such as corn or cassava, hence it would not cause food security risks. In 2014, he founded Bioface, a company that commercialized such bioplastics, with various products including drinking straws, cutlery, and containers. These bioplastic straws were claimed to degrade after 240 days of being buried in the ground unlike their fossil-based plastic counterparts which may take up to 100 years to degrade or disseminate.

Biofase's bioplastic straw made from avocado seeds (*Source*: Credit: 4eco Limited).

(Continued)

Table 10.2 Examples of recent innovative application of bioplastic (*Continued*).

Case VIII: First house on Mars will be made from bioplastics, U.S. (Barrett, 2018c; SpaceFactory, n.d.)

The NASA's Centennial Challenges program approved a Marsha Project led by a U.S. research firm called AI SpaceFactory to build a 3D-printed bioplastic habitat for deep space exploration, including the agency's journey to Mars. The 3D print technology utilizes recyclable biopolymer composite which has been proven to outperform concrete in strength, durability and impact resistance. It was ASTM certified for safety and feasibility in multi-planetary application.

The conceptual bioplastic housing on Mars (*Source*: Credit: AI SpaceFactory).

Case IX: Bioplastic PPEs to combat COVID-19

Bioplastic face shield designed by Alice Potts (*Source*: Credit: dezeen.com).

Since December 2019, the world has been affected by a pandemic originated from a novel coronavirus (SARS-CoV-2) responsible for a severe respiratory syndrome known as COVID-19. The Covid-19 pandemic has revealed the vulnerability of global systems to protect the environment, health and economy. Dependence on plastic personal protective equipment (PPE), masks, food packaging and online shopping has increased at an exponential rate thus creating plastic waste exceeding the capacity of incineration, landfill and municipalities to handle. In such scenarios, bioplastic becomes an innovative application to decouple human safety and harmful impacts of plastics. Several bioplastic innovations during the COVID pandemic are summarized below.

(Continued)

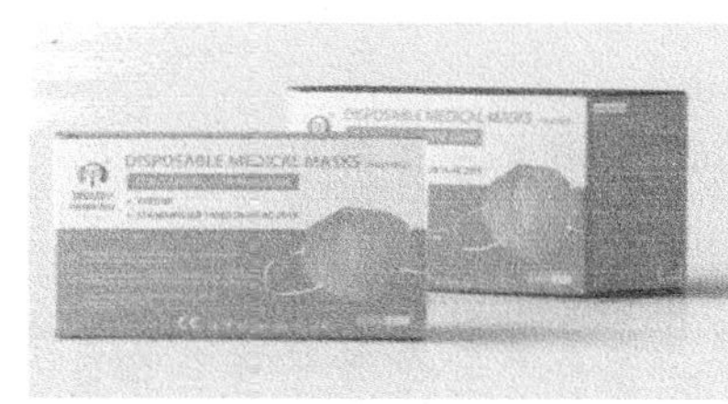

The biodegradable mask of Health Box (*Source*: Credit: Health Box Technology Co., Ltd.).

- A UK designer Alice Potts used food waste and flowers to create bioplastic face shields for the National Gallery of Victoria Triennial 2020 in Australia. The face shield was composed of a 3D-printed top section combined with a biodegradable, bioplastic shield. The designer believed that it can serve as an alternative to many items of personal protective equipment (PPE) that are made from single-use plastic during the global pandemic (Carlson, 2020).

- A Chinese mask-maker named 'Health Box' began the production of polylactide-based masks which are sourced from cornstarch and straw and are fully biodegradable. The cost of such biodegradable mask is approximately twice of a traditional mask made of non-woven or melt-blown cloth. Nonetheless, due to its eco-friendly nature, high interest and inquiries have been received from business partners from Japan, Spain and Saudi Arabia among others (Ye, 2021).

- The 2 in 1 virus protection face mask/ bioplastic shield project initiated by Mindanao State University, Philippines, aimed at creating bioplastics from locally available materials to make virus/dust protection face masks and face shields. The project converts water hyacinth into face masks and disposable pads and banana peel to bioplastic shields. These materials are found to be abundant and useless in the local area. The innovation is also believed to be valuable in supporting the local economy through establishing small business and providing job opportunities (Enactus Plus, n.d.).

10.4 BIODEGRADATION AND STANDARDIZATION OF BIOPLASTICS

Biodegradability is an attractive feature of bioplastics as it is possible to help addressing the persistent accumulation of waste plastic litter in the environment. Degradation of plastic polymers can be triggered by different mechanisms and processed in the natural environment and industrial treatment facilities. Understanding these degradation processes and their mechanisms will help to make a proper treatment and disposal decision. It is valuable to properly assess the environmental and ecosystem risks attributing to plastic mismanagement and littering. Biodegradability of bioplastics is often a misused and abused term. Since green consumerism is on the rise, manufacturers and companies can often be seen labelling their products in eco-friendly terms such as biodegradable, bioplastic, oxodegradable and compostable without specifying particular conditions or environments that the product will need to actually biodegrade. Bioplastics have become a buzzword in the market to promote the green brands to the customers. Just because a product is bioplastics doesn't necessarily mean it is eco-friendly. The results of biodegradability largely depend on their testing conditions such as humidity, environment, temperature and microorganism. Although bioplastics degrade faster than most conventional plastics, it still possibly leads to issues like microplastic leakage, contamination during recycling and confusion to the current plastic recycling system. Therefore, the standardization of bioplastic waste management is necessary to better incorporate bioplastics into the conventional plastic waste management systems and further expand their advantages and market potential.

10.4.1 Polymer degradation

Various environmental factors such as mechanical force, light, heat, chemicals and microorganisms influence the physical properties (e.g., integrity, shape, and morphology) and chemical structures and/or compositions of polymers, and result in their degradation. Depending on the driving forces of polymer transformation, the primary types of environmental degradation include mechanical degradation, chemical degradation, thermal degradation, photo degradation and biological degradation (Fotopoulou & Karapanagioti, 2019; Luyt & Malik, 2019). Figure 10.16 shows these main degradation routes of polymers in the natural environment. Mechanical degradation usually causes the initial morphology change and bulk fragmentation of polymers through cracking, embrittlement and flaking, resulting in the leaching of plastic litter in the environment (Chamas *et al.*, 2020). It can be caused by various physical processes such as wind erosion, wave action and any machinery force in nature.

Chemical degradation refers to the changes at the molecular level of the polymers such as bond cleavage or oxidation of long chains to create new molecule typically with shorter chain lengths. Chemical degradation of plastic polymers usually involves two reactions oxidation or hydrolysis which occurs at near-ambient temperatures in the environment. These chemical reactions can simultaneously occur with the other processes such as thermal, photo and biological degradation which accelerate the deterioration of polymers (Chamas *et al.*, 2020).

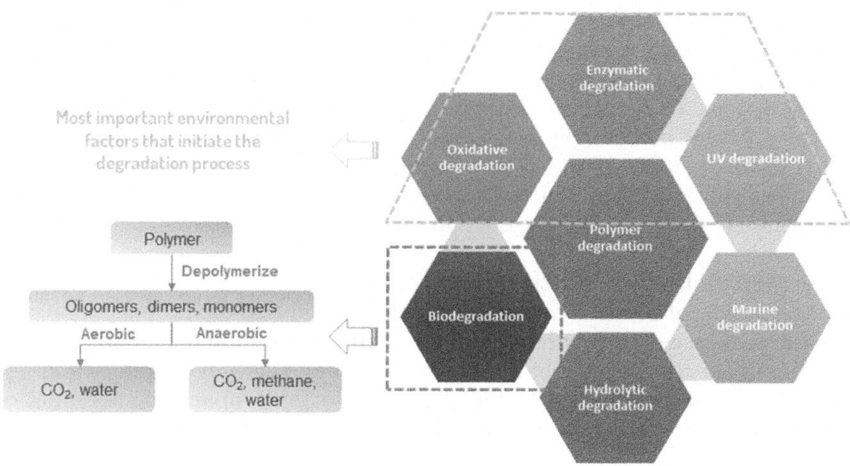

Figure 10.16 Polymer degradation in natural environment.

Thermal degradation refers to the deterioration of a polymer molecule due to overheating (Luyt & Malik, 2019). The mechanism of thermal degradation of polymers is complex in nature. It involves the occurrence of molecular scission which leads to the changes in molecular weight distribution of the material (Das & Tiwari, 2017). Heating results in many physio-chemical changes of plastic polymers such as melting, sublimation, polymorphic transformation, and degradation. Thermal degradation of plastic polymers typically occurs in different thermal treatment processes such as pyrolysis and gasification, or during the downcycling process which involved thermal molding and extrusion (Lamberti *et al.*, 2020).

Photo degradation refers to the alternation of material through photolysis, which is initiated by the sunlight, for example, ultraviolet radiation, gamma, and X-rays. The majority of organic polymers tends to absorb the low wavelength band radiations such as UVA (~315–400 nm) and UVB (~295–315 nm), which activate the electrons to transit to a higher energy level (Shah *et al.*, 2008). This process causes the bond cleavage of the polymeric chains through oxidation and/ or hydrolysis, resulting in lower molecular weight components. Photo degradation is one of the most important processes which initiates the degradation of plastic polymers in the natural environment (Luyt & Malik, 2019).

Biodegradation refers to the process that utilize microorganisms, such as bacteria and fungi, to decompose the long-chain polymers into short chains, monomers, or even the end products such as carbon dioxide, water, methane and minerals in proper environment. Biodegradation of polymers can occur in natural environment such as in soils, sediments, rivers, lakes and oceans. It can be achieved in artificial environment such as wastewater treatment facilities, waste composting facilities and landfills.

Notably, the abovementioned degradation processes are not solely occurring in the environment. Instead, they are functioning simultaneously or in combinations which accelerate the degradation process. Despite various degradation mechanisms available, it takes a long time to completely decompose the conventional plastic polymers due to their stable molecular structure and inter-physicochemical characteristics. Chamas *et al.* (2020) compared the degradation rates of different plastic polymers under variable natural environment (Figure 10.17). Biodegradable plastics are believed to have significantly enhanced degradation performance which significantly shortens the duration needed for their complete decomposition in the natural environment.

10.4.2 Mechanisms of plastic biodegradation

Biodegradation of plastics can occur under aerobic and anaerobic conditions. With the presence of molecular oxygen, aerobic degradation will be triggered by microorganisms resulting in biomass production and carbon dioxide, water and other mineral salts. Whereas in the absence of oxygen, anaerobic degradation will occur that converts polymers to carbon disulfide, methane, carbon dioxide, water and mineral salts.

The biodegradation process by microorganisms consists of three steps: bio-deterioration, bio-fragmentation and mineralization (see Figure 10.18). Bio-deterioration involves the process whereby microorganisms irreversibly attach to and grow on the surface and internal pores of polymers and produce extracellular substances to facilitate the attachment and biofilm matrix. In this process, the physical, mechanical and chemical properties of the polymers are possibly changed because of the growing microorganisms. In bio-fragmentation, extracellular enzymes are secreted and transported to the surface of the polymers. The action of extracellular enzymes cleaves the long-chain polymers into monomers and oligomers. As the macromolecules of polymers are not able to pass through the cell membrane and absorbed by microorganisms directly, the decomposition of polymer through extracellular enzymes is essential for the whole biodegradation process. The small monomers and oligomers can enter the microbial cell and be metabolized to produce energy and new biomass. The end product of microbial metabolization varies depending on the chemical composition of polymers and the presence of oxygen (Ahmed *et al.*, 2018; Luyt & Malik, 2019). Table 10.3 summarizes bioplastic-depredating microorganisms.

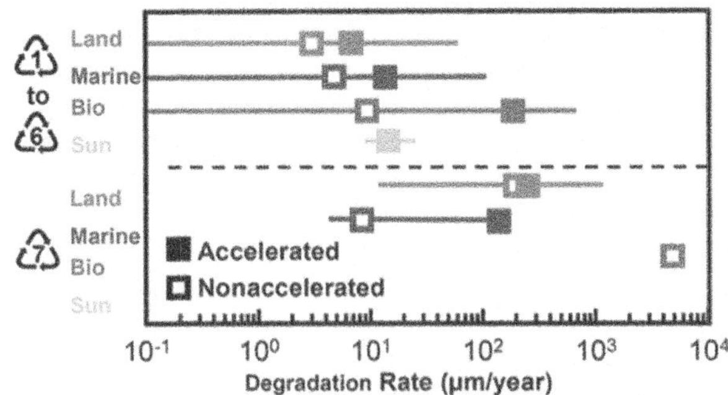

Figure 10.17 Factors affecting degradability of plastics (*Source*: Chamas *et al.*, 2020).

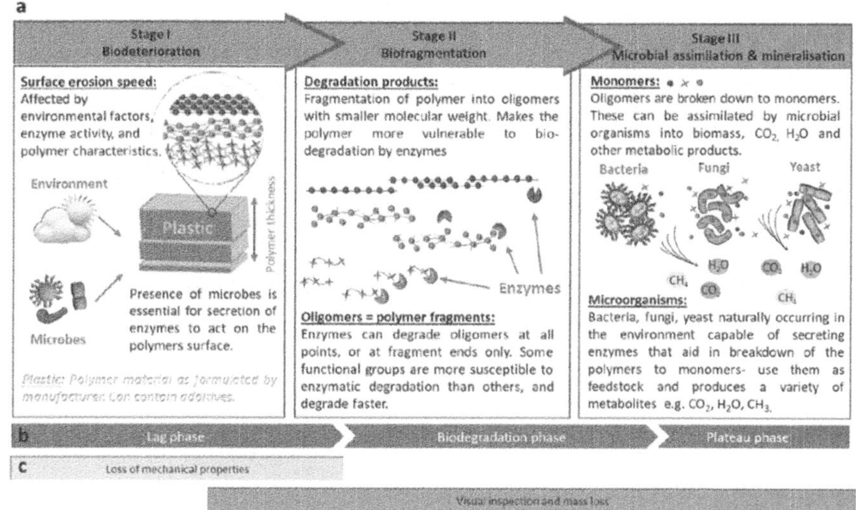

Figure 10.18 Biodegradation process of plastics (*Source*: Kjeldsen *et al.*, 2018).

Factors affecting the biodegradation of bioplastics can be classified into environmental factors and polymer characteristics (see Figure 10.19). Environmental factors such as temperature and pH are essential for microbial growth and reproduction. Salinity and moisture can influence the degradation rates of bioplastics in the environment. Typically, a high moisture content aids in accelerating the degradation of biodegradable polymers, while, many polymers do not degrade if the salinity in the environment is below a necessary level (Emadian *et al.*, 2017). In addition, polymer characteristics including their flexibility, chemical structure, molecular weight, morphology and functional groups play dominant roles in the biodegradability of polymers.

10.4.3 Biodegradation of bioplastic in different environments

Biodegradability in the natural environment is a valuable property of bioplastics. One of the key reasons for development of bioplastics was the possibility of avoiding plastic waste accumulation in the environment, especially in the oceans. However, the degradation of bioplastics is a slow process and takes place under only certain conditions and often, in ambient conditions.

- **Aerobic biodegradation of bioplastics**

The biodegradation of bioplastics under aerobic condition is an environmentally friendly approach which can contribute to accelerated decomposition of heterogeneous waste via the enzymatic action of a high diversity microbial population in a moist and warm environment under controlled conditions. The common environment for the aerobic degradation of the bioplastics is soil and composting. In general, PLA and PCL degrade more efficiently in compost compared to soil and liquid environment as the

Table 10.3 Isolated microorganisms degrading bioplastics.

Source of Bioplastic	Type of Bioplastic	Microbial Type	Microorganism	Source	References
Bio-based	PLA-based	Bacteria	*Amycolatopsis* sp., *Amycolatopsis thailandensis*, *Thermomactinomyces* sp., *Laceyella* sp., *Nonomuraea* sp., *Bacillus licheniformis*, *Actinomadura keratinilytica*, *Micromonospora* sp., *Streptomyces* sp., *Bordetella petrii*, *Paenibacillus amylolyticus*, *Paenibacillus* sp., *Amycolatopsis* sp., *Saccharaothrix* sp., *Lentzea* sp., *Kibdelosporangium* sp., *Streptoalloteichus* sp., *Burkholderia capacia*, *Thermomyces*	Soil	Jarerat *et al.* (2002), Teeraphatpornchai *et al.* (2003), Sukkhum *et al.* (2009), Wu (2009), Kim and Park (2010), Chomchoei *et al.* (2011), Penkhrue *et al.* (2015)
		Fungi	*Fennellomyces linderi*, *Fusarium solani*, *Purpureocillium* sp., *Cladosporium* sp., *Verticillium* sp., *Lecanicillium saksenae*, *Cladosporium* sp., *Aspergillus ustus*, *Penicillium verrucosum*, *Aspergillus fumigatus*, *Aspergillus Sydowii*, *Paecilomyces*, *Paecilomyces lilacinus*	Soil, compost	Szumigaj *et al.* (2008), Karamanlioglu *et al.* (2014), Penkhrue *et al.* (2015)
	PHA-based	Bacteria	*Streptomyces* sp., *Burkholderia capacia*, *Bacillus* sp., *Cupriavidus* sp., *Mycobacterium* sp., *Nocardiopsis* sp., *Streptomyces bangladeshensis*, *Pseudomonas aerogusina*, *Bacillus subtilis*, *Pseudomonas putida*, *Leptothrix* sp., *Variovorax* sp., *Pseudomonas fluorescens*, *Pseudomonas putida*, *Pseudomonas aeruginosa*, *Pseudomonas* sp., *Soil Candida albicans*, *Fusarium oxysporum*, *Pseudomonas lemoignei*, *Entrobacter* sp., *Bacillus sp.*, *Gracilibacillus* sp., *Actinomadura* sp., *Microcossus*	Soil, river sediment, river water, marine water, active sludge	Colak and Güner (2004), Hoang *et al.* (2007), Shah *et al.* (2007), Volova *et al.* (2007), Bhatt *et al.* (2008), Kumaravel *et al.* (2010), Shah *et al.* (2010), Volova *et al.* (2010), Phukon *et al.* (2012), Boyandin *et al.* (2013)

(Continued)

		Fungi	*Aspergillus niger, Penicillium* sp., *Aspergillus* sp., *Penicillium* sp., *Trichoderma pseudokoningii, Paecilomyces lilacinus, Cogronella* sp., *Acremonium recifei*	Soil, compost	Lee *et al.* (2005), Kumaravel *et al.* (2010), Phukon *et al.* (2012), Boyandin *et al.* (2013)
	Starch-based	Bacteria	*Clostridium acetobutylicum, Laceyella sacchari*	Soil	Yoshida *et al.* (2013), Lomthong *et al.* (2015)
		Fungi	*Aspergillus* sp., *Aspergillus niger*	Soil, compost	Accinelli *et al.* (2012), Li *et al.* (2015)
Fossil-based	PCL-based	Bacteria	*Amycolatopsis* sp., *Streptomyces* sp., *Streptomyces thermovioaceus, Paenibacillus* sp., *Streptomyces thermonitrificans, Streptomyces* sp., *Bacillus pumilus, Leptothrix* sp., *Pseudomonas* sp., *Tenacibaculum* sp., *Alcanivorax* sp., *Psychrobacter* sp., *Pseudomonas* sp., *Moritella* sp., *Shewanella* sp., *Paenibacillus amylolyticus*	Soil, compost, river sediment, fresh water, sea water, sea sediment	Teeraphatpornchai *et al.* (2003), Tezuka *et al.* (2004), Nakasaki *et al.* (2006), Hoang *et al.* (2007), Nakajima-Kambe *et al.* (2009), Sekiguchi *et al.* (2011b), Sekiguchi *et al.* (2011a), Chua *et al.* (2013), Penkhrue *et al.* (2015)
		Fungi	*Purpureocillium* sp., *Cladosporium* sp.	Soil	Penkhrue *et al.* (2015)
	PBS-based	Bacteria	*Amycolatopsis* sp., *Streptomyces* sp., *Paenibacillus* sp., *Paenibacillus amylolyticus, Azospirillum brasilense, Paenibacillus amylolyticus, Pseudomonas aeruginosa, Burkholderia capacia, Bacillus pumilus, Leptothrix* sp., *Pseudomonas aeruginosa, Azospirillum brasilense*	Soil, fresh water, activated sludge	Teeraphatpornchai *et al.* (2003), Hayase *et al.* (2004), Tezuka *et al.* (2004), Nakajima-Kambe *et al.* (2009), Lee and Kim (2010), Wu (2012), Penkhrue *et al.* (2015)
		Fungi	*Purpureocillium* sp., *Cladosporium* sp., *Aspergillus fumigatus, Aspergillus niger, Fusarium solani, Aspergillus oryzae, Rhizopus oryzae, Aspergillus oryzae, Rhizopus oryzae*	Soil	Maeda *et al.* (2005), Ishii *et al.* (2008), Abe *et al.* (2010), Li *et al.* (2011), Wu (2011), Penkhrue *et al.* (2015)

Source: Modified from Emadian *et al.* (2017).

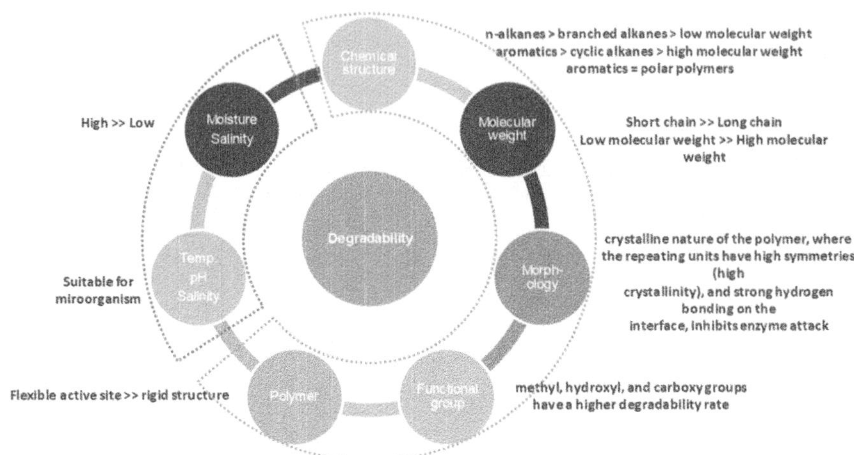

Figure 10.19 Factors affecting the biodegradation of biopolymers.

process conditions (pH, temperature, humidity) are favorable during controlled composting (Cho *et al.*, 2011).

- **Anaerobic degradation of bioplastics**

Under anaerobic conditions, the biodegradation process results in the production of biogas, water, hydrogen sulfide, ammonia and digestate via microbial metabolism. Anaerobic decomposition of bioplastics offers some notable advantages over the aerobic composting such as limitations of odor emissions, utilization of the produced methane as an energy source, and the nutrient-rich digestable residue, which can be used as a fertilizer (Bátori *et al.*, 2018).

In anaerobic digestors (AD), bioplastics degrade mainly due to the enzymatic hydrolysis by micro-organisms where long-chain polymers are converted into oligomers and monomers. The oligomers and monomers then diffuse through the cell wall and are utilized as carbon and energy sources by microorganisms. The waste stream of AD is typically biowaste, such as kitchen or food waste, which are rich in nitrogen. A C/N ratio below 15 is not ideal for AD process since it results in the generation of toxic ammonia. However, adding bioplastics to the mixture can increase the carbon-rich substrate and thus C/N ratio to optimal values.

- **Biodegradation in soil**

The rate of biodegradation in soil depends on various intrinsic and extrinsic factors like population of the microorganisms in the soil, type of the soil, pH, temperature and moisture. Biodegradation of bioplastics can be undertaken by several bacteria found in soil such as *Pseudomonas* sp, *Streptococcus* sp, *Staphylococcus* sp, *Bacillus* sp and *Moraxella* sp. Optimal aerobic biodegradation requires soil humidity of 50–60% (Siracusa, 2019), while, cellulose-based bioplastics have shown the highest biodegradability (from 80% to 100%) in a period over 100 days (Koch & Mihalyi, 2018).

The impact of degraded plastics to soil needs to be assessed. Seed germination index is usually applied, a value over 95% indicating the absence of toxic effects to the soil. During the biodegradation of several natural substances (i.e., polysaccharides), phyto-toxicity may occur as the biodegradation process can deplete soil oxygen and produce metabolic intermediates. This might cause a short-term disturbance and unfavorable conditions to the soil microorganisms and the plants. Such issues must be taken into considerations when evaluating biodegradation of the bioplastics in the soil.

- **Biodegradation in aquatic environment**

A few studies have focused on the biodegradation of bioplastics, under aquatic environment when compared to the much more widely investigated terrestrial systems. Different sea waters may contain different bioplastic degrading microorganisms, which are a key factor in the biodegradation process and efficiency. This is due to the various parameters like temperature, pH, which vary according to different habitats. Biodegradation rates have been found to be higher in sea water, and particularly at the interface of the water sediment, rather than in fresh water ponds and rivers.

10.4.4 Standards for plastic biodegradability

As biodegradability is a highly concerned property of plastic polymers, and the standardized criteria used to identify and test biodegradability is of great significance to their production, consumption, and waste management, several industrial standards regulating the biodegradability of plastics have been formulated and applied. Figure 10.20 presents several existing standards implemented at international, regional and national levels. ISO is an

Figure 10.20 Biodegradability standards for plastic.

internationally accepted standardization system. CEN and OECD standards are accepted and practiced regionally among the member countries in EU and the Organization for Economic Co-operation and Development (OECD), respectively. In addition, several developed countries such as the US, Japan, Germany, UK, and France have introduced their own national standards within the countries. These standards typically define the terminology, testing methods, and/or industrial composting related to biodegradable plastics. Table 10.4 lists the terminology related to biodegradable plastics and biodegradability in different standards. Slight differences in the definition of biodegradable plastics may appear among the different standards. For instance, ISO standard defines biodegradation of biodegradable plastics based on significant chemical structure change caused by microorganisms, Japanese standard recognizes the plastic as biodegradable as long as one step (e.g., chain cleavage) of the biodegradation process is enabled, whereas, CEN and ASTM require the completion of the entire biodegradation process or results of metabolic end products. These differences in terminology and definitions among the standards will directly link to varied standard testing methods and the way to certify biodegradable plastic products.

The biodegradability and compostability of plastic packaging are two concepts that can easily create confusion in the market. Hence, the CEN standard has clearly defined these two terms separately with mentioning that a plastic product can be recognized compostable only when its biodegradation has been demonstrated in a composting system following the standard methods. Typically, the customers can only identify a biodegradable plastic product if it is marked with a recognized label. Several international organizations such as TUV AUSTRIA, DIN Certro and EU are authorized to certify such labels (Filiciotto & Rothenberg, 2021). Figure 10.21 illustrates several labels for bio-based, biodegradable and compostable plastics in the market.

The OECD testing guidelines were launched in 1980s with various testing methods and were offered free of charge online to users. However, these guidelines are not recognized worldwide and typically act as self-certification (Filiciotto & Rothenberg, 2021). The ISO offers a series of standard testing procedures for biodegradability in various environments (e.g., soil, activated sludge, seawater and marine sediment). They are widely applied and recognized as biodegradable plastic testing and certificating by many manufacturers and companies around the world. In the US, the ASTM standards are applied to test and determine biodegradability in various specifications. Table 10.5 lists several biodegradable plastic testing methods provided by different standards.

Principally, the testing methods can be generally classified into three categories: field tests, simulated tests and laboratory tests (see Figure 10.22). A field test provides the closest results to the actual biodegradation of plastics. Field tests can be performed by placing plastics in lakes or rivers, burying plastic samples in soil, or performing a full-scale composting/landfill process. However, field tests usually suffer several constrictions such as difficult experimental control, and limited on-site analysis approaches and accuracy. To overcome those shortcomings, a simulation test, which mimics various defined environmental conditions (e.g., compost, soil, and seawater) in a laboratory-scale reactor, may

Table 10.4 Terminology regarding biodegradable plastics and biodegradability in different standards.

Standard	Terminology	Definition
ISO 472	Biodegradable plastics	A plastic designed to undergo a significant change in its chemical structure under specific environmental conditions resulting in a loss of some properties that may vary as measured by standard test methods appropriate to the plastic and the application in a period of time that determines its classification. The change in the chemical structure results from the action of naturally occurring microorganisms.
CEN	Biodegradable plastics	A degradable material in which the degradation results from the action of microorganisms and ultimately the material is converted to water, carbon dioxide and/or methane, and a new cell biomass.
	Biodegradation	Biodegradation is a degradation caused by biological activity, especially by enzymatic action, leading to a significant change in the chemical structure of a material.
	Inherent biodegradability	The potential of a material to be biodegraded, established under laboratory conditions.
	Ultimate biodegradability	The breakdown of an organic chemical compound by microorganisms in the presence of oxygen to biodegradable carbon dioxide, water and mineral salts of any other elements present (mineralization), and new biomass or in the absence of oxygen to carbon dioxide, methane, mineral salts, and new biomass.
	Compostability	Compostability is a property of a packaging to be biodegraded in a composting process. To claim compostability, it must have been demonstrated that a packaging can be biodegraded in a composting system as can be shown by standard methods. The end product must meet the relevant compost quality criteria.
ASTM sub-committee D20-96	Biodegradable plastics	A degradable plastic in which the degradation results from the action of naturally occurring microorganisms such as bacteria, fungi, and algae.
DIN	Biodegradable plastics	A plastic material is called biodegradable if all its organic compounds undergo a complete biodegradation process. Environmental conditions and rates of biodegradation are to be determined by standardized test methods.
	Biodegradation	Biodegradation is a process, caused by biological activity, which leads to change of the chemical structure and result in naturally occurring metabolic products.
Japanese Biodegradable Plastics Society	Biodegradable plastics	Polymeric materials which are changed into lower molecular weight compounds where at least one step in the degradation process is through metabolism in the presence of naturally occurring organisms.

Source: Ashter (2016b).

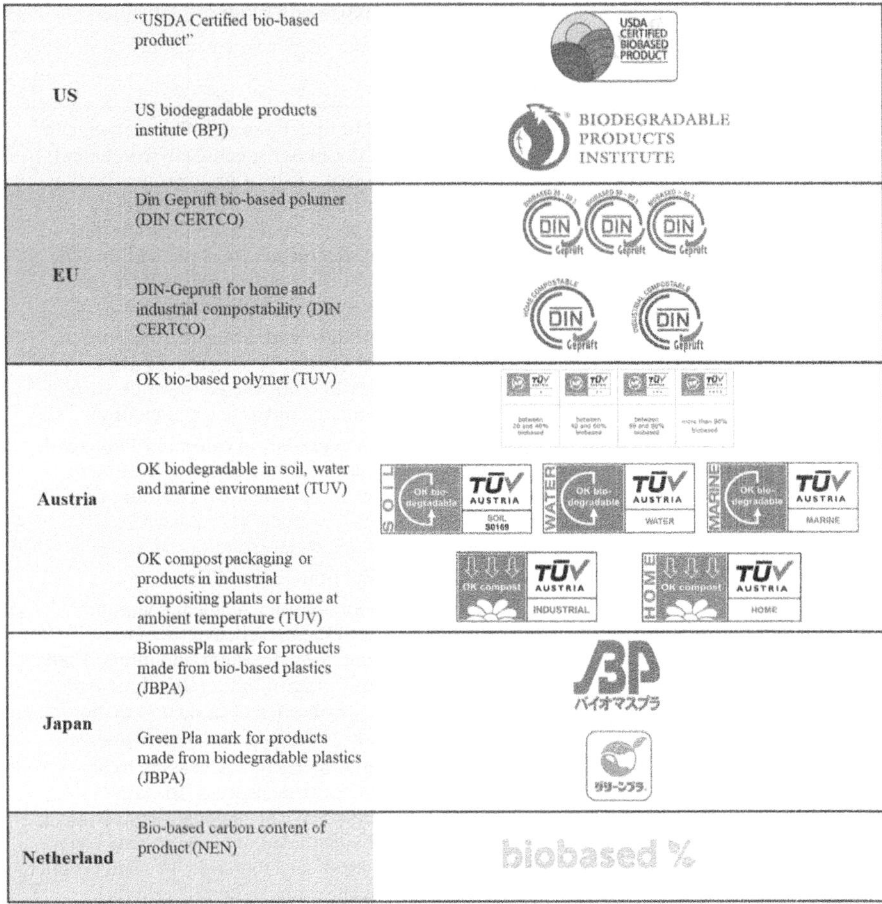

Figure 10.21 Labels for bio-based, biodegradable, and compostable plastics (*Source*: modified from Niaounakis, 2019).

be carried out. In this case, the testing conditions such as temperature, pH and humidity can be well controlled, and advanced measurements and analyses such as gas formation and chemical analysis for biodegradation intermediate and residuals are possible. To study the basic mechanisms and kinetics of the biodegradation process, laboratory tests may be performed using designed media and screened microbial strain or mixture, and/or selected enzyme for special polymer. Such optimized reaction conditions will typically result in a higher degradation rate compared with that found in the natural environment. This method is usually the most reproducible and efficient to study the fundamental theory and mechanisms of polymer biodegradation. However, the results obtained from laboratory tests usually represent ideal biodegradation condition, which may be far from the actual environment.

Table 10.5 Several biodegradability testing standards and methods.

Standard	Testing Method
ISO standard	ISO 9439 – Evaluation of ultimate aerobic biodegradability of organic compounds in aqueous medium – carbon dioxide evolution test ISO 14593 – Evaluation of ultimate aerobic biodegradability of organic compounds in aqueous medium – method by analysis of inorganic carbon in sealed vessels (CO_2 headspace test) ISO 14852 – Determination of the ultimate aerobic biodegradability of plastic materials in an aqueous medium – method by analysis of evolved carbon dioxide ISO 14855 – Determination of the ultimate aerobic biodegradability of plastic materials under controlled composting conditions – method by analysis of evolved carbon dioxide
OECD standard	OECD 301 Testing – Ready/Ultimate Biodegradability (includes OECD 310). OECD 301 series (OECD 301A, OECD 301B, OECD 301C, OECD 301D, OECD 301E, OECD 301F) allows for direct, explicit certification of a material's biodegradability. OECD 311 – Anaerobic biodegradability of organic compounds in digested sludge
ASTM standard	ASTM D 5210 – Standard Test Method for Determining the Anaerobic Biodegradation of Plastic Materials in the Presence of Municipal Sewage Sludge ASTM D 5338 – Standard Test Method for Determining Aerobic Biodegradation of Plastic Materials Under Controlled Composting Conditions, Incorporating Thermophilic Temperatures ASTM D 5864 – Standard Test Method for Determining Aerobic Aquatic Biodegradation of Lubricants or Their Components ASTM D 5988 – Standard Test Method for Determining Aerobic Biodegradation in Soil of Plastic Materials or Residual Plastic Materials After Composting ASTM D 6400 – Standard Specification for Compostable Plastics

Sources: Ashter (2016b), Filiciotto and Rothenberg (2021), Müller (2005), Pagga (1997) and RespirTek (2021).

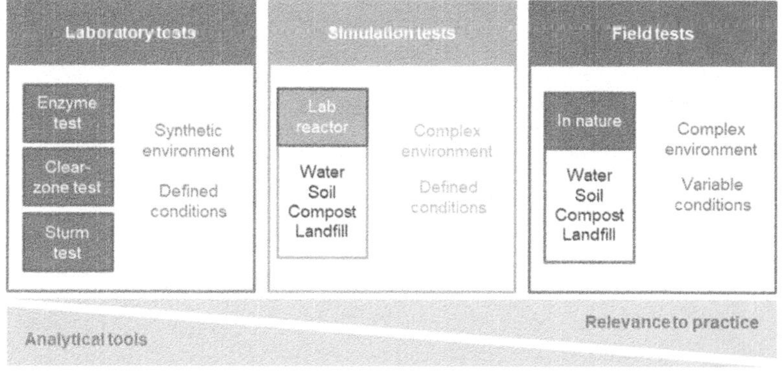

Figure 10.22 Typical testing methods for biodegradability of plastic (*Source*: Müller, 2005).

10.5 WASTE MANAGEMENT OPTIONS FOR BIOPLASTICS

Like any solid waste, the management options for bioplastics generally follow the classic waste management hierarchy (see Figure 10.23). Prevention and reuse rank as the top priorities before the bioplastic products are disposed of as waste. These are followed by several types of material and energy recycling. Due to their unique biodegradability, there are various techniques used to recycle the materials and recover energy from bioplastic wastes through biological treatment processes such as composting and anaerobic digestion. However, the environmental impact taking place at the end-of-life disposal (e.g., landfill) is typically considered less serious than that observed with fossil-based plastic wastes. The discussion in this section will mainly focus on the postconsumer management and will not include prevention and reuse.

10.5.1 Recycling

Despite the outstanding biodegradability of most bioplastics in comparison with traditional fossil-based plastics, most commercial biodegradable polymers such as PLA degrade slowly under ordinary conditions, even in the presence of microorganism. Large amounts of bioplastic consumption may cause the accumulation to far exceed the biodegradation capacity in the environment, and eventually create environmental problems in the future. Furthermore, biodegradable plastics disposed of in landfills can lead to methane emissions and consequently negative climate impact. On the other hand, the recycling of

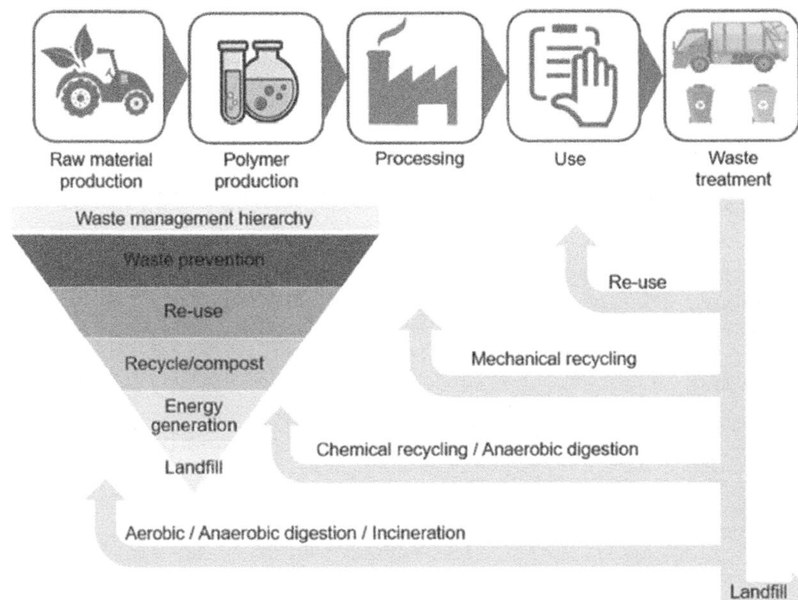

Figure 10.23 Bioplastic waste management options within the waste management hierarchy (*Source*: Saalah *et al.*, 2020).

bioplastics is crucial in reducing the consumption of reusable and/or renewable resources needed for new plastic products or synthesis of the corresponding monomers. Hence, the recycling of biopolymers complies with the requirements of circular economy and is a preferable method for bioplastic waste management among all available options.

Similar to fossil-based plastic wastes, waste bioplastic materials can be recycled in three possible routes after being properly collected, sorted and cleaned. (1) Primary recycling or re-extrusion is a closed-loop recycling method which is simple and low cost. It can only be performed with high-quality plastic scrap, hence rarely applied among the recyclers. This type of recycling can either be the reuse of the material or the closed-loop mechanical recycling of scrap plastic to produce products with the original structure (Al-Salem *et al.*, 2009; Grigore, 2017; Lamberti *et al.*, 2020). (2) Secondary recycling or mechanical recycling refers to the transformation or downgrading of waste plastic into a less-demanding product via mechanical means (screw extrusion, injection molding, blow molding, etc.) (Al-Salem *et al.*, 2009; Grigore, 2017; Lamberti *et al.*, 2020; Singh *et al.*, 2017). This method can only be practiced with thermoplastics as it involves the re-melting and re-processing of the polymers. Mass loss may occur during the secondary recycling depending on the quality and purity of plastic waste. The disadvantages of this method are the heterogeneity of the solid waste and the deterioration of the product's properties in each cycle which occurs due to the low molecular weight of the recycled resin. However, it has certain advantages over the chemical recycling, including a lower processing cost, lower global warming potential, less non-renewable energy use and less acidification and eutrophication (Shen *et al.*, 2010). (3) Tertiary recycling or chemical recycling involves a wide range of thermo-chemical processes such as hydrogenation, glycolysis, gasification, hydrolysis, pyrolysis and enzymatic cracking which depolymerize the plastic waste into biofuels, monomers, or other useful feedstock materials. Chemical recycling offers advantages over mechanical recycling including the opportunity to produce value-added materials, and the potential for a circular polymer production economy since recovered virgin monomers can be re-polymerized for an indefinite amount of recycles (Hopewell *et al.*, 2009; Lamberti *et al.*, 2020; Payne *et al.*, 2019).

The quality and purity of plastic waste is a crucial factor that determines the selection of the recycle strategies and the values of recycled products. Hence, the success of bioplastics recycling is greatly depends on the proper collection and sorting technologies and systems. One main concern related to bioplastic recycling is the compatibility with the current existing plastic recycling systems. Bio-based non-biodegradable polymers, such as bio-PET, bio-PE and bio-PA can be mixed with their fossil-based counterparts and recycled in existing recycling facilities. However, several new biopolymers which are not present in the conventional fossil-based plastics may make the recycling process more complicated or even fail without proper sorting technologies. Niaounakis (2019) has summarized several innovative approaches for biopolymer identification and sorting, including manual sorting via markers and labels, density separation, identification using optical or near-infrared systems, fluorescent and coloring dyes, and separation by dissolution (see Table 10.6).

Table 10.6 Innovative technologies for bioplastic identification and sorting.

Technology	Description	Application Example
Marks and labels	Product labels provided by authorized international standards help to identify the biodegradable, compostable, and bio-based plastics	See Figure 10.21
Density separation	PLA is heavier than most PE and can be separated from other plastic polymers by density difference using float-sink technology	
NIR	Sensor-based NIR sorting process offers an innovative and highly potential option to automatically separate biopolymers and their blends out of the plastic waste stream.	The Nature Works applied NIR to separate PLA (Ingeo™) from plastic wastes containing PET bottles. A sorting accuracy up to 97.5% was obtained in testings (NatureWorks LLC, 2008).
Fluorescent and coloring dye	Fluorescent dyes such as the fluorescent whitening dyes, green dye, and edible dye, and so on have been used for the identification and separation of various aliphatic polyesters from a plastic waste. The dyes can be either used as fluorescent inks, which are printed to form optically readable barcode patterns, or compounded with the aliphatic polyester to enable differentiation of the aliphatic polyesters from the plastics waste at the time of recovery.	The Mitsui Toatsu Chem Inc. uses fluorescent whitening dyes to mark biodegradable aliphatic polyesters. The fluorescence labeling facilitates sorting in a compost mill of the biodegradable polymer from the unlabeled non-degradable polymers.
Dissolution	Chemical dissolution using selective solvents (e.g., hexafluoroisopropanol/chloroform, lactic acid ester, etc.) combined with separation, such as filtration, precipitation, has been used to dissolve PLA and simultaneously separate solid polymers and undissolved impurities (e.g., PE. PP, PVC, PET, etc.) from waste mixture, laminated packaging, and PLA blends.	Fraunhofer Institute for Process Engineering and Packaging IVV showed that PLA can be recovered from a waste with 30% impurities by implementing the CreaSolv® process (Fraunhofer IVV, 2016; Hiebel *et al.*, 2017).

Source: Niaounakis (2019.).

10.5.2 Incineration (with energy recovery)

Incineration refers to the thermal decomposition of a substance through combustion. This technology allows the energy content to be recovered from low-quality plastic wastes or the mixture of plastics with municipal solid wastes. In recent years, incineration has become a popular waste management option compared with landfill, as it can significantly save the footprint of waste

Table 10.7 Heating values of various fossil-based polymers, biopolymers, and solid fuels.

Fossil-Based Polymer	Calorific Value (MJ/kg)	Biopolymer	Calorific Value (MJ/kg)	Solid Fuel	Calorific Value (MJ/kg)
PE	45	Bio-PE	44	Coke	30
PP	43	PCL	32	Coal	26–29
PS	40	Polyester	26	Firewood	14–19
PS	31	PHAs	24	Paper	19
PC	31	PLA and blend	19–21	Charcoal	15
PVC	18	Starch blend	21		
		Cellulose derivative	18		

Source: Modified from Saalah *et al.* (2020).

end-of-life facilities. Several literatures classified incineration as quaternary recycling as well (Al-Salem *et al.*, 2009; Grigore, 2017; Lamberti *et al.*, 2020). Many fossil-based polymers contain high heating values which result in high-energy yield during combustion comparing with several solid fuels like coals and charcoals. The energy contents of biopolymers are equal to or slightly lower than fossil-based polymers depending on the chemical structure, while, their heat values are still comparable with several solid fuels like coal, wood and charcoal (see Table 10.7). The recovered energy through incineration can be used to generate steam and/or electricity. Different from conventional polymers, the CO_2 generated by bio-based polymer incineration is considered neutral for the global warming potential. However, this approach should only be selected when material recycling is not applicable, as not only the inherent energy of the polymer will be lost, but also harmful chemicals and dioxins will be released into the atmosphere (Al-Salem *et al.*, 2009; Lamberti *et al.*, 2020).

10.5.3 Composting and anaerobic digestion

Preferable biopolymers should remain stable during manufacture and use, and break down rapidly after disposal and/or be converted into biomass within an acceptable duration. Microbial degradation is one crucial method of waste management as it increases environmental safety and economic value. Unlike many shared or similar recycling systems and technologies for bioplastics and fossil-based plastics, composting and anaerobic digestion are only applicable to biodegradable polymers.

Composting is a self-heating process which degrades organic wastes to compost and by-products, for example, carbon dioxide and heat (Cerda *et al.*, 2018). Compost is a nutrient-rich organic which can provide N, P, K and organics for soil amendment (Wang & Zeng, 2018). Composting can save the primary resources contained in organic waste which can be recycled as fertilizers in horticulture and agriculture. Industrial aerobic composting is composed of consecutive phases (Figure 10.24): (1) mechanical pretreatment and preparation of the waste bulk; (2) the mesophilic (25–40°C) phase with a duration of several days for biological process start-up; (3) the thermophilic

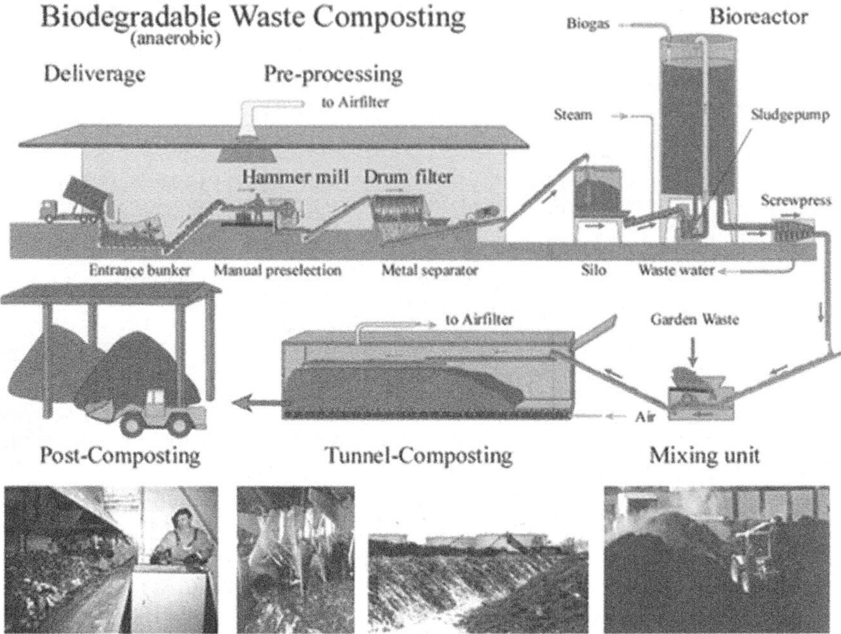

Figure 10.24 Industrial composting process (*Source*: Yip & Madl, 2003).

(55–65°C) phase which lasts approximately three weeks; (4) maturation phase lasting one to two months with the compost cooling down to ambient temperature; and (5) final refining sieving the compost to obtain acceptable compost quality (Gioia *et al.*, 2021). In addition to operating composting in industrial facilities, home composting has become popular in recent years, particularly in rural areas (Cucina *et al.*, 2021). However, home composting usually cannot guarantee the optimum composting condition (especially temperature) and duration, and it is difficult to ensure adequate compost qualities. Several relevant international and national standards used to identify compostable plastics are listed in Table 10.8.

Recent literatures have reported the performance of various biodegradable polymers in composting processes (Accinelli *et al.*, 2012; Arrieta *et al.*, 2014; Balaguer *et al.*, 2016; Báreková *et al.*, 2021; Cafiero *et al.*, 2021; Gómez & Michel, 2013; Kale *et al.*, 2007; Kalita *et al.*, 2021; Mihai *et al.*, 2014; Mohee *et al.*, 2008; Ruggero *et al.*, 2021; Sun *et al.*, 2021; Tabasi & Ajji, 2015; Weng *et al.*, 2011). Most of the tested biodegradable plastics showed quick biodegradation during composting, with the kinetic coefficients of degradation following the order of starch-based blends < PHAs blends ≤ PLA blends. As a conclusion, PLA blends consume the shortest duration to complete biodegradation in composting of approximated 84 ± 47 days, followed by PHAs and starch blends that take 119 ± 43 days and 124 ± 83 days, respectively.

Table 10.8 Relevant standards for biodegradable plastic composting.

Composting Environment	Standard	Description
Industrial composting	EN 13432: 2000	Requirements for packaging to be considered recoverable through composting and biodegradation.
	EN 14995: 2006	Evaluation of compostability of plastics.
	ISO17088: 2021	Specifications for compostable plastics.
	AS4736: 2006	Biodegradable plastics suitable for composting and other microbial treatment.
	ASTM D 6400	Labeling of plastics designed to be aerobically composted in municipal or industrial facilities.
Home composting	AS 5810-2010	Biodegradable plastics suitable for home composting.
	NF T51-800	Specifications for plastics suitable for home composting.

Source: Modified from Gioia *et al.* (2021).

Although composting has proven to be a valuable strategy for treatment of biodegradable plastics, it is worth noting that composting still owns the drawbacks such as, long time consumption, high land-use intensity and high GHG emission, and so on. Taking this into consideration, anaerobic digestion or anaerobic digestion coupled to digestate composting may represent preferred treatments for bioplastics' end-of-life treatment (Cucina *et al.*, 2021; Edwards *et al.*, 2018; Wainaina *et al.*, 2020).

Anaerobic digestion is one classic biological treatment technology for organic waste and wastewater. It converted organics anaerobically into biogas through a four-step biological process as hydrolysis, acidogenesis, acetogenesis, and methanogenesis. The biogas containing methane and hydrogen can be used to generate bioenergy or be converted into high-value products (Patel *et al.*, 2020; Wu *et al.*, 2016). Anaerobic digestion of several biodegradable polymers have been intensively studied in recent years (Calabro' *et al.*, 2020; Cazaudehore *et al.*, 2021; Ren *et al.*, 2018; Tseng *et al.*, 2019; Yagi *et al.*, 2013; Zhang *et al.*, 2018). It was concluded that PHA blends degraded much faster than PLA and starch blends under both thermophilic and mesophilic conditions. However, anaerobic digestion usually takes longer time to complete the mineralization of biopolymers comparing with aerobic composting.

10.5.4 Landfilling

Although being a popular waste management system because of its low cost and simplicity of operation, landfilling is the least favorable option for the waste management of bioplastics. One major challenge regarding waste landfill is the limitation on suitable sites and space over the time, which is faced especially by many developing countries. In addition, concerns have been raised by the public on the impact of landfill on neighboring environment and public heath due to the release of methane and toxic chemicals. As one of the most impactful GHG, methane has a global warming potential 25–36 times of CO_2. In conclusion, landfill of bioplastics especially biodegradable plastics is relatively beneficial than fossil-based plastics due to the degradability and carbon neutral feature. However, this approach should be only selected as a last resort, just like any other municipal solid wastes.

10.6 CHALLENGES AND FUTURE OUTLOOK OF BIOPLASTICS

Owning to their renewable feedstock sources and biodegradability in different environments, bioplastics are widely considered to be a more sustainable substitute of fossil-based plastics, and are currently used to produce a wide range of products. Whereas, bioplastics have not yet been produced in a larger amount or consumed in a wider field because of a number of limitations and challenges such as the cost constraints, performance limitations, and lack of knowledge about available substitutes in the market, and so on. Although bioplastic is deemed as more sustainable in some aspects, it might have some environmental and occupational health and safety hazards during its production, and may be a threat as a potential source of marine microplastics if they are not properly managed and disposed in the end of life. A wiser development and application of bioplastics are imperative in order to enable a brighter future of these attractive materials.

10.6.1 Feedstock production of bioplastics

Potential environmental and occupational health hazard may present during the production of bioplastics. It has been pointed out that the cultivation of starch, corn zein and soy protein may consume different persistent bio-accumulative and toxic chemicals such as pesticides and herbicides. The feedstock growth for bio-based plastics may also involve genetically modified organisms (GMOs). Although GMOs have been allowed for commercial/industrial agricultural production, limited knowledge is available for their potential hazards to the water, air and soil while being applied and leached into the environment through bioplastic supply chains. Álvarez-Chávez *et al.* (2012) have recommended the spectra for occupational health and environmental impact of bioplastics according to the sustainability criteria in Figure 10.25. Further efforts are needed to gain better understanding on the potential hazards associated with bioplastic feedstock production and to minimize them in line with the principles of sustainable materials (The Sustainable Biomaterials Collaborative, 2009).

Furthermore, there is a concern about excess land use for polymer production, instead of using it for growing crops, which may exacerbate the global food crisis. The European Bioplastics (2020d) has estimated the global land use for bioplastic feedstock production in 2020 and 2025. It shows that approximately 0.7–1.1 million ha of arable land will be used for bioplastic productions, which accounts for only around 0.015–0.020% of global arable land. On the other hand, agricultural by-products (e.g. rice and corn straw), non-crop biomass (e.g., lingo-cellulose), marine sourced biomass (e.g., seaweed and microalgae), and organic waste and wastewater become attractive alternatives as the feedstock for bioplastic production (Bhatia *et al.*, 2021; Brodin *et al.*, 2017; Chong *et al.*,

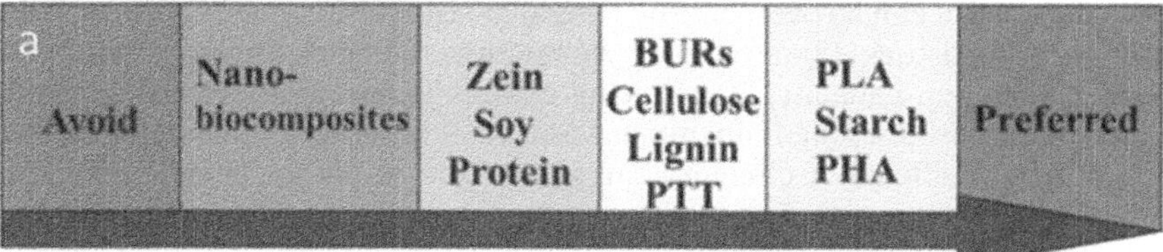

Preferred: Feedstock grown sustainably; avoid GMOs, hazardous chemicals, untested materials, or petroleum based co-polymers.
Avoid: Feedstock grown unsustainably; GMOs, hazardous chemicals, untested materials, or petroleum based co-polymers are used.

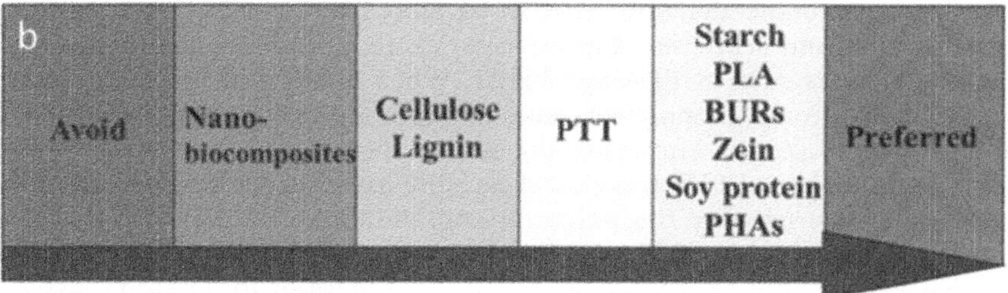

Preferred: Feedstock grown sustainably; avoid GMOs, PBT chemicals, untested materials, or petroleum based co-polymers; flexibility in environmentally friendly disposal options; production is energy and water efficient and does not impact food supply.
Avoid: Feedstock grown unsustainably; GMOs, PBT chemicals, untested materials, or petroleum based co-polymers are used; environmentally unfriendly disposal options, production is energy and water inefficient, impact food supply.

Figure 10.25 The bioplastics spectra for comparative occupational health and safety impacts (a) and comparative environmental impacts (b) (*Source*: Álvarez-Chávez *et al.*, 2012).

2021; Zhang *et al.*, 2019). The cascade use of agricultural feedstock also provides a new aspect to maximize resource efficiency, which helps to relieve the stress on food crisis due to bioplastic production.

10.6.2 Weak competitiveness in the market

The production cost of bioplastics is relatively high compared with fossil based plastics. To enable competitive prices and devise a level-playing field for bioplastics and fossil-based plastics, economies of scale effects are necessary. It might be challenging considering the long supply chain of bioplastics and the low price of the crude oil. Feedstock selection can be one of the challenges considering that (1) substrate cost typically accounts for 40–60% of the total cost for bioplastic product (Demain, 2007); (2) the overall yield or production capacity of bio-based monomers is significantly lower than fossil-based monomers, and (3) the limited scale of bioplastic manufactures may cause relatively high unit cost of bioplastic products.

The high production cost will ultimately lead to high prices of bioplastics in the market, thus shifting the cost burden to consumers and resulting in a low market competitiveness. Low competitiveness due to the current immature bioplastic market can be overcome by several measures: (1) selecting low cost and eco-friendly feedstock, (2) developing the manufacturing technology and scaling up the production capacity, (3) ensuring comprehensive and diversified supply chains for bioplastics, (4) enhancing the consumer acceptance through extensive publicizing the effects of climate change, and the needs to detach from fossil resources, and so on and (5) introducing effective policy and financial support.

10.6.3 Limited recycling options

A debate is ongoing over the pros and cons of the recycling of biopolymers. The opponents of recycling assert that biopolymers pose a serious problem when they enter either conventional plastics recycling or green-waste composting streams. Despite the dynamic growth of biopolymers, their waste streams are still small and scattered, and no separate recycling stream yet exists for new biopolymers such as PLA and PHAs. When blended in the conventional plastic waste stream, biopolymers may increase the risk of contamination in the recycling process and influence the quality and values of recycled products. A growing number of PET recyclers, recycling associations and officials have expressed concerns over the potential contamination of the PET recycling stream by PLA bottles, yield loss, increased cost for sorting, and impact on recycled PET quality and processing. A critical mass of PLA (of at least 200 million kg produced annually) is necessary for independent rigid packaging recycling to be profitable (Niaounakis, 2019).

Lack of appropriated collection, sorting and recycling technologies, facilities, and systems for bioplastics especially biodegradable plastics remain as one of the main obstacles hindering the extensive application and sustainable management of bioplastics. Intensive research and development to ensure high recyclability of bioplastics will become the key for the future of this industry.

10.6.4 Potential contribution to marine microplastics

The impact of microplastics derived from non-biodegradable plastics is obvious. The question is whether biodegradable plastics also pose such impact to marine ecosystem. In fact, the degradation of many commercialized degradable plastics which are marked as 'green' or 'eco-friendly' in the market tends to be incomplete, leading to the accumulation of partially degraded residuals in the nature environment. This process is likely to cause the release of microparticles into the environment, even though such microplastics may not be as harmful as conventional fossil-based microplastics. It should be noted that the leakage and dispersion of bioplastics in the natural environment must be avoided, despite the often widespread and incorrect idea that a biodegradable material can easily disappear into the environment without creating pollution.

10.6.5 Insufficient policies, legislations and standards

Despite the dynamic growth of global bioplastic market, only a few counties and regions have related policies, legislations and standards to promote, incentivize and regulate the production, consumption and waste management of bioplastics. Such situation is specially faced by the developing countries in

Asia and Africa which contribute most to global mismanagement and leakage of plastic litters.

EU as a global pioneer towards circular economy has accelerated the momentum of growth of the bioplastic industry and continued to make substantial investment in bioplastic R&D and new production in Europe through a series of top-level strategies and policies including the Circular Economy Package, Bio-economy Strategy, Europe 2020/Innovation Union, Lead Markets Initiative for Bio-based Products, and Resource Efficiency Strategy. At the beginning of 2018, the European institutions adopted a revised EU waste package that encourages member states to support the use of bio-based materials for the production of packaging and to improve market conditions for such products. In addition, the mandatory separate collection of bio-waste will be ensured across Europe, facilitated by certified collection tools such as compostable bio-waste bags. In order to build a supportive legislative framework, regulatory measures such as extended producer responsivity (EPR) or product design rules, or carbon pricing mechanism, would help to boost the bioplastics production and development in Europe. Meanwhile, regulation measures to support biodegradable and compostable plastics in specific applications such as agricultural mulch films, fresh food packaging, bio-waste bags and disposable catering items will help to improve the efficiency of plastic waste management and reduce the environmental burden due to plastic leakage (European Bioplastics, 2020e). Such proactive efforts and actions are expected by more and more countries and regions in order to create wider collaboration to address the global plastic issues through further development and applications of bio-economy in the future.

In terms of bioplastic-related standards, though a number of standards and test methods for measuring the biodegradability of polymers within specified environments have been developed, several issues may still limit their reliability: (1) undefined approaches to inoculum preparation and test conditions, (2) a lack of specific guidelines for employing different test materials, (3) insufficient statistical replication, (4) a lack of suitable procedures for unmanaged aquatic environments, and (5) deficiencies in toxicity testing and understanding of the wider impacts of plastic litter on aquatic ecosystems (Harrison *et al.*, 2021).

Despite a number of limitations and challenges faced by the bioplastic industry, it is no doubt that bioplastics is one of the potential routes towards plastic sustainability. Researchers and engineers are committed to finding the optimal ways to substitute the functions of fossil-based plastics with innovative materials. Rationally designing the bioplastics to impart desired functionality and recyclability, and utilizing unaccounted biomass as a valuable resource would together establish a sustainable production value chain for bioplastics. Despite the fact that some of the bioplastics production technologies are lacking the scalability and productivity comparable to petroleum-based routes, governmental regulations and consumer pressure have been fostering the bioplastics industry to adopt and implement sustainable production routes. Circular bio-economy is also gaining global momentum, which in turn triggers a wide range of stakeholders to leverage the synergistic potential of bioplastics manufacturing and upscaling/recycling strategies. Innovations in fundamental redesigning of bioplastics with improved economics for recycling will pave a way for the next generation of sustainable bioplastics (Brockhaus *et al.*, 2016; Garcia & Robertson, 2017; Hong & Chen, 2017; RameshKumar *et al.*, 2020).

REFERENCES

Abe M., Kobayashi K., Honma N. and Nakasaki K. (2010). Microbial degradation of poly(butylenesuccinate)by*Fusariumsolani*insoilenvironments.*PolymerDegradation and Stability*, **95**, 138–143, https://doi.org/10.1016/j.polymdegradstab.2009.11.042

Accinelli C., Saccà M. L., Mencarelli M. and Vicari A. (2012). Deterioration of bioplastic carrier bags in the environment and assessment of a new recycling alternative. *Chemosphere*, **89**, 136–143, https://doi.org/10.1016/j.chemosphere.2012.05.028

Ahmed T., Shahid M., Azeem F., Rasul I., Shah A. A., Noman M., Hameed A., Manzoor N., Manzoor I. and Muhammad S. (2018). Biodegradation of plastics: current scenario and future prospects for environmental safety. *Environmental Science and Pollution Research*, **25**, 7287–7298, https://doi.org/10.1007/s11356-018-1234-9

Al-Salem S. M., Lettieri P. and Baeyens J. (2009). Recycling and recovery routes of plastic solid waste (PSW): A review. *Waste Management*, **29**, 2625–2643, https://doi.org/10.1016/J.WASMAN.2009.06.004

Álvarez-Chávez C. R., Edwards S., Moure-Eraso R. and Geiser K. (2012). Sustainability of bio-based plastics: general comparative analysis and recommendations for improvement. *Journal of Cleaner Production*, **23**, 47–56, https://doi.org/10.1016/J.JCLEPRO.2011.10.003

Anjum A., Zuber M., Zia K. M., Noreen A., Anjum M. N. and Tabasum S. (2016). Microbial production of polyhydroxyalkanoates (PHAs) and its copolymers: a review of recent advancements. *International Journal of Biological Macromolecules*, **89**, 161–174, https://doi.org/10.1016/J.IJBIOMAC.2016.04.069

Anugrahwidya R., Armynah B. and Tahir D. (2021). Bioplastics starch-based with additional fiber and nanoparticle: characteristics and biodegradation performance: a review. *Journal of Polymers and the Environment*, **29**, 3459–3476, https://doi.org/10.1007/s10924-021-02152-z

Arca H. C., Mosquera-Giraldo L. I., Bi V., Xu D., Taylor L. S. and Edgar K. J. (2018). Pharmaceutical applications of cellulose ethers and cellulose ether esters. *Biomacromolecules*, **19**, 2351–2376, https://doi.org/10.1021/acs.biomac.8b00517

Arrieta M. P., López J., Rayón E. and Jiménez A. (2014). Disintegrability under composting conditions of plasticized PLA–PHB blends. *Polymer Degradation and Stability*, **108**, 307–318, https://doi.org/10.1016/j.polymdegradstab.2014.01.034

Ashter S. A. (2016a). Introduction. In: Introduction to Bioplastics Engineering, S. A. Ashter (ed.), Elsevier Inc., Amsterdam, The Netherlands., pp. 1–17.

Ashter S. A. (2016b). Fundamentals on biodegradability. In: Introduction to Bioplastics Engineering, S. A. Ashter (ed.), Elsevier Inc., Amsterdam, The Netherlands., pp. 61–80.

ASTM International (2019) Standard Specification for Labeling of Plastics Designed to be Aerobically Composted in Municipal or Industrial Facilities [WWW Document]. URL https://www.astm.org/Standards/D6400.htm (accessed 10.6.21)

Atiwesh G., Mikhael A., Parrish C. C., Banoub J. and Le T.-A. T. (2021). Environmental impact of bioplastic use: a review. *Heliyon*, **7**, e07918, https://doi.org/10.1016/J.HELIYON.2021.E07918

Avella M., De Vlieger J. J., Errico M. E., Fischer S., Vacca P. and Volpe M. G. (2005). Biodegradable starch/clay nanocomposite films for food packaging applications. *Food Chemistry*, **93**, 467–474, https://doi.org/10.1016/j.foodchem.2004.10.024

Babu R. P., O'Connor K. and Seeram R. (2013). Current progress on bio-based polymers and their future trends. *Progress in Biomaterials*, **2**, 8, https://doi.org/10.1186/2194-0517-2-8

Balaguer M. P., Aliaga C., Fito C. and Hortal M. (2016). Compostability assessment of nano-reinforced poly(lactic acid) films. *Waste Management*, **48**, 143–155, https://doi.org/10.1016/j.wasman.2015.10.030

Báreková A., Demovičová M., Tátošová L., Danišová L., Medlenová E. and Hlaváčiková S. (2021). Decomposition of single-use products made of bioplastic under real conditions of urban composting facility. *Journal of Ecological Engineering*, **22**, 265–272, https://doi.org/10.12911/22998993/134040

Barrett A. (2018a). Vegan, Biodegradable and Compostable Glitter. Bioplastics News.

Barrett A. (2018b). Lactips is a Milk-Based Biodegradable and Water-Soluble Packaging. Bioplastics News.

Barrett A. (2018c). First House on Mars will be Made from Bioplastics.

Bátori V., Åkesson D., Zamani A., Taherzadeh M. J. and Sárvári Horváth I. (2018). Anaerobic degradation of bioplastics: A review. *Waste Management*, **80**, 406–413, https://doi.org/10.1016/j.wasman.2018.09.040

Benselfelt T., Engström J. and Wågberg L. (2018). Supramolecular double networks of cellulose nanofibrils and algal polysaccharides with excellent wet mechanical properties. *Green Chemistry*, **20**, 2558–2570, https://doi.org/10.1039/C8GC00590G

Benson Ford Research Center (n.d.). Soybean Car [WWW Document]. URL https://www.thehenryford.org/collections-and-research/digital-resources/popular-topics/soy-bean-car/# (accessed 1.22.22).

Bhatia S. K., Otari S. V., Jeon J. M., Gurav R., Choi Y. K., Bhatia R. K., Pugazhendhi A., Kumar V., Rajesh Banu J., Yoon J. J., Choi K. Y. and Yang Y. H. (2021). Biowaste-to-bioplastic (polyhydroxyalkanoates): conversion technologies, strategies, challenges, and perspective. *Bioresource Technology*, **326**, 124733, https://doi.org/10.1016/J.BIORTECH.2021.124733

Bhatt R., Shah D., Patel K. C. and Trivedi U. (2008). PHA–rubber blends: synthesis, characterization and biodegradation. *Bioresource Technology*, **99**, 4615–4620, https://doi.org/10.1016/j.biortech.2007.06.054

Bi S., Li F., Qin D., Wang M., Yuan S., Cheng X. and Chen X. (2021). Construction of chitin functional materials based on a "green" alkali/urea solvent and their applications in biomedicine: recent advances. *Applied Materials Today*, **23**, 101030, https://doi.org/10.1016/j.apmt.2021.101030

Boyandin A. N., Prudnikova S. V., Karpov V. A., Ivonin V. N., Đỗ N. L., Nguyễn T. H., Lê T. M. H., Filichev N. L., Levin A. L., Filipenko M. L., Volova T. G. and Gitelson I. I. (2013). Microbial degradation of polyhydroxyalkanoates in tropical soils. *International Biodeterioration & Biodegradation*, **83**, 77–84, https://doi.org/10.1016/j.ibiod.2013.04.014

BPI (2015). Confused by the terms Biodegradable & Biobased.

Brockhaus S., Petersen M. and Kersten W. (2016). A crossroad for bioplastics: exploring product developers' challenges to move beyond petroleum-based plastics. *Journal of Cleaner Production*, **127**, 84–95, https://doi.org/10.1016/j.jclepro.2016.04.003

Brodin M., Vallejos M., Opedal M. T., Area M. C. and Chinga-Carrasco G. (2017). Lignocellulosics as sustainable resources for production of bioplastics – a review. *Journal of Cleaner Production*, **162**, 646–664, https://doi.org/10.1016/j.jclepro.2017.05.209

Cafiero L. M., Canditelli M., Musmeci F., Sagnotti G. and Tuffi R. (2021). Assessment of disintegration of compostable bioplastic bags by management of electromechanical and static home composters. *Sustainability*, **13**, 263, https://doi.org/10.3390/su13010263

Calabro' P. S., Folino A., Fazzino F. and Komilis D. (2020). Preliminary evaluation of the anaerobic biodegradability of three biobased materials used for the production of disposable plastics. *Journal of Hazardous Materials*, **390**, 121653, https://doi.org/10.1016/j.jhazmat.2019.121653

Campbell R. and Hotchkiss S. (2017). Carrageenan industry market overview BT – tropical seaweed farming trends, problems and opportunities: focus on *Kappaphycus* and *Eucheuma* of commerce. In: A. Q. Hurtado, A. T. Critchley and I. C. Neish (eds.), Springer International Publishing, Cham, pp. 193–205, https://doi.org/10.1007/978-3-319-63498-2_13

Carlson C. (2020). Alice Potts Makes Bioplastic Face Shields From Food Waste. Dezeen, London, UK.

Castilho L. R., Mitchell D. A. and Freire D. M. G. (2009). Production of polyhydroxyalkanoates (PHAs) from waste materials and by-products by submerged and solid-state fermentation. *Bioresource Technology*, **100**, 5996–6009, https://doi.org/10.1016/J.BIORTECH.2009.03.088

Cazaudehore G., Monlau F., Gassie C., Lallement A. and Guyoneaud R. (2021). Methane production and active microbial communities during anaerobic digestion of three commercial biodegradable coffee capsules under mesophilic and thermophilic conditions. *Science of The Total Environment*, **784**, 146972, https://doi.org/10.1016/j.scitotenv.2021.146972

Cazón P., Velazquez G., Ramírez J. A. and Vázquez M. (2017). Polysaccharide-based films and coatings for food packaging: A review. *Food Hydrocolloids*, **68**, 136–148, https://doi.org/10.1016/j.foodhyd.2016.09.009

Cerda A., Artola A., Font X., Barrena R., Gea T. and Sánchez A. (2018). Composting of food wastes: Status and challenges. *Bioresource Technology*, **248**, 57–67, https://doi.org/10.1016/j.biortech.2017.06.133

Chaireh S., Ngasatool P. and Kaewtatip K. (2020). Novel composite foam made from starch and water hyacinth with beeswax coating for food packaging applications. *International Journal of Biological Macromolecules*, **165**, 1382–1391, https://doi.org/10.1016/j.ijbiomac.2020.10.007

Chamas A., Moon H., Zheng J., Qiu Y., Tabassum T., Jang J. H., Abu-Omar M., Scott S. L. and Suh S. (2020). Degradation rates of plastics in the environment. *ACS Sustainable Chemistry & Engineering*, **8**, 3494–3511, https://doi.org/10.1021/acssuschemeng.9b06635

Chen G., Li S., Jiao F. and Yuan Q. (2007). Catalytic dehydration of bioethanol to ethylene over TiO2/γ-Al$_2$O$_3$ catalysts in microchannel reactors. *Catalysis Today*, **125**, 111–119, https://doi.org/10.1016/j.cattod.2007.01.071

Cho H. S., Moon H. S., Kim M., Nam K. and Kim J. Y. (2011). Biodegradability and biodegradation rate of poly(caprolactone)-starch blend and poly(butylene succinate) biodegradable polymer under aerobic and anaerobic environment. *Waste Management*, **31**, 475–480, https://doi.org/10.1016/j.wasman.2010.10.029

Chomchoei A., Pathom-aree W., Yokota A., Kanongnuch C. and Lumyong S. (2011). *Amycolatopsis thailandensis* sp. nov., a poly(l-lactic acid)-degrading actinomycete, isolated from soil. *International Journal of Systematic and Evolutionary Microbiology*, **61**, 839–843, https://doi.org/10.1099/ijs.0.023564-0

Chong J. W. R., Khoo K. S., Yew G. Y., Leong W. H., Lim J. W., Lam M. K., Ho Y.-C., Ng H. S., Munawaroh H. S. H. and Show P. L. (2021). Advances in production of bioplastics by microalgae using food waste hydrolysate and wastewater: A review. *Bioresource Technology*, **342**, 125947, https://doi.org/10.1016/J.BIORTECH.2021.125947

Chua T.-K., Tseng M. and Yang M.-K. (2013). Degradation of poly(ε-caprolactone) by thermophilic *Streptomyces thermoviolaceus* subsp. *thermoviolaceus* 76T-2. *AMB Express*, **3**, 8, https://doi.org/10.1186/2191-0855-3-8

Colak A. and Güner S. (2004). Polyhydroxyalkanoate degrading hydrolase-like activities by *Pseudomonas* sp. isolated from soil. *International Biodeterioration & Biodegradation*, **53**, 103–109, https://doi.org/10.1016/j.ibiod.2003.10.006

Cucina M., de Nisi P., Tambone F. and Adani F. (2021). The role of waste management in reducing bioplastics' leakage into the environment: A review. *Bioresource Technology*, **337**, 125459, https://doi.org/10.1016/J.BIORTECH.2021.125459

Dai L., Cheng T., Duan C., Zhao W., Zhang W., Zou X., Aspler J. and Ni Y. (2019). 3D printing using plant-derived cellulose and its derivatives: A review. *Carbohydrate Polymers*, **203**, 71–86, https://doi.org/10.1016/j.carbpol.2018.09.027

Das P. and Tiwari P. (2017). Thermal degradation kinetics of plastics and model selection. *Thermochimica Acta*, **654**, 191–202, https://doi.org/10.1016/j.tca.2017.06.001

Demain A. L. (2007). Reviews: The business of biotechnology. *Industrial Biotechnology*, **3**, 269–283, https://doi.org/10.1089/ind.2007.3.269

Duan B., Huang Y., Lu A. and Zhang L. (2018). Recent advances in chitin based materials constructed via physical methods. *Progress in Polymer Science*, **82**, 1–33, https://doi.org/10.1016/j.progpolymsci.2018.04.001

Edgar K. J., Buchanan C. M., Debenham J. S., Rundquist P. A., Seiler B. D., Shelton M. C. and Tindall D. (2001). Advances in cellulose ester performance and application. *Progress in Polymer Science*, **26**, 1605–1688, https://doi.org/10.1016/S0079-6700(01)00027-2

Edwards J., Othman M., Crossin E. and Burn S. (2018). Life cycle assessment to compare the environmental impact of seven contemporary food waste management systems. *Bioresource Technology*, **248**, 156–173, https://doi.org/10.1016/j.biortech.2017.06.070

Elsawy M. A., Kim K. H., Park J. W. and Deep A. (2017). Hydrolytic degradation of polylactic acid (PLA) and its composites. *Renewable and Sustainable Energy Reviews*, **79**, 1346–1352, https://doi.org/10.1016/J.RSER.2017.05.143

Emadian S. M., Onay T. T. and Demirel B. (2017). Biodegradation of bioplastics in natural environments. *Waste Management*, **59**, 526–536, https://doi.org/10.1016/J.WASMAN.2016.10.006

Enactus Plus (n.d.). 2 in 1 Virus Protection Face Mask/Bioplastic Shield [WWW Document]. URL https://plus.enactus.org/s/project/a0d3n000006ZAPAAA4/2-in-1-virus-protection-face-maskbioplastic-shield?language=en_US (accessed 1.23.22)

European Bioplastics (2020a). What are bioplastics? [WWW Document]. URL https://www.european-bioplastics.org/bioplastics/ (accessed 10.6.21)

European Bioplastics (2020b). Environmental benefits of bioplastics [WWW Document]. Eur. Bioplastics. URL https://www.european-bioplastics.org/bioplastics/environment/ (accessed 9.25.21)

European Bioplastics (2020c). Bioplastics market data [WWW Document].

European Bioplastics (2020d). Renewable feedstock [WWW Document]. URL https://www.european-bioplastics.org/bioplastics/feedstock/ (accessed 10.7.21)

European Bioplastics (2020e). Relevant EU policies [WWW Document]. URL https://www.european-bioplastics.org/policy/ (accessed 10.6.21)

European Commission (2021). Bio-based, biodegradable and compostable plastics [WWW Document]. URL https://ec.europa.eu/environment/topics/plastics/bio-based-biodegradable-and-compostable-plastics_en (accessed 10.29.21)

Fang J. (2014). Turning Shrimp Shells Into Plastic. ZDNet, San Francisco, California.

Fernandez J. G. and Ingber D. E. (2014). Manufacturing of large-scale functional objects using biodegradable chitosan bioplastic. *Macromolecular Materials and Engineering*, **299**, 932–938, https://doi.org/10.1002/mame.201300426

Filiciotto L. and Rothenberg G. (2021). Biodegradable plastics: standards, policies, and impacts. *ChemSusChem*, **14**, 56–72, https://doi.org/10.1002/cssc.202002044

Fotopoulou K. N. and Karapanagioti H. K. (2019). Degradation of various plastics in the environment. In: Hazardous Chemicals Associated with Plastics in the Marine Environment, H. Takada and H. K. Karapanagioti (eds.), Springer International Publishing, Cham, pp. 71–92, https://doi.org/10.1007/698_2017_11

Fraunhofer IVV (2016). Sustainable recycling strategies for products and waste streams containing biobased plastics – SustRecPLA [WWW Document]. URL https://www.ivv.fraunhofer.de/en/recycling-environment/packaging-recycling/sustrecpla.html (accessed 10.3.21)

Futerro (2021). Who's Futerro [WWW Document]. URL http://www.futerro.com/index.html (accessed 9.29.21)

Garcia J. M. and Robertson M. L. (2017). The future of plastics recycling. *Science*, **358**, 870–872, https://doi.org/10.1126/science.aaq0324

Gigante V., Cinelli P., Seggiani M., Alavarez V. A. and Lazzeri A. (2020). Processing and thermomechanical properties of PHA. In: The Handbook of Polyhydroxyalkanoates M. Koller (ed.), CRC Press, Boca Raton, Florida, USA. p. 28.

Gioia C., Giacobazzi G., Vannini M., Totaro G., Sisti L., Colonna M., Marchese P. and Celli A. (2021). End of life of biodegradable plastics: composting versus re/upcycling[GT1]. *ChemSusChem*, **14**, 4167–4175, https://doi.org/10.1002/cssc.202101226

Giri B. R., Poudel S. and Kim D. W. (2021). Cellulose and its derivatives for application in 3D printing of pharmaceuticals. *Journal of Pharmaceutical Investigation*, **51**, 1–22, https://doi.org/10.1007/s40005-020-00498-5

Gómez E. F. and Michel F. C. (2013). Biodegradability of conventional and bio-based plastics and natural fiber composites during composting, anaerobic digestion and long-term soil incubation. *Polymer Degradation and Stability*, **98**, 2583–2591, https://doi.org/10.1016/j.polymdegradstab.2013.09.018

Grigore M. E. (2017). Methods of recycling, properties and applications of recycled thermoplastic polymers. *Recycling*, **2**, 24, https://doi.org/10.3390/recycling2040024

Harrison J. P., Boardman C., O'Callaghan K., Delort A.-M. and Song J. (2021). Biodegradability standards for carrier bags and plastic films in aquatic environments: a critical review. *Royal Society Open Science*, **5**, 171792, https://doi.org/10.1098/rsos.171792

Hayase N., Yano H., Kudoh E., Tsutsumi C., Ushio K., Miyahara Y., Tanaka S. and Nakagawa K. (2004). Isolation and characterization of poly(butylene succinate-co-butylene adipate)-degrading microorganism. *Journal of Bioscience and Bioengineering*, **97**, 131–133, https://doi.org/10.1016/S1389-1723(04)70180-2

He M., Wang X., Wang Z., Chen L., Lu Y., Zhang X., Li M., Liu Z., Zhang Y., Xia H. and Zhang L. (2017). Biocompatible and biodegradable bioplastics constructed from chitin via a "green" pathway for bone repair. *ACS Sustainable Chemistry & Engineering*, **5**, 9126–9135, https://doi.org/10.1021/acssuschemeng.7b02051

Hiebel M., Maga D., Jesse A., Wesphalen C., Bauer J., Kroll L., Rinberg R., Harmann T. H.-J., Kötter-Gribber S., Mäuer A. and Dörgens A. (2017). PLA in the waste stream.

HISUN (2021). Development Innovation to The World [WWW Document]. URL http://en.hisunplas.com/about/&i=3&comContentId=3.html (accessed 9.29.21)

Hoang K.-C., Lee C.-Y., Tseng M. and Chu W. S. (2007). Polyester-degrading *Actinomycetes* isolated from the Touchien River of Taiwan. *World Journal of Microbiology and Biotechnology*, **23**, 201–205, https://doi.org/10.1007/s11274-006-9212-7

Hong M. and Chen E. Y.-X. (2017). Chemically recyclable polymers: a circular economy approach to sustainability. *Green Chemistry*, **19**, 3692–3706, https://doi.org/10.1039/C7GC01496A

Hong L. G., Yuhana N. Y. and Zawawi E. Z. E. (2021). Review of bioplastics as food packaging materials. *AIMS Materials Science*, **8**, 166–184, https://doi.org/10.3934/matersci.2021012

Hopewell J., Dvorak R. and Kosior E. (2009). Plastics recycling: challenges and opportunities. *Philosophical Transactions of the Royal Society B: Biological Sciences*, **364**, 2115–2126, https://doi.org/10.1098/rstb.2008.0311

Ishii N., Inoue Y., Tagaya T., Mitomo H., Nagai D. and Kasuya K. (2008). Isolation and characterization of poly(butylene succinate)-degrading fungi. *Polymer Degradation and Stability*, **93**, 883–888, https://doi.org/10.1016/j.polymdegradstab.2008.02.005

ISWA (2017). The Impact of the 4th Industrial Revolution on the Waste Management Sector.

Jarerat A., Pranamuda H. and Tokiwa Y. (2002). Poly(L-lactide)-degrading activity in various actinomycetes. *Macromolecular Bioscience*, **2**, 420–428, https://doi.org/10.1002/mabi.200290001

Jerez A., Partal P., Martínez I., Gallegos C. and Guerrero A. (2007). Protein-based bioplastics: effect of thermo-mechanical processing. *Rheologica Acta*, **46**, 711–720, https://doi.org/10.1007/s00397-007-0165-z

Jiménez-Rosado M., Zarate-Ramírez L. S., Romero A., Bengoechea C., Partal P. and Guerrero A. (2019). Bioplastics based on wheat gluten processed by extrusion. *Journal of Cleaner Production*, **239**, 117994, https://doi.org/10.1016/j.jclepro.2019.117994

Jiménez-Rosado M., Bouroudian E., Perez-Puyana V., Guerrero A. and Romero A. (2020). Evaluation of different strengthening methods in the mechanical and functional properties of soy protein-based bioplastics. *Journal of Cleaner Production*, **262**, 121517, https://doi.org/10.1016/j.jclepro.2020.121517

Jin T., Liu T., Lam E. and Moores A. (2021). Chitin and chitosan on the nanoscale. *Nanoscale Horizons*, **6**, 505–542, https://doi.org/10.1039/D0NH00696C

Kale G., Auras R., Singh S. P. and Narayan R. (2007). Biodegradability of polylactide bottles in real and simulated composting conditions. *Polymer Testing*, **26**, 1049–1061, https://doi.org/10.1016/j.polymertesting.2007.07.006

Kalita N. K., Hazarika D., Kalamdhad A. and Katiyar V. (2021). Biodegradation of biopolymeric composites and blends under different environmental conditions: approach towards end-of-life panacea for crop sustainability. *Bioresource Technology Reports*, **15**, 100705, https://doi.org/10.1016/j.biteb.2021.100705

Karamanlioglu M., Houlden A. and Robson G. D. (2014). Isolation and characterisation of fungal communities associated with degradation and growth on the surface of poly(lactic) acid (PLA) in soil and compost. *International Biodeterioration & Biodegradation*, **95**, 301–310, https://doi.org/10.1016/j.ibiod.2014.09.006

Kim M. N. and Park S. T. (2010). Degradation of poly(L-lactide) by a mesophilic bacterium. *Journal of Applied Polymer Science*, **117**, 67–74, https://doi.org/10.1002/app.31950

Kjeldsen A., Price M., Lilley C. and Guzniczak E. (2018). A Review of Standards for Biodegradable Plastics.

Koch D. and Mihalyi B. (2018). Assessing the change in environmental impact categories when replacing conventional plastic with bioplastic in chosen application fields. **70**, 853–; SE-Research Articles, https://doi.org/10.3303/CET1870143

Kumaravel S., Hema R. and Lakshmi R. (2010). Production of polyhydroxybutyrate (bioplastic) and its biodegradation by *Pseudomonas lemoignei* and *Aspergillus niger*. *E-Journal of Chemistry*, **7**, 148547, https://doi.org/10.1155/2010/148547

Lamberti F. M., Román-Ramírez L. A. and Wood J. (2020). Recycling of bioplastics: routes and benefits. *Journal of Polymers and the Environment*, **28**, 2551–2571, https://doi.org/10.1007/s10924-020-01795-8

Laycock B., Halley P., Pratt S., Werker A. and Lant P. (2013). The chemomechanical properties of microbial polyhydroxyalkanoates. *Progress in Polymer Science*, **38**, 536–583, https://doi.org/10.1016/J.PROGPOLYMSCI.2012.06.003

Lee S.-H. and Kim M.-N. (2010). Isolation of bacteria degrading poly(butylene succinate-co-butylene adipate) and their lip A gene. *International Biodeterioration & Biodegradation*, **64**, 184–190, https://doi.org/10.1016/j.ibiod.2010.01.002

Lee K.-M., Gimore D. F. and Huss M. J. (2005). Fungal degradation of the bioplastic PHB (poly-3-hydroxy-butyric acid). *Journal of Polymers and the Environment*, **13**, 213–219, https://doi.org/10.1007/s10924-005-4756-4

Li F., Hu X., Guo Z., Wang Z., Wang Y., Liu D., Xia H. and Chen S. (2011). Purification and characterization of a novel poly(butylene succinate)-degrading enzyme from *Aspergillus* sp. XH0501-a. *World Journal of Microbiology and Biotechnology*, **27**, 2591–2596, https://doi.org/10.1007/s11274-011-0731-5

Li M., Witt T., Xie F., Warren F. J., Halley P. J. and Gilbert R. G. (2015). Biodegradation of starch films: The roles of molecular and crystalline structure. *Carbohydrate Polymers*, **122**, 115–122, https://doi.org/10.1016/j.carbpol.2015.01.011

Lim C., Yusoff S., Ng C. G., Lim P. E. and Ching Y. C. (2021). Bioplastic made from seaweed polysaccharides with green production methods. *Journal of Environmental Chemical Engineering*, **9**, 105895, https://doi.org/10.1016/j.jece.2021.105895

Loliware (2020). Meet the Loliware Straw [WWW Document]. URL https://www.loliware.com/the-straw (accessed 9.28.21)

Lomthong T., Chotineeranat S. and Kitpreechavanich V. (2015). Production and characterization of raw starch degrading enzyme from a newly isolated thermophilic filamentous bacterium, *Laceyella sacchari* LP175. *Starch – Stärke*, **67**, 255–266, https://doi.org/10.1002/star.201400150

Lu J., Tappel R. C. and Nomura C. T. (2009). Mini-review: biosynthesis of poly(hydroxyalkanoates). *Polymer Reviews*, **49**, 226–248, https://doi.org/10.1080/15583720903048243

Luyt A. S. and Malik S. S. (2019). Can biodegradable plastics solve plastic solid waste accumulation? In: Plastics Design Library. Al-Salem, S.M.B.T.-P. to E. (ed.), William Andrew Publishing, Norwich, New York, USA, pp. 403–423, https://doi.org/10.1016/B978-0-12-813140-4.00016-9

Maeda H., Yamagata Y., Abe K., Hasegawa F., Machida M., Ishioka R., Gomi K. and Nakajima T. (2005). Purification and characterization of a biodegradable plastic-degrading enzyme from *Aspergillus oryzae*. *Applied Microbiology and Biotechnology*, **67**, 778–788, https://doi.org/10.1007/s00253-004-1853-6

Mannina G., Presti D., Montiel-Jarillo G., Carrera J. and Suárez-Ojeda M. E. (2020). Recovery of polyhydroxyalkanoates (PHAs) from wastewater: a review. *Bioresource Technology*, **297**, 122478, https://doi.org/10.1016/J.BIORTECH.2019.122478

Mihai M., Legros N. and Alemdar A. (2014). Formulation-properties versatility of wood fiber biocomposites based on polylactide and polylactide/thermoplastic starch blends. *Polymer Engineering & Science*, **54**, 1325–1340, https://doi.org/10.1002/pen.23681

Mohee R., Unmar G. D., Mudhoo A. and Khadoo P. (2008). Biodegradability of biodegradable/degradable plastic materials under aerobic and anaerobic conditions. *Waste Management*, **28**, 1624–1629, https://doi.org/10.1016/j.wasman.2007.07.003

Mordor Intelligence (2021). Bio-polylactic acid (PLA) market – Growth, trends, COVID-19 impact, and forecast (2021–2026).

Mujtaba M., Morsi R. E., Kerch G., Elsabee M. Z., Kaya M., Labidi J. and Khawar K. M. (2019). Current advancements in chitosan-based film production for food technology; a review. *International Journal of Biological Macromolecules*, **121**, 889–904, https://doi.org/10.1016/j.ijbiomac.2018.10.109

Müller R.-J. (2005). Biodegradability of polymers: regulations and methods for testing. In Biopolymers Online, A. Steinbüchel (ed.), Wiley Online Library, New Jersey, U.S., pp. 366–388. https://doi.org/10.1002/3527600035.bpola012

Muneer F., Rasul I., Azeem F., Siddique M. H., Zubair M. and Nadeem H. (2020). Microbial polyhydroxyalkanoates (PHAs): efficient replacement of synthetic polymers. *Journal of Polymers and the Environment*, **28**, 2301–2323, https://doi.org/10.1007/s10924-020-01772-1

Nakajima-Kambe T., Toyoshima K., Saito C., Takaguchi H., Akutsu-Shigeno Y., Sato M., Miyama K., Nomura N. and Uchiyama H. (2009). Rapid monomerization of poly(butylene succinate)-co-(butylene adipate) by *Leptothrix* sp. *Journal of Bioscience and Bioengineering*, **108**, 513–516, https://doi.org/10.1016/j.jbiosc.2009.05.018

Nakasaki K., Matsuura H., Tanaka H. and Sakai T. (2006). Synergy of two thermophiles enables decomposition of poly-ε-caprolactone under composting conditions. *FEMS Microbiology Ecology*, 58, 373–383, https://doi.org/10.1111/j.1574-6941.2006.00189.x

NatureWorks (2021). NatureWorks [WWW Document]. URL https://www.natureworksllc.com/ (accessed 9.29.21)

NatureWorks LLC (2008). Using near-infrared sorting to recycle PLA bottles [WWW Document]. URL https://www.natureworksllc.com/What-is-Ingeo/Where-it-Goes/Recycling (accessed 10.3.21)

Niaounakis M. (2017). 1 – The problem of marine plastic debris. In: Plastics Design Library, Niaounakis, M.B.T.-M. of M.P.D. (ed.), William Andrew Publishing, Norwich, New York, USA, pp. 1–55, https://doi.org/10.1016/B978-0-323-44354-8.00001-X

Niaounakis M. (2019). Recycling of biopolymers – The patent perspective. *European Polymer Journal*, 114, 464–475, https://doi.org/10.1016/J.EURPOLYMJ.2019.02.027

nova-Institut GmbH (2010). Oculus3D – Cereplast Offer Bioplastic 3D Glasses to Movie Theaters. Nova-Institut GmbH, Hürth, Nordrhein-Westfalen, Germany.

Padil V. V. T., Senan C., Wacławek S., Černík M., Agarwal S. and Varma R. S. (2019). Bioplastic fibers from gum Arabic for greener food wrapping applications. *ACS Sustainable Chemistry & Engineering*, **7**, 5900–5911, https://doi.org/10.1021/acssuschemeng.8b05896

Pagga U. (1997). Testing biodegradability with standardized methods. *Chemosphere*, **35**, 2953–2972, https://doi.org/10.1016/S0045-6535(97)00262-2

Patel P. (2020). The time is now for edible packaging. Chemical News, **98** (4). https://cen.acs.org/food/food-science/time-edible-packaging/98/i4 (accessed 20 June 2021).

Patel S. K. S., Gupta R. K., Kondaveeti S., Otari S. V., Kumar A., Kalia V. C. and Lee J.-K. (2020). Conversion of biogas to methanol by methanotrophs immobilized on chemically modified chitosan. *Bioresource Technology*, 315, 123791, https://doi.org/10.1016/j.biortech.2020.123791

Pavon C., Aldas M., López-Martínez J. and Ferrándiz S. (2020). New materials for 3D-printing based on polycaprolactone with gum rosin and beeswax as additives. *Polymers*, **12**, 334, https://doi.org/10.3390/polym12020334

Payne J., McKeown P. and Jones M. D. (2019). A circular economy approach to plastic waste. *Polymer Degradation and Stability*, **165**, 170–181, https://doi.org/10.1016/J.POLYMDEGRADSTAB.2019.05.014

Penkhrue W., Khanongnuch C., Masaki K., Pathom-aree W., Punyodom W. and Lumyong S. (2015). Isolation and screening of biopolymer-degrading microorganisms from northern Thailand. *World Journal of Microbiology and Biotechnology*, **31**, 1431–1442, https://doi.org/10.1007/s11274-015-1895-1

Philp J. C., Ritchie R. J. and Guy K. (2013). Biobased plastics in a bioeconomy. *Trends in Biotechnology*, **31**, 65–67, https://doi.org/10.1016/j.tibtech.2012.11.009

Phukon P., Saikia J. P. and Konwar B. K. (2012). Bio-plastic (P-3HB-co-3HV) from *Bacillus circulans* (MTCC 8167) and its biodegradation. *Colloids and Surfaces B: Biointerfaces*, **92**, 30–34, https://doi.org/10.1016/j.colsurfb.2011.11.011

Ramakrishnan N., Sharma S., Gupta A. and Alashwal B. Y. (2018). Keratin based bioplastic film from chicken feathers and its characterization. *International Journal of Biological Macromolecules*, **111**, 352–358, https://doi.org/10.1016/j.ijbiomac.2018.01.037

RameshKumar S., Shaiju P., O'Connor K. E. and Ramesh Babu P. (2020). Bio-based and biodegradable polymers – state-of-the-art, challenges and emerging trends. *Current Opinion in Green and Sustainable Chemistry*, **21**, 75–81, https://doi.org/10.1016/J.COGSC.2019.12.005

Ren Y., Yu M., Wu C., Wang Q., Gao M., Huang Q. and Liu Y. (2018). A comprehensive review on food waste anaerobic digestion: research updates and tendencies. *Bioresource Technology*, **247**, 1069–1076, https://doi.org/10.1016/j.biortech.2017.09.109

RespirTek (2021). Biodegradability Testing [WWW Document]. URL https://www.respirtek.com/biodegradability-testing/?gclid=CjwKCAjw49qKBhAoEiwAHQVTo1LoOlKdFggqUMZEeWb2tliItZ6axDomE9rPoNSfFjrKlDAqFyhu_BoCxgUQAvD_BwE (accessed 10.2.21)

Rodriguez-Perez S., Serrano A., Pantión A. A. and Alonso-Fariñas B. (2018). Challenges of scaling-up PHA production from waste streams. A review. *Journal of Environmental Management*, **205**, 215–230, https://doi.org/10.1016/J.JENVMAN.2017.09.083

Roy A. (2017). Starch based bioplastics market.

Ruggero F., Onderwater R. C. A., Carretti E., Roosa S., Benali S., Raquez J.-M., Gori R., Lubello C. and Wattiez R. (2021). Degradation of film and rigid bioplastics during the thermophilic phase and the maturation phase of simulated composting. *Journal of Polymers and the Environment*, **29**, 3015–3028, https://doi.org/10.1007/s10924-021-02098-2

Saalah S., Saallah S., Rajin M. and Yaser A. Z. (2020). Management of biodegradable plastic waste: a review BT – advances in waste processing technology. In: A. Z. Yaser (ed.), Springer Singapore, Singapore, pp. 127–143, https://doi.org/10.1007/978-981-15-4821-5_8

Sagong H.-Y., Son H. F., Choi S. Y., Lee S. Y. and Kim K.-J. (2018). Structural insights into polyhydroxyalkanoates biosynthesis. *Trends in Biochemical Sciences*, **43**, 790–805, https://doi.org/10.1016/j.tibs.2018.08.005

Savenkova L., Gercberga Z., Nikolaeva V., Dzene A., Bibers I. and Kalnin M. (2000). Mechanical properties and biodegradation characteristics of PHB-based films. *Process Biochemistry*, **35**, 573–579, https://doi.org/10.1016/S0032-9592(99)00107-7

Sekiguchi T., Saika A., Nomura K., Watanabe T., Watanabe T., Fujimoto Y., Enoki M., Sato T., Kato C. and Kanehiro H. (2011a). Biodegradation of aliphatic polyesters soaked in deep seawaters and isolation of poly(ε-caprolactone)-degrading bacteria. *Polymer Degradation and Stability*, **96**, 1397–1403, https://doi.org/10.1016/j.polymdegradstab.2011.03.004

Sekiguchi T., Sato T., Enoki M., Kanehiro H., Uematsu K. and Kato C. (2011b). Isolation and characterization of biodegradable plastic degrading bacteria from deep-sea environments. *JAMSTEC Report of Research and Development*, **11**, 33–41, https://doi.org/10.5918/jamstecr.11.33

Serizawa S., Inoue K. and Iji M. (2006). Kenaf-fiber-reinforced poly(lactic acid) used for electronic products. *Journal of Applied Polymer Science*, **100**, 618–624, https://doi.org/10.1002/app.23377

Shah A. A., Hasan F., Hameed A. and Ahmed S. (2007). Isolation and characterisation of poly(3-hydroxybutyrate-co-3-hydroxyvalerate) degrading actinomycetes and purification of PHBV depolymerase from newly isolated *Streptoverticillium kashmirense* AF1. *Annals of Microbiology*, **57**, 583–588, https://doi.org/10.1007/BF03175359

Shah A. A., Hasan F., Hameed A. and Ahmed S. (2008). Biological degradation of plastics: A comprehensive review. *Biotechnology Advances*, **26**, 246–265, https://doi.org/10.1016/j.biotechadv.2007.12.005

Shah A. A., Hasan F. and Hameed A. (2010). Degradation of poly(3-hydroxybutyrate-co-3-hydroxyvalerate) by a newly isolated *Actinomadura* sp. AF-555, from soil. *International Biodeterioration & Biodegradation*, **64**, 281–285, https://doi.org/10.1016/j.ibiod.2009.10.012

Shen L., Worrell E. and Patel M. K. (2010). Open-loop recycling: A LCA case study of PET bottle-to-fibre recycling. *Resources, Conservation and Recycling*, **55**, 34–52, https://doi.org/10.1016/J.RESCONREC.2010.06.014

Shukla S. K., Mishra A. K., Arotiba O. A. and Mamba B. B. (2013). Chitosan-based nanomaterials: A state-of-the-art review. *International Journal of Biological Macromolecules*, **59**, 46–58, https://doi.org/10.1016/j.ijbiomac.2013.04.043

Singh N., Hui D., Singh R., Ahuja I. P. S., Feo L. and Fraternali F. (2017). Recycling of plastic solid waste: A state of art review and future applications. *Composites Part B: Engineering*, **115**, 409–422, https://doi.org/10.1016/J.COMPOSITESB.2016.09.013

Siracusa V. (2019). Microbial degradation of synthetic biopolymers waste. *Polymers*, **11**, 1066, https://doi.org/10.3390/polym11061066

SpaceFactory, A. (n.d.). Advanced Construction Technology for earth and space [WWW Document]. URL https://www.aispacefactory.com/ (accessed 1.22.22)

Spierling S., Knüpffer E., Behnsen H., Mudersbach M., Krieg H., Springer S., Albrecht S., Herrmann C. and Endres H. J. (2018). Bio-based plastics – A review of environmental, social and economic impact assessments. *Journal of Cleaner Production*, **185**, 476–491, https://doi.org/10.1016/J.JCLEPRO.2018.03.014

Storz H. and Vorlop K.-D. (2013). Bio-based plastics: status, challenges and trends. *Landbauforschung Applied Agricultural and Forestry Research*, **4**, 321–, https://doi.org/10.3220/LBF_2013_321-332

Sukkhum S., Tokuyama S., Tamura T. and Kitpreechavanich V. (2009). A novel poly (L-lactide) degrading actinomycetes isolated from Thai forest soil, phylogenic relationship and the enzyme characterization. *The Journal of General and Applied Microbiology*, **55**, 459–467, https://doi.org/10.2323/jgam.55.459

Sun J.-Y., Zhao X., Illeperuma W. R. K., Chaudhuri O., Oh K. H., Mooney D. J., Vlassak J. J. and Suo Z. (2012). Highly stretchable and tough hydrogels. *Nature*, **489**, 133–136, https://doi.org/10.1038/nature11409

Sun Q., Xi T., Li Y. and Xiong L. (2014). Characterization of corn starch films reinforced with CaCO₃ nanoparticles. *PLoS One*, **9**, e106727, https://doi.org/10.1371/journal.pone.0106727

Sun Y., Ren X., Rene E. R., Wang Z., Zhou L., Zhang Z. and Wang Q. (2021). The degradation performance of different microplastics and their effect on microbial community during composting process. *Bioresource Technology*, **332**, 125133, https://doi.org/10.1016/j.biortech.2021.125133

Szumigaj J., Żakowska Z., Klimek L., Rosicka-Kaczmarek J. and Bartkowiak A. (2008). Assessment of polylactide foil degradation as a result of filamentous fungi activity. *Polish Journal of Environmental Studies*, **17**, 335–341.

Tabasi R. Y. and Ajji A. (2015). Selective degradation of biodegradable blends in simulated laboratory composting. *Polymer Degradation and Stability*, **120**, 435–442, https://doi.org/10.1016/j.polymdegradstab.2015.07.020

Tedeschi G., Benitez J. J., Ceseracciu L., Dastmalchi K., Itin B., Stark R. E., Heredia A., Athanassiou A. and Heredia-Guerrero J. A. (2018). Sustainable fabrication of plant cuticle-like packaging films from tomato pomace agro-waste, beeswax, and alginate. *ACS Sustainable Chemistry & Engineering*, 6, 14955–14966, https://doi.org/10.1021/acssuschemeng.8b03450

Teeraphatpornchai T., Nakajima-Kambe T., Shigeno-Akutsu Y., Nakayama M., Nomura N., Nakahara T. and Uchiyama H. (2003). Isolation and characterization of a bacterium that degrades various polyester-based biodegradable plastics. *Biotechnology Letters*, 25, 23–28, https://doi.org/10.1023/A:1021713711160

Tezuka Y., Ishii N., Kasuya K. and Mitomo H. (2004). Degradation of poly(ethylene succinate) by mesophilic bacteria. *Polymer Degradation and Stability*, 84, 115–121, https://doi.org/10.1016/j.polymdegradstab.2003.09.018

The New York Times (1941). Ford shows auto built of plastic; Strong Material Derived From Soy Beans, Wheat, Corn is Used for Body and Fenders Saving of steel is cited Car Is 1,000 Pounds Lighter Than Metal Ones – 12 Years of Research Developed It. New York Times.

The Sustainable Biomaterials Collaborative (2009). Guidelines for Sustainable Bioplastics.

Thiruchelvi R., Das A. and Sikdar E. (2021). Bioplastics as better alternative to petro plastic. *Materials Today: Proceedings*, 37, 1634–1639, https://doi.org/10.1016/j.matpr.2020.07.176

Total Corbion (2021). Total Corbion PLA announces the first world-scale PLA plant in Europe [WWW Document]. URL https://www.total-corbion.com/ (accessed 9.29.21)

Tseng H.-C., Fujimoto N. and Ohnishi A. (2019). Biodegradability and methane fermentability of polylactic acid by thermophilic methane fermentation. *Bioresource Technology Reports*, 8, 100327, https://doi.org/10.1016/j.biteb.2019.100327

UNEP (2019). Made in Latin America – four innovations for sustainable living that could change the world.

Villaluz K. (2017). Meet Ooho! A Sustainable Water Bottle You Can Eat and Drink.

Volova T. G., Gladyshev M. I., Trusova M. Y. and Zhila N. O. (2007). Degradation of polyhydroxyalkanoates in eutrophic reservoir. *Polymer Degradation and Stability*, 92, 580–586, https://doi.org/10.1016/j.polymdegradstab.2007.01.011

Volova T. G., Boyandin A. N., Vasiliev A. D., Karpov V. A., Prudnikova S. V., Mishukova O. V., Boyarskikh U. A., Filipenko M. L., Rudnev V. P., Bá Xuân B., Việt Dũng V. and Gitelson I. I. (2010). Biodegradation of polyhydroxyalkanoates (PHAs) in tropical coastal waters and identification of PHA-degrading bacteria. *Polymer Degradation and Stability*, 95, 2350–2359, https://doi.org/10.1016/j.polymdegradstab.2010.08.023

Wainaina S., Awasthi M. K., Sarsaiya S., Chen H., Singh E., Kumar A., Ravindran B., Awasthi S. K., Liu T., Duan Y., Kumar S., Zhang Z. and Taherzadeh M. J. (2020). Resource recovery and circular economy from organic solid waste using aerobic and anaerobic digestion technologies. *Bioresource Technology*, 301, 122778, https://doi.org/10.1016/j.biortech.2020.122778

Wang S. and Zeng Y. (2018). Ammonia emission mitigation in food waste composting: A review. *Bioresource Technology*, 248, 13–19, https://doi.org/10.1016/j.biortech.2017.07.050

Wang H., Qian J. and Ding F. (2018). Emerging chitosan-based films for food packaging applications. *Journal of Agricultural and Food Chemistry*, 66, 395–413, https://doi.org/10.1021/acs.jafc.7b04528

Weng Y.-X., Wang X.-L. and Wang Y.-Z. (2011). Biodegradation behavior of PHAs with different chemical structures under controlled composting conditions. *Polymer Testing*, 30, 372–380, https://doi.org/10.1016/j.polymertesting.2011.02.001

Wu C.-S. (2009). Renewable resource-based composites of recycled natural fibers and maleated polylactide bioplastic: characterization and biodegradability. *Polymer Degradation and Stability*, 94, 1076–1084, https://doi.org/10.1016/j.polymdegradstab.2009.04.002

Wu C.-S. (2011). Performance and biodegradability of a maleated polyester bioplastic/recycled sugarcane bagasse system. *Journal of Applied Polymer Science*, **121**, 427–435, https://doi.org/10.1002/app.33713

Wu C.-S. (2012). Characterization and biodegradability of polyester bioplastic-based green renewable composites from agricultural residues. *Polymer Degradation and Stability*, **97**, 64–, https://doi.org/10.1016/j.polymdegradstab.2011.10.012

Wu B., Zhang X., Shang D., Bao D., Zhang S. and Zheng T. (2016). Energetic-environmental-economic assessment of the biogas system with three utilization pathways: combined heat and power, biomethane and fuel cell. *Bioresource Technology*, **214**, 722–728, https://doi.org/10.1016/j.biortech.2016.05.026

Xie F., Pollet E., Halley P. J. and Avérous L. (2013). Starch-based nano-biocomposites. *Progress in Polymer Science*, **38**, 1590–1628, https://doi.org/10.1016/j.progpolymsci.2013.05.002

Xue J., Wu Y., Shi K., Xiao X., Gao Y., Li L. and Qiao Y. (2019). Study on the degradation performance and kinetics of immobilized cells in straw-alginate beads in marine environment. *Bioresource Technology*, **280**, 88–94, https://doi.org/10.1016/j.biortech.2019.02.019

Yagi H., Ninomiya F., Funabashi M. and Kunioka M. (2013). Thermophilic anaerobic biodegradation test and analysis of eubacteria involved in anaerobic biodegradation of four specified biodegradable polyesters. *Polymer Degradation and Stability*, **98**, 1182–1187, https://doi.org/10.1016/j.polymdegradstab.2013.03.010

Yakovleva I. S., Banzaraktsaeva S. P., Ovchinnikova E. V., Chumachenko V. A. and Isupova L. A. (2016). Catalytic dehydration of bioethanol to ethylene. *Catalysis in Industry*, **8**, 152–167, https://doi.org/10.1134/S2070050416020148

Yang J., Ching Y. C., Chuah C. H. and Liou N.-S. (2021). Preparation and characterization of starch/empty fruit bunch-based bioplastic composites reinforced with epoxidized oils. *Polymers*, **13**, 94, https://doi.org/10.3390/polym13010094

Ye Y. (2021). Chinese PPE makers become greener, trying to limit pollution. Global Times.

Ye C., Voet V. S. D., Folkersma R. and Loos K. (2021). Robust superamphiphilic membrane with a closed-loop life cycle. *Advanced Materials*, **33**, 2008460, https://doi.org/10.1002/adma.202008460

Yip M. and Madl P. (2003). Waste Management [WWW Document]. URL https://biophysics.sbg.ac.at/waste/organic.htm (accessed 9.30.21)

Yoshida N., Ye L., Liu F., Li Z. and Katayama A. (2013). Evaluation of biodegradable plastics as solid hydrogen donors for the reductive dechlorination of fthalide by *Dehalobacter* species. *Bioresource Technology*, **130**, 478–485, https://doi.org/10.1016/j.biortech.2012.11.139

Zargar V., Asghari M. and Dashti A. (2015). A review on chitin and chitosan polymers: structure, chemistry, solubility, derivatives, and applications. *ChemBioEng Reviews*, **2**, 204–226, https://doi.org/10.1002/cben.201400025

Zhang Y., Rempel C. and Liu Q. (2014). Thermoplastic starch processing and characteristics – a review. *Critical Reviews in Food Science and Nutrition*, **54**, 1353–1370, https://doi.org/10.1080/10408398.2011.636156

Zhang Y., Cui L., Che X., Zhang H., Shi N., Li C., Chen Y. and Kong W. (2015). Zein-based films and their usage for controlled delivery: origin, classes and current landscape. *Journal of Controlled Release*, **206**, 206–219, https://doi.org/10.1016/j.jconrel.2015.03.030

Zhang W., Heaven S. and Banks C. J. (2018). Degradation of some EN13432 compliant plastics in simulated mesophilic anaerobic digestion of food waste. *Polymer Degradation and Stability*, **147**, 76–88, https://doi.org/10.1016/j.polymdegradstab.2017.11.005

Zhang C., Show P.-L. and Ho S.-H. (2019). Progress and perspective on algal plastics – A critical review. *Bioresource Technology*, **289**, 121700, https://doi.org/10.1016/j.biortech.2019.121700

doi: 10.2166/9781789063448_0437

Chapter 11

Innovations and future trends in plastic waste management

Wenchao Xue

Table of Contents

11.1 PRINCIPLES AND MEASURES OF INNOVATIONS FOR SUSTAINABLE PLASTIC WASTE MANAGEMENT

11.1.1 What is innovation?

There are many ways to define innovation. In this chapter, we consider innovation as the generation of a new idea and its implementation into a new product, process or service, leading to the dynamic growth of the national economy and the increase of employment as well as to a creation of pure profit for the innovative business enterprise (Urabe *et al.*, 2018).

11.1.1.1 *Innovation vs. invention vs. creativity*

The terminologies of innovation, invention, and creativity may be used interchangeably in our daily lives. Specifically, creativity is used to represent an opportunity to create new appearance, content or process by combining existing inputs of factors of production; invention refers to a process of creating something new, which contributes to the level of mankind knowledge; innovation is linked to the definitive marketing of the new product, service or technology, which is the result of the inventiveness. Figure 11.1 illustrates the relationships among science, creativity, invention and innovation.

11.1.1.2 *Classifications of innovation*

Innovations can be classified using various classification systems (Table 11.1). For instance, based on the applications, innovations can be categorized into agricultural innovation, industrial innovation, social innovation and organizational innovation. The Science-Technology-Process (STP) model is used to identify the different stages in an innovation life cycle (Figure 11.2).

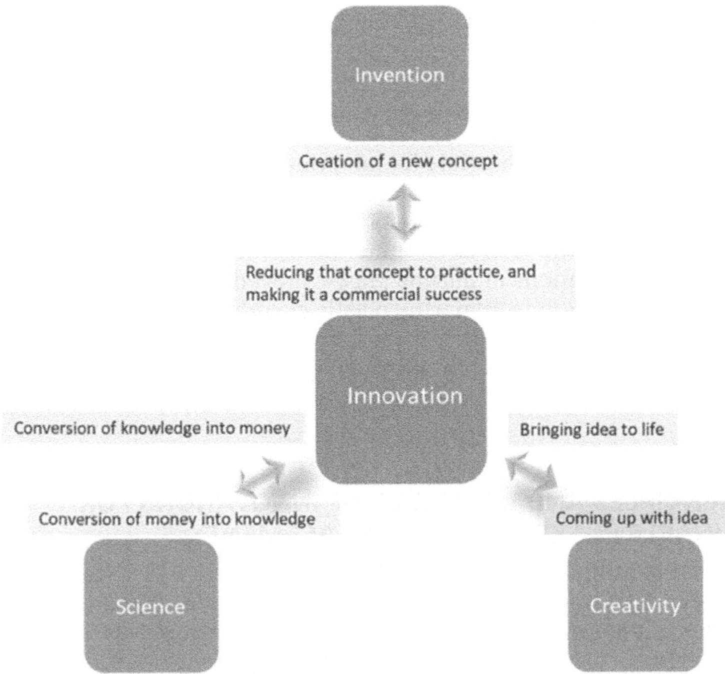

Figure 11.1 Relationships among science, creativity, invention, and innovation (*Source*: Kogabayev & Maziliauskas, 2017).

Table 11.1 Classification systems of innovation.

Classification Sign	Categories of Innovation
Application of innovation	Managerial, organizational, social, industrial, agricultural, and so on.
STP stages, resulting in innovation	Scientific, technological, engineering, manufacturing, information
Intensity of innovation	Incremental, radical/breakthrough
Pace of implementation of innovation	Fast, slow, decaying, growing, uniform, abrupt
Scope of innovation	Transcontinental, transnational, regional, large, medium, small
Effectiveness of innovation	High, stable, low
Target of innovation	Economic, social, ecological, integrated

Source: Kogabayev & Maziliauskas (2017).

The life cycle of a typical technological innovation is usually composed of three key stages, that is, scientific discovery stage, technical invention stage, and engineering prototype stage. These stages are used to classify the level and degree of an innovation in its life cycle.

Intensity of innovation is another frequently used system to classify innovations. According to the framework suggested by Henderson and Clark (1990), innovations can be divided into four categories, namely incremental innovation, architectural innovation, disruptive innovation and radical innovation based on their technological improvements and market impacts (see Figure 11.3). Incremental innovation is considered the most common form of innovation. It is a 'small-scale' innovation that adds little advancement on the existing technology, which results in increasing value for the customer within

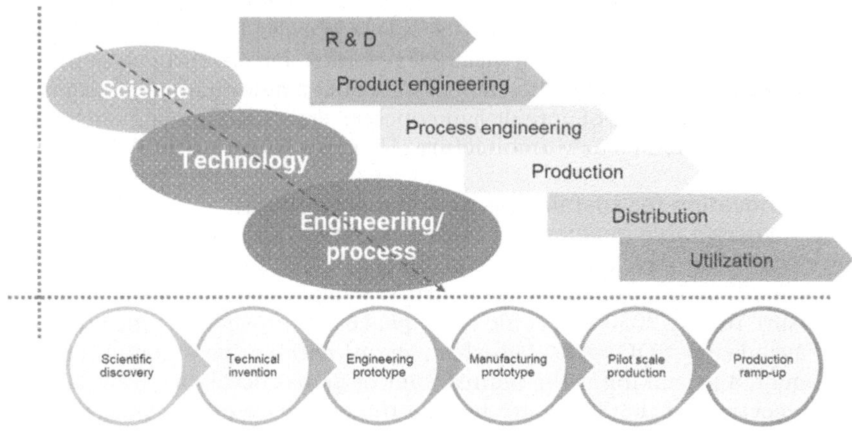

Figure 11.2 A typical life cycle of technological innovation (*Source:* Kogabayev & Maziliauskas, 2017).

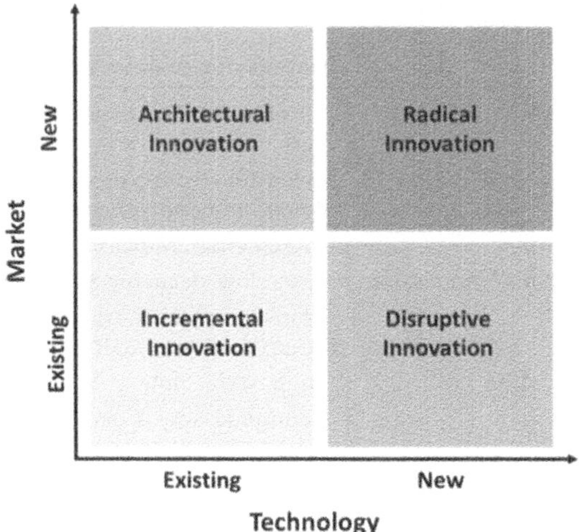

Figure 11.3 Intensity of innovation (*Source*: Henderson & Clark, 1990).

the same market. Architectural innovation, as the name suggests, refers to the innovation of an architecture of any product. Through this innovation, various components of the system can be changed or modified the way it links or relates to each other, that is, lighter in weight or smaller in size. However, the main technology of the component stays intact (Procto, 2019). Disruptive innovation, also known as stealth innovation, involves application of new technology or processes to one's company's current market. This newer technology is usually difficult to use and expensive. After few iterations, this newer technology surpasses the older ones and disrupts all the existing companies. For example, take the case of Apple's iPhone in the market. Before iPhone, the market used phones with buttons, keypads or scroll wheels (Lopez, 2015). Radical innovation is considered to bring the highest market impacts through significant technological improvements. It usually establishes a new dominant design and, consequently, a new set of core design concepts embodied in components that are linked together in a new architecture (Henderson & Clark, 1990).

11.1.2 Innovation for sustainable plastic management
11.1.2.1 Sustainable development goals, plastics, and innovation
The sustainable development goals (SDGs) were adopted by the United Nations in 2015 as a universal call to action to end poverty, protect the planet and ensure that by 2030 all people enjoy peace and prosperity. Since the UN's 2030 agenda on SDGs was launched, many nations have committed and implemented it, making it the mainstream of global development today. The main objectives of the SDGs are to take the collective decisions against the most critical multi-dimensional socio-economic-environmental global threats that humans face. Plastics, as one of the most amazing industrial products, become a geological indicator for the Anthropocene era. Its appearance has

contributed to economic growth, benefited social life, and at the same time caused adverse effects on the environment and human health. Hence, the concern on plastic issues is closely associated with sustainable development. The life cycle of plastics including their raw material extraction, production, consumption, end-of-life management and environmental disposal/leakage have been closely interlinked with the 17 SDGs (Kumar *et al.*, 2021). As presented in Figure 11.4, the global concerns on plastic pollution and its negative impact on the environment, socio-economy and health are directly or indirectly related to SDGs 1, 2, 3, 6, 11, 13, 14 and 15, covering all aspects of food safety, human well-being, water security, climate change, and life below water and on land. Addressing these sustainability issues and challenges resulting from plastic pollution entails diverse innovations. Figure 11.5 provides several innovation options throughout the life cycle of plastics that potentially address sustainable development issues.

On the other hand, several principles of SDGs such as quality education and gender-based roles (SDGs 4 and 5), industrial innovation and new economic growth (SDG 8 and 9), circular economy (SDG 12), and global institution and partnership (SDG 17) become key drivers and approaches for innovations in plastic management. In fact, improvements in industrial processes, social practices and market value help achieve all SDGs by providing transformational solutions to numerous challenges that go above incremental innovation through (1) adopting new visions for the product value and design; (2) building and trying new business models, such as reuse and closed-loop business models, and (3) developing disruptive technologies that enhance plastic waste management and recycling. These innovative actions aim to boost organizational position in overcoming market disruptions and pursuing growth opportunities, while

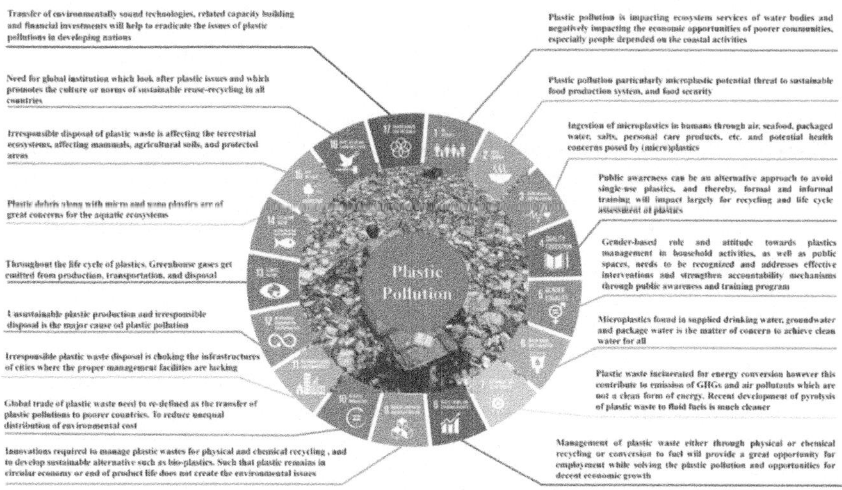

Figure 11.4 Perspectives on interlinkages and relationships between plastic pollution and management with SDGs (*Source*: Kumar *et al.*, 2021).

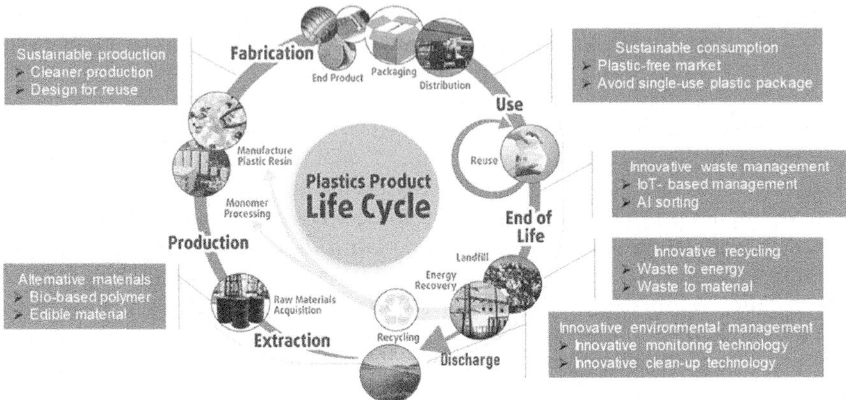

Figure 11.5 Innovation options in the life cycle of plastic (*Source*: Modified from Devasahayam *et al.*, 2019).

at the same time helping to achieve the SDGs and tackling plastic pollution. Table 11.2 presents several examples of how sustainable plastic innovations can address the SDGs.

11.1.2.2 Emerging innovation approaches for plastic sustainability

To achieve the broad and ambitious agenda of SDGs, a new perspective on innovation, referred to 'innovation for sustainability' is necessary. Different from conventional market-orientated innovations, new approaches of innovation should be socially inclusive and environmentally benign. A recent report by UNCTAD (2017) has proposed several emerging approaches to achieve innovations for sustainability. Descriptions of these emerging innovation approaches and their application cases in conducting innovations for sustainable plastic management are provided below.

Mission-oriented innovation: Organizing networked research programs at national and international levels, as well as the incentive structures that can direct innovation towards the achievement of specific technological, environmental or social goals. Mission-oriented innovation is based on the conventional approach but reallocates the innovation resources from market-orientated to mission-orientated in order to resolve the social and environmental challenges such as global climate change, environmental management and infectious diseases. Contemporary mission-oriented innovation programs are initiated and funded by variable sectors or organizations, typically with substantial scales. Examples of this innovation approach include various state-funded programs, large multilateral cooperation initiatives, philanthropic organization-funded programs, state investment bank mission-oriented programs, and different public–private initiatives. Unlike traditional research programs, such new initiatives typically seek to move beyond research and development, to actively support prototyping, scaling up and the commercial or public diffusion of new

Table 11.2 Several SDGs addressed by adopting sustainable plastic innovation.

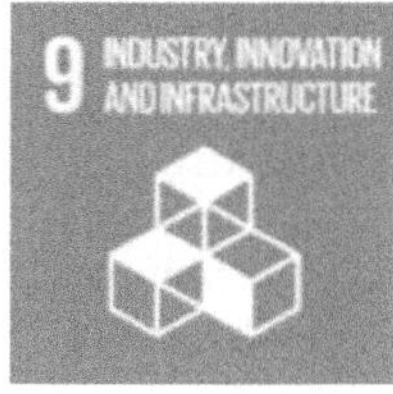

Different requirements throughout the plastic lifecycle make it highly energy intensive and expensive. One of the main challenges to tackle when attempting to achieve SDG 9 is to transform the conventional processes that start from the extraction of raw materials to the end-of-life management, into more sustainable processes, that is, developing innovative technologies for recycling non-recyclable plastic waste, or for enhancing outputs from waste-to-energy solutions. Innovations in this sense will significantly improve the economic and environmental footprint of the plastic sector.

Preventing the creation of unnecessary plastic in the first place in packaging, or single-use plastic products, and reusing plastic that is already created will definitely shape both the consumers' and the producers' behaviors to lean toward sustainability, and thus contributes to achieving SDG 12. Also, producing innovative alternatives to plastic materials that are bio-based and biodegradable can be a path to achieve responsible consumption and production.

Decreasing plastic demand and production while increasing plastic recycling and recovery helps cutting down GHGs emitted by the fossil fuel-based industries, and thus contributes to the goal of achieving SDG 13.

Innovations in tackling plastic waste will lead to the reduction of the amount of leakage into the environment and into the ocean, which would otherwise disrupt the environment and threaten marine ecosystems that SDG 14 aims to conserve.

tcchnologics; it also seeks to involve, and provide incentives to a more diverse range of innovators than researchers, across public, private and civil society sectors. An example of applying the mission-oriented innovation approach to developing a supplementary material for plastic replacement is provided in Box 11.1.

Box 11.1: The CounterMEASURE Project for plastic-free rivers

Jointly funded by the UNEP and Japanese government, and managed by the UNEP's Regional Office for Asia and the Pacific, the CounterMEASURE Project identifies sources and pathways of plastic pollution in river

> systems in Asia, particularly the Mekong and the Ganges. The project has developed plastic leakage models for localities in six different countries using an innovative and replicable approach. Deploying technologies like GIS, machine learning and drones has allowed the CounterMEASURE team to augment ground-level research in an efficient and scalable way. This scientific knowledge can then be used to inform policy decisions and actions to defeat plastic pollution and ensure rivers are free of plastic waste (CounterMEASURE, 2021).

Pro-poor and inclusive innovation: Aiming to actively include and involve poor people in mainstream processes of technology-related development, either as consumers in new products and service markets or, more ambitiously, as participants in innovation processes themselves. This approach focuses on innovating low-cost products that can serve untapped markets with new commercialization and distribution strategies. Initiatives using this innovation approach may often associate with ideas related to frugal innovation, which focuses on informal forms of innovation with resource constraints. Low-cost, affordable, or locally developed solutions are identified as the key properties of these innovations. These innovations have potential to address many of the SDGs with creative and resilient solutions especially for the marginalized communities (Foster & Heeks, 2013). An example of applying the pro-poor and inclusive innovation approach in recycling plastic waste in remote areas is provided in Box 11.2.

Box 11.2: Upcycling plastic waste for road construction in rural areas

One of the innovative ways India is addressing the challenges of road development in rural areas is the use of plastic waste as an alternative material for road construction. This does not only contribute to the reduction of the amount of plastic waste ending up in landfills, but also benefits the communities living in remote and rural areas (Heriawan, 2020).

Grass-roots innovation: Aiming to broaden the innovation process to involve social movements and networks of academics, activists and practitioners experimenting with alternative forms of knowledge creation and innovation (Fressoli *et al.*, 2014). In contrast to the mainstream innovation process that are led by firms and operated in formal markets, grass-root initiatives operate in civil society arenas, typically within a social economy of community and voluntary activities and social enterprises, rather than the formal business sector. They are often driven by practical social and environmental needs and empowered by cooperation and community. Funding resources for such innovations may come from development aid, community finance, donations or state funding. Grass-roots innovation occurs in the sectors such as water and sanitation, housing, food and agriculture, energy, public health, and education, which can directly contribute to the corresponding SDGs particularly at local scales. An example of grass-roots innovation aiming to recycle PET is provided in Box 11.3.

Box 11.3: Everdrop's dissolvable cleaning tablets as an alternative to liquid detergents

These cleaning products are obtained by mixing the tablets with water in reusable spray bottles at home. By using the tablets, waste from single-use plastic bottles is avoided. The tablets are also convenient and cost effective (the refill tablets retailing for EUR 1 per refill). The reusable spray bottles are made from 100% recycled polyethylene terephthalate. As a result, transport volume is decreased by 80–90%, which contributes to reduction in CO_2 emissions (Ellen Macarthur Foundation, 2020).

Social innovation: Referring to the design and implementation of new solutions that imply conceptual, process, structural, or organizational change, which ultimately aim to improve the welfare and wellbeing of individuals and communities (van der Have & Rubalcaba, 2016). Shifting beyond technological innovation to social innovation can address SDGs that emphasize inclusion and social equity, especially in areas of education, health, work and poverty reduction. Social innovation may be partially overlapping with grass-roots innovation in terms of their implementation approaches. An example of the social innovation approach used to support plastic management is described in Box 11.4.

Box 11.4: Conceptos Plasticos and UNICEF's building classrooms using recycled plastic bricks

UNICEF and Conceptos Plásticos, a Colombian social enterprise, partnered to transform plastic waste into affordable and sustainable building materials. The plastic waste comes from the Ivory Coast, with its majority collected by poor women workers from their local communities. The bricks are cheap, durable and easy to assemble. All types of plastic waste, except for PVC, are used to produce the bricks. As a result, construction cost is 30% lower than that done with the conventional process as a lot of savings can be made on material and labor, and more children from poor communities can have access to school. The factory has the capacity to recycle 9200 tons of plastic waste per year, and it takes 5 tons of plastic to build a classroom of an area 68 m² (Lufadeju & Chavanel, 2019).

Digitally enabled open and collaborative innovation: Collaborative innovation recently attracts increasing attention owing to the extensive development of internet connection, data science and digital technologies. The highly developed technology-based social networks help foster joint learning, enable co-creation of knowledge and provide widespread access of tools, data and resources. Two key requirements for enabling collaborative innovation are open access to knowledge and wide participation in the process of developing ideas, products and technologies. An example of digital mapping used to support plastic management is described in Box 11.5.

Box 11.5: Trashmap Visualizing plastic waste

Created by OpenMap Development Tanzania in 2019, this interactive map portrays informal dumpsites and waste hotspots along Dar es Salaam's polluted rivers, as well as informal settlements across the city. Over 35 drone flights, coupled with spatial analysis, helped produce the map giving an overview of the dumpsites which normally cause rivers to be clogged, thus increasing flooding risks. This spatial data is used to develop more efficient plastic recycling and recovery (OpenMap Development Tanzania, 2017).

11.1.2.3 Sustainability-driven innovation in business to end plastic pollution

Conventional innovation methodologies follow a linear structure in which scientific studies are carried out, followed by technological improvements, engineering developments and finally marketing commercialization. This approach targets the market and aims primarily to generate economic benefits. As per innovation for sustainability, social, economic and environmental benefits are all targeted equally as a multi-goal approach. It is unique for its interactive structure that considers multiple stakeholders and open collaborations to achieve sustainability. For instance, limiting plastic footprints can be a sustainability driver for organizations that seek to achieve it by working with different stakeholders, including the public. Organizations innovate in order to achieve sustainability throughout the plastic lifecycle in their products and services. Moreover, the circular economy principle is also an important player as it supports the redesign of the entire plastic value chain to overcome the plastic pollution challenge. Accordingly, the two types of innovations are as follows (Ellen Macarthur Foundation, 2020).

Upstream innovation: In a circular economy, upstream innovation means that rather than working out how to deal with a pile of waste, we work out how to prevent the waste from being created in the first place (Ellen Macarthur Foundation, 2020). The products and services are considered starting at the design stage to prevent waste from ever being generated in the first place. This can include developing new materials, product designs or business models.

Treating the root cause of a problem, rather than the symptoms, is fundamental to finding a solution that truly tackles the issue.

According to the 'upstream innovation' guide to packaging solutions developed by Ellen McArthur Foundation, 'upstream innovation requires a shift in mindset'. It is considered necessary to unlock new opportunities and move beyond incremental innovations. It supports delivering products and services the best way possible not only packaging-wise but also the product/ service itself and the whole business model. Therefore, upstream innovation supports phasing out waste while delivering the same or a better value to end users in the following ways.

- **Rethinking the packaging** (concept, design, components, amount of material, type of material) to provide the same essential function of conventional packaging while phasing out waste, some examples of which are edible or dissolvable packaging material, naked packaging, biodegradable, compostable or recyclable packaging.
- **Rethinking the product** by innovating at the design stage of the product (formulation, concept, weight, size etc.) to decrease the potentially generated waste while delivering the same or improved user experience, some examples of which are switching from physical services to digital services, or from heavy liquid formulas to concentrated solid formulas.
- **Rethinking the business model** including production, supply chain, storage, selling methods, revenue streams, delivery or take back schemes, an example of which is cutting down single-use plastic packaging through refill machines.

Since businesses contribute heavily to the high percentage of plastic packaging waste generation, the following three upstream innovation strategies should be adopted by the packaging industry to help limit their plastic footprint.

- **Elimination**: elimination of plastic waste can be achieved through the naked packaging concept that eliminates the need for packaging or by designing packaging that is edible, dissolvable, naked, biodegradable, compostable, or recyclable.
- **Reuse**: reuse of plastic packages, rather than discarding them after one single use, creating value for both users and businesses based on the following four main models.
  - Refill at home: Reusable containers which are refilled at home (refills delivered through a subscription service).
  - Return from home: A dedicated collection services which pick up the used packaging from the customers' homes.
  - Return on the go: Packaging returned by customers to a store or a drop-off point.
  - Refill on the go: Reusable container refilled at an in-store dispensing system.
- **Material circulation**: Packaging is designed so that its core materials can be recycled or composted. This may include plastics recycling, composting or substitution to a non-plastic material.

Downstream innovation: This type of innovation is initiated after the product is used and becomes waste. This can come in the form of new technologies or innovative measures for collecting, sorting, recycling, energy recovery or upcycling.

Plastic waste management refers to the activities and actions that handle plastic waste. An innovation strategy for plastic waste management can tackle plastics waste prevention, reuse, collection, transportation, recycling, and recovery with the aim of ensuring proper end-of-life management that conserves natural resources and saves on labor, cost and energy. The need for innovations for plastic waste management emanate from the following challenges.

- Environmental impacts associated with the conventional end-of-life options for plastic waste such as open dumps/burns and uncontrolled landfills are becoming increasingly severe.
- Single-use products are used within a very short lifetime and thrown away to be decomposed to microplastics, which potentially endangers marine life and human health.
- Plastic products may contain hazardous additives and contaminants that make recycling complex and costly.
- Segregation is poor or inexistent in the majority of the countries around the world.
- Recycling rates of plastic waste are very low globally.
- There is lack of adequate technology for plastic recycling and energy recovery.

According to the plastic waste management hierarchy as shown in Figure 11.6, accelerating progress on plastic waste management depends on giving priority to plastic prevention and reduction. However, reuse, recycling and energy recovery can also be supported by innovative approaches through improved design, sustainable business models and efficient supply chains, and

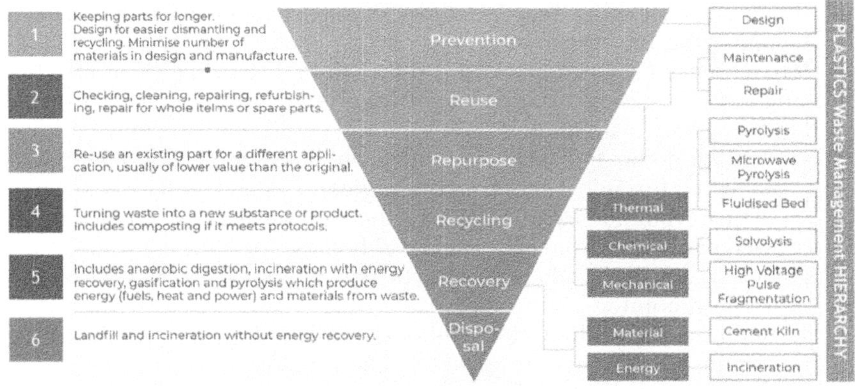

Figure 11.6 Plastic waste management hierarchy (*Source*: Modified from BAX & Company, 2021).

by supporting take-back schemes, recycling and energy recovery systems, thus promoting efficient and sustainable utilization of resources.

11.1.2.4 *Innovations to limit the plastic footprint: case studies*

Circular economy: Circular economy seeks a sustainable way to develop, with the resources being used more efficiently and retained in the economic loop as long as possible. This concept has been widely accepted and gained a global momentum as a sustainable tool to manage the modern production and consumption, including that for plastics (Hahladakis *et al.*, 2020). Today, various innovative circular economy models have been developed and adopted in the industries which have been intensively engaged in plastic production and consumption, such as the fashion industry and packaging industry to limit the plastic footprint in their business and process.

- ***Case I:*** Rothy's circular economy model

Rothy's is a US-based fashion company specialized in shoes and bags founded in 2012. It aims to close the loop of its circular economy model from materials manufacturing, products and recycling (Rothy's, 2021) as shown in Figure 11.7.

- ***Case II:*** Girlfriend Collective

Girlfriend Collective is a 'guilt-free luxury' fashion brand that recycles old water bottles and fishing nets into sportswear, which is certified Standard 100 by Oeko-Tex, the world's leader in testing fabrics to regulate harmful substances. In 2021, Girlfriend Collective has recycled 5 089 471 water bottles, prevented 4 906 822 lbs of CO_2, and saved 9 279 262 gallons of water. With their slogan 'Trash looks better on you than it does polluting the planet' (Girlfriend Collective, 2021), their circular economy model is based on the following principles.

  - Usage of recycled fabric with proven moisture-wicking abilities, that is, recycled polyester from PET bottles and nylon obtained from recycled fishing nets, which is suitable for sportswear

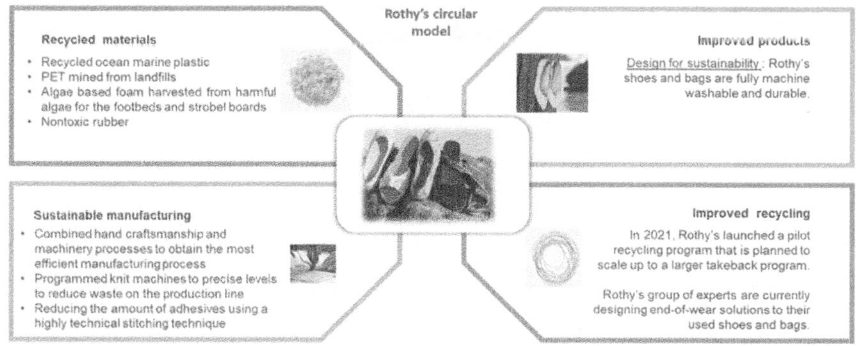

Figure 11.7 Rothy's circular economy model (*Source*: Modified from Rothy's Inc., 2021).

○ Usage of safe dyed colors, OEKO-certified for safety with the dye mud donated to a local pavement facility where it's recycled into sidewalks and roads and the wastewater sent to a wastewater treatment plant
○ 100% recycled and recyclable packaging

• ***Case III:*** Refill machine for cleaning products

The city of Medellin in Colombia is taking a circular economy approach to selling cleaning products (Figure 11.8). The project was initiated to reduce plastic waste by reusing and recycling containers for conventional household detergents. The venture enables people to refill their cleaning products and allows companies to repackage them for distribution. The goal is to generate a conscious and ecological purchase and reduce the environmental impact of consumption.

The project began in 2018. Its objective is to formalize a real commitment to the care of the planet by creating a circular economy. Such a system will benefit both the citizens and producers. Quality cleaning products are easily accessible at a low price for families and small businesses. The cost is less for everyone because producers do not need to pay for packaging, and the consumers do not need to pay for a container (Steffen, 2019).

Digital innovations: Digital technologies play an essential role in enhancing efficiency of downstream innovations by facilitating consistent supply of high-quality recycled plastic pellets. Consequently, this leads to boosting upstream innovations through products made out of recycled plastic content, and designed for recycling and dismantling. Digital technologies, such as apps, artificial intelligence (AI) and connected devices, improve public engagement in responsible consumption and waste segregation practices. For instance, apps can be developed to raise awareness around proper waste segregation practices, enable identification of hotspots, and increase opportunities to access recycled materials.

Figure 11.8 Refill machines for cleaning products in Colombia (*Source*: Steffen, 2019).

Such technologies can also act as drivers for social and economic benefits, for instance, by matching waste pickers from the informal sector with the supply of plastic waste (from households and businesses) to other relevant stakeholders in the plastic value chain. As such, digital innovations can also enhance collection and recycling activities, such as the use of AI and the internet of things (IoT) plastic segregation and bin monitoring.

The high level of transparency these solutions bring to the supply chain is an important requirement for global brands that aim to increase their use of ethically sourced recycled plastics in their manufacturing processes (Wilson *et al.*, 2021). Hence, allowing the tracing of provenance and quality of plastic to contribute to the emergence of an ethical plastics sector.

- ***Case I:*** Mobile app for plastic waste collection and recycling

Coliba in Côte d'Ivoire provides a mobile phone app that offers plastic collection and recycling services by connecting waste pickers to homes and businesses. It was launched in 2017 with the aim of helping waste pickers collect plastic bottles in exchange for points. After being segregated and cleaned, the collected plastic waste is turned into pellets in Coliba's local factory and resold to local or international companies to produce repurposed and recycled products. In 2018, the app deployed door-to-door plastic bottle collection service by rewarding users with MTN data credit (Coliba, 2021; Wilson *et al.*, 2021). The Regenize is a similar mobile app which provides residential recycling collection services in South Africa. The app allows customers to drop their plastic waste at collection points free of charge, and provides door-to-door collection service with certain payment. Customers who recycle through this app will be rewarded virtual money to encourage such environmentally sustainable practices (Regenize, 2021).

Partnerships-based innovations

- ***Case I:*** Loop – reusable containers initiative partnering with numerous brands

Loop is a global reuse platform, e-commerce and in store system offering products in reusable packaging (Loop, 2021) This revolutionary initiative was developed and piloted by TerraCycle, a global waste management and recycling company, the logistics giant UPS and some of the world's leading retailers (e.g., Procter & Gamble, Mondelēz, PepsiCo, Coca-Cola, Unilever, Nestlé and Danone). According to Loop, the enhanced design and functionality of the packaging are the major drivers of this reuse business model's success as 97% of reusable packaging is returned within the following 90 days of purchase (Ellen Macarthur Foundation, 2020).

Convenience is the major factor contributing to customer satisfaction for Loop as users do not have to clean or sort the used container. Loop's reusable packaging can be either picked up from the customer's home, or dropped off at a partner store. It also takes care of thoroughly cleaning, refilling and using the packaging for the same or a different product for another customer.

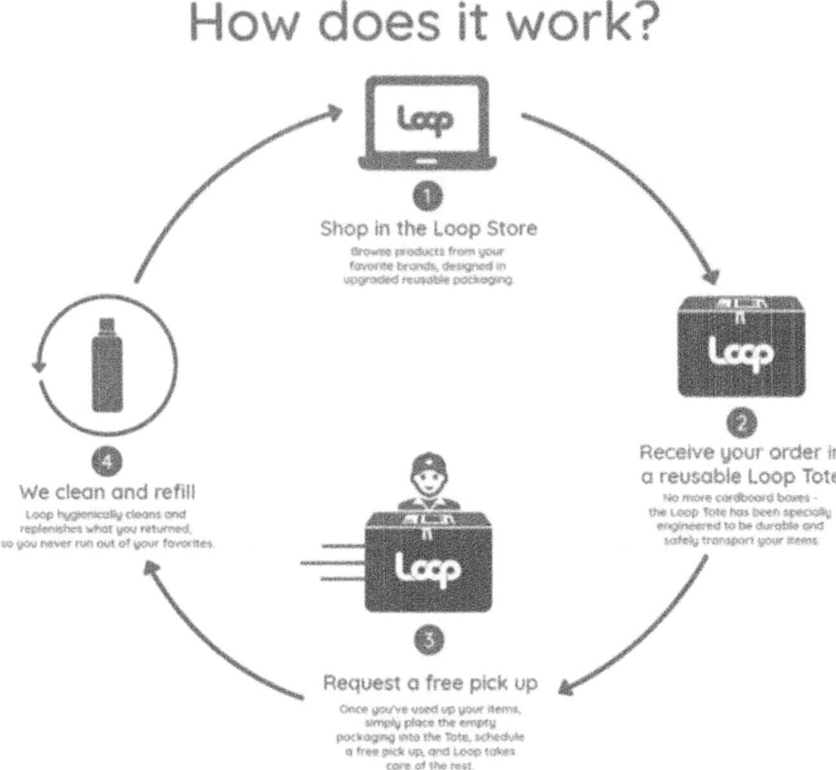

Figure 11.9 The loop model (*Source*: World Economic Forum, 2019).

This reuse model offers very efficient and eco-friendly alternative to single-use packaging. As such, Loop promotes responsible consumption and waste prevention by introducing a convenient and durable alternative for customers. The Loop model is represented in Figure 11.9.

- ***Case II:*** ReSource: Plastic – an WWF action hub to prevent global plastic waste

This initiative represents WWF's activation hub for companies that commit to limiting their plastic footprint by supporting them in developing their action plan through an 'innovative measurement framework that tracks corporate action against ReSource's three-pronged approach to leveraging business as a catalyst for systems change' (McCarthy, 2019) by

- Eliminating unnecessary plastic through business model innovation, reduction and substitution,
- For plastic that is necessary, shifting from virgin plastic sourcing to sustainable inputs, including recycled content, sustainably sourced bio-based content and advanced materials and

- Doubling global collection, recycling and composting of plastic so that the plastic going into the system is circulated back.

Launched in 2019 with five participating member companies, ReSource aims to prevent 50 million metric tons of plastic pollution from entering our oceans and other ecosystems by 2030. In order to achieve this ambitious goal, ReSource works with its member companies to track and report on their plastic footprint annually using a ReSource footprint tracker system as well as implementing recommended actions.

The pilot version of the ReSource footprint tracker consists of three components.

- A survey tool that companies fill out with information on the plastics they use and sell, which provides summary outputs related to the composition of their portfolio.
- A plastic waste management model that uses the survey data as an input, and estimates the share of plastic that is recycled, landfilled, incinerated and mismanaged based on country-level waste management data and the polymer and format of the item.
- A beyond supply chain survey where companies are asked to provide detailed information about any projects or investments, they are involved in, and focused on, reducing plastic pollution beyond their own supply chain, as well as any measured impacts to date

More importantly, ReSource aims to multiply the impact by establishing more collaborations with other companies and stakeholders to boost investments for systems change and scale critical solutions in waste management.

11.1.2.5 *Challenges facing innovations to end plastic pollution*

Although the importance of innovation to address the plastic pollution issue has been widely recognized, and increasing innovative activities are on-going with various emerging approaches, a number of practical challenges are still faced especially by the less developed regions and countries to further advance the achievement. Table 11.3 summarizes these challenges in 3Rs circumstance and provides recommendations to better address them.

11.2 DIGITIZATION AND ITS APPLICATION IN PLASTIC WASTE MANAGEMENT

The emergence of modern wireless technology coupled with development of revolutionary ubiquitous computing, has resulted in the intricate integration of digitalization in every paradigm of society. With the plethora of the advantages and the convenience, digitalization has been embraced in every aspect of the development and management. The plastic waste management is no exception to this novel trend. Plastic-associated organizations are using digital technologies to improve citizen engagement, to drive operational efficiencies, and to increase the plastic value supply chain.

Table 11.3 Challenges and recommendations for 3Rs innovation to end plastic pollution.

Challenge	Recommendation
Reduction	
Lack of the know-how	• Developing capacity building and knowledge transfer opportunities • Increasing cooperation between relevant stakeholders
Difficulty to break habits of the 'take–make–dispose' system (e.g., reliance on single-use cutlery and straws)	Raising awareness around the potential human health and environmental impacts of the 'take–make–dispose' in terms of overexpolitation of natural resources and plastic pollution
Communicating without a label for naked packaging (from listing ingredients to providing customers with information on the product)	Implementing item recognition through an app that enables users to scan the product and receive all information about it
Reuse	
Preference of single-use plastic items due to their convenience for the public	Implementing policies banning single-use plastic
Insufficient reuse of plastic products	• Incentivizing the creation of repair centers, rental services, tool libraries and reuse business models • Increasing cooperation across supply chains, retailers and industries.
Covid 19 pandemic encouraging single-use plastic consumption (diposable containers, cutlery, masks, gloves) for fear of infection	Implementing proper sanitizing measures for reusable items
Recycling	
Plastic recycling bins are often contaminated because of: • Dirty waste • Wishcycling: Putting non-recyclables in recycling bins by good intentions • Compostable items Contaminated waste also damage recycling equipments and endanger workers	• Indicating nearby the recycling bins that disposed items should be rinsed or cleaned • Raising awareness about non-recyclables and compostable items • Implementing cleaning pre-treatment processes and advanced sorting technologies
Lack of plastic waste input for recycling	Implementing deposit-refund systems as an incentive to increase the amounts of plastic waste sent to recycling streams
Lack of take back schemes	Implementing extended producers responsibility strategies to hold producers responsible for the collection and recycling of the waste their products and services generate

(*Continued*)

Table 11.3 Challenges and recommendations for 3Rs innovation to end plastic pollution. (*Continued*)

Challenge	Recommendation
Limited budgets for sustainability-driven innovations Low demand for recycled products	• Starting with small-scale pilot projects • Supporting startups and SMEs • Developing product standards imposing a minimum level of recycled content as part of the green purchasing policies • Implementing sustainable public to accelerate the use of products made out of recycled content • Disincentivizing virgin plastic
Technical challenges related to identification of the types of materials and polymers in waste	Considering the following during the design stage of a product: • Labeling for disassembly • Incorporation of stamps to indicate plastic types
High cost of processing recycled plastic	Innovating for automation of some processing steps of plastic recycling to optimize operations and associated costs
Non-recyclable plastic waste as a dominant component in the household generated wastes	Implementing waste-to-energy technologies, such as pyrolysis to convert non-recyclable plastics to oil and fuel

According to European commission (EU), digital transformation is the fusion of advanced technologies with physical and digital systems integration that enables creation of smart products and services (European Commission, 2018). It started with the popularity of the internet, followed by mobile technology, and then various smart devices. Technologies related to IoT, AI and blockchain are creating disruptive solutions that ensure not only the economic growth but also the environmental sustainability. Plastic waste management requires an integrated system to overlook the cradle-to-grave process, the generation, storage, collection, transportation, treatment and disposal of waste materials in a way that is sustainable to society, economy and environment. Adoption of digital technologies in plastic waste management allows closing of the loop on plastic material flow by providing real-time information, collecting and sorting plastic wastes smarter, and recycling them more efficiently. Driven by different factors such as increased environmental burden caused by population growth and waste generation, enhanced ability of global manufacturers to save costs through recycling the cheaper alternatives, and customers' expectation on environmentally resilient urbanization, many aspects of plastic waste management, especially logistics, communication, collection and sorting, treatment and handling, recycling, data analyzing and policy interventions regarding impactful waste management are considered and are to be transformed by digitalization. It is expected that digital transformation in plastic waste management section will create more opportunities to promote the plastic

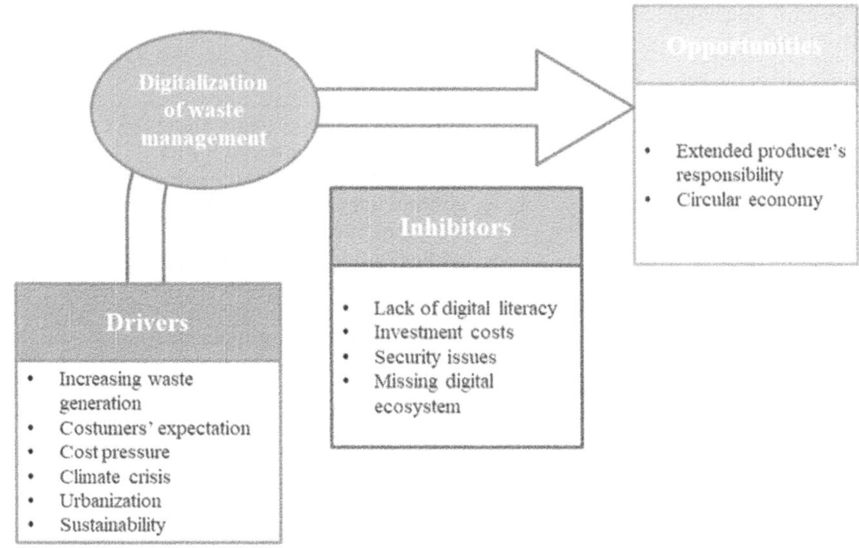

Figure 11.10 Diagrammatic representation of the drivers, inhibitors and opportunities in digitalization of plastic waste management (*Source*: Berg *et al.*, 2020).

circular economy and to advance the implementation of extended producer's responsibility (EPR). Some of the drivers, inhibitors and opportunities of digitization in plastic waste management are mentioned in Figure 11.10. This section will focus on innovations that apply digital technologies to address the problems of plastic waste.

11.2.1 Digital technologies for smart plastic waste management

Digital solutions provide a number of opportunities to shift toward a circular economy for plastic management and to strive toward the zero-emission target for marine plastic pollution. Through applying variable digital technologies, a more efficient plastic recycling scheme which results in less material that end up as non-recyclable waste and leaking into the environment is reachable. As the signature innovation products of Industry 4.0, digital technologies such as IoTs, AI, blockchains, automation and robotics, as well as data analytic and cloud computing technologies, provide various possibilities for sustainable plastic waste management (Figure 11.11). These digital solutions are used synergistically to complete the high-efficiency sustainable management of plastic waste.

11.2.1.1 Internet of things

IoT is defined by the Global Standards Initiative (GSI) as the inter-networking of physical devices, vehicles (also referred to as 'connected devices' and 'smart devices'), buildings and other items embedded with electronics, software, sensors, actuators and network connectivity which enable these objects

Figure 11.11 Digital technologies for smart plastic waste management.

to collect and exchange data (Kiran, 2019). It is an emerging paradigm that makes rapid progress in the scenario of modern wireless telecommunications. IoT enables communications and interactions among the networks of objects without human interference, and makes things smarter. An IoT-based smart system is basically composed of three layers of architecture.

- **Sensor and actuator layer**

The lowest layer of an IoT system is the smart objects that are integrated with sensors and actuators. Sensing is required to collect the desired information from the targeted physical objects and to transform the off-line signals into communicable information through sending it back to the server or cloud. Nowadays, various sensors, actuators, and sensing devices are available to measure the physical properties and convert them into signals that can be understood by instruments, functioning as a digital nervous system for example, GPS sensors for locating data, cameras and microphones for vision and audience information, and different meters to record the information such as temperature, moisture, pressure and velocity (see Figure 11.12). Sensors are grouped according to their unique application purpose such as environmental sensors, body sensors, home appliance sensors and vehicle telematics sensors (Patel & Patel, 2016).

- **Connectivity layer**

Connectivity in an IoT system refers to the medium platform that connects devices and sensors with the network. All the data that are collected by sensors are transferred, exchanged or stored in the cloud networks through connectivity. To process the massive data produced by sensors, it requires a robust and high-performance network infrastructure. Connectivity can be wired network such as IPv4 and IPv6, as well as wireless network such as the internet, cellular network, Wi-Fi, Bluetooth or satellite. With demand needed to serve a wider range of IoT services and applications such as high-speed transactional services, context-aware applications, multiple networks with various technologies and access protocols are needed to work with each other in a heterogeneous configuration. These networks can be in the form of

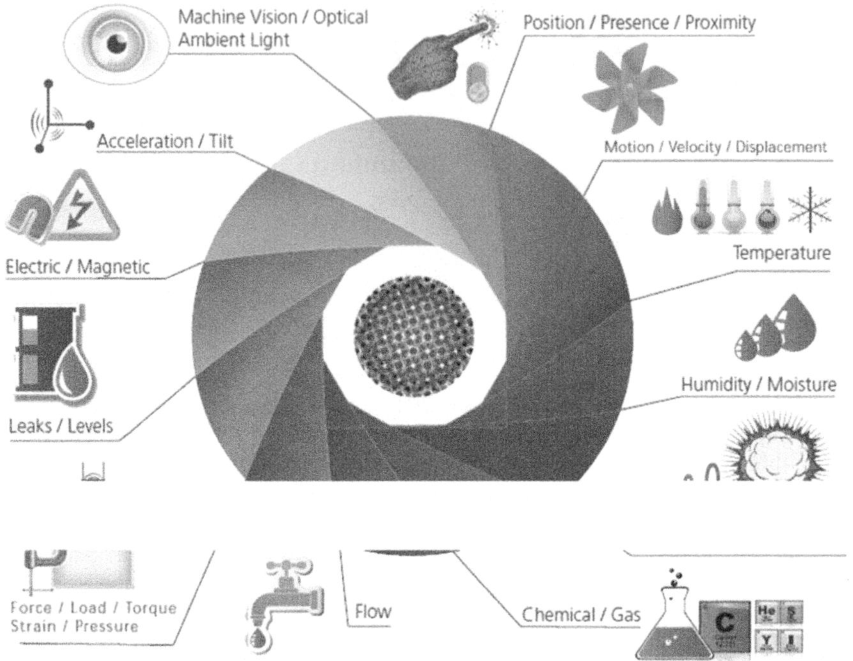

Figure 11.12 Sensors and actuators in IoT (*Source*: Postecapes, 2015).

a private, public or hybrid models and are built to support the communication requirements for latency, bandwidth or security (see Figure 11.13).

- **Processes and service layer**

The processing of networked data is carried out through process models, analytics, security controls and data management with an integration of data, people, process and systems for better decision making and service providing (see Figure 11.14). One of the important features of this layer is the process rule engines. These rule engines support the formulation of decision logics and trigger interactive and automated process to enable a responsive IoT system. Big data analytic tools are used to extract relevant information from massive amount of raw data and to process them at a fast rate. The power of data analytic incorporating in this layer enables the users to carry out agile decisions despite the huge amount of information. Data management functions allow efficient access to, integration of and control of the collected data. It can also filter out the unnecessary information and report only the essential data for efficient service based on specific users' need. Furthermore, security management in the system prevents system hacking and compromises by unauthorized personnel, thus reducing the possibility of risks. Once necessary processing is completed, outputs can be integrated to the user applications and user interfaces (web or

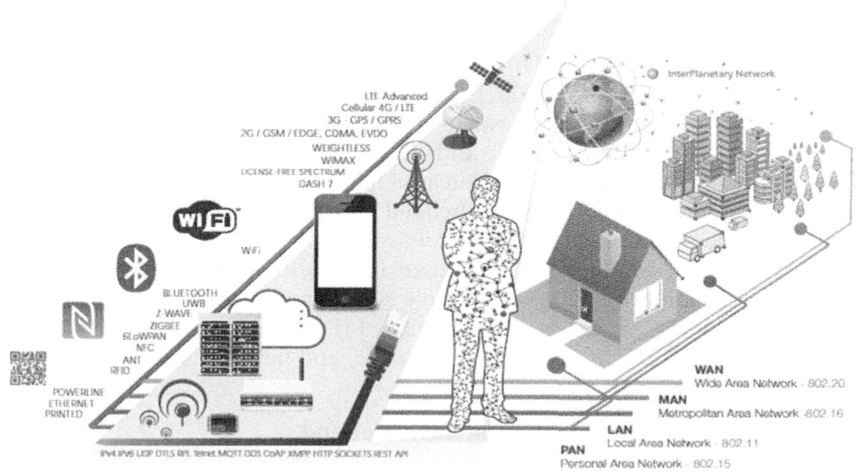

Figure 11.13 Connectivity of IoT (*Source*: Postecapes, 2015).

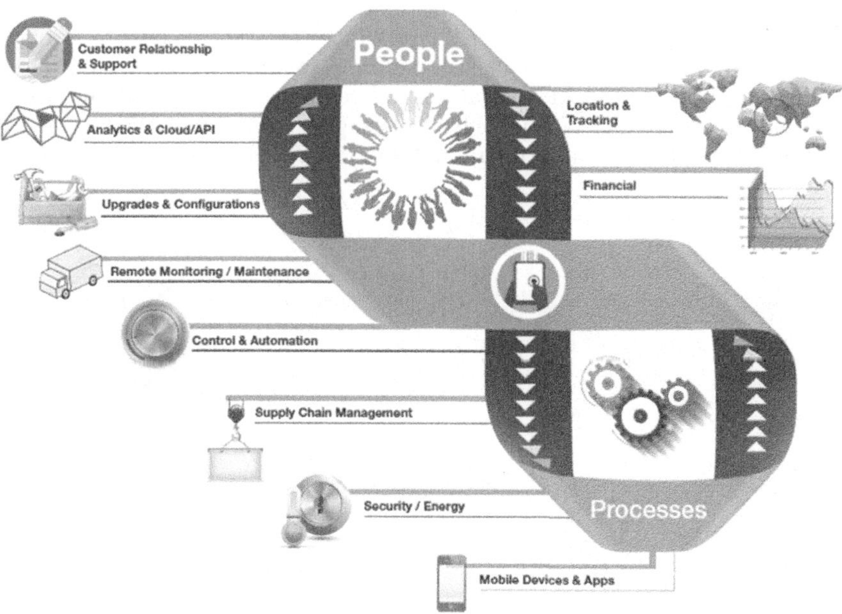

Figure 11.14 Process and services provided by IoT (*Source*: Postecapes, 2015).

mobile) to provide them various smart services covering wide domains such as transportation, building, city, lifestyle, healthcare, agriculture, industry, supply chain, emergency, culture and tourism, environment, and energy management and decision making (Patel & Patel, 2016).

The opportunities of IoT in waste management systems are diverse and significant. IoT has made it possible and affordable to integrate connectivity and functionality into various devices. The anticipated applications of IoT in plastic waste management lies in sensor-embedded plastic products, sensor-supported trash containers and networking of waste collection vehicles to improve logistics as shown in Figure 11.15. It can help to integrate the disseminated waste management procedures as a whole system and helps to develop holistic approach to addressing the waste recycling. IoT can help the reverse logistics of the plastic be more simplified and scientific, and help in execution of policy intervention.

- **IoT for waste identification at customer end**

Collection of plastics after use is not sufficient if the collected plastics cannot then be sorted properly. Improved sorting of plastics can improve both the quality and the quantity of recycled plastics on the market. One way that more efficient sorting can be achieved is to embed digital tags in various plastic products that can help facilitate the identification of product features and exchange of data between stakeholders and machines. Tagging the plastic products can potentially increase the efficiency of EPR systems, improve customer engagement and market surveillance, as well as ensure a better recycling of plastic wastes. A classic example of utilizing IoT-based digital tagging for plastic waste identification and recycling at the customer end is the application of RFID tags in plastic packages (Amritkar, 2017). RFID is a small tag that contains datasets and automatically identifies objects. It stands

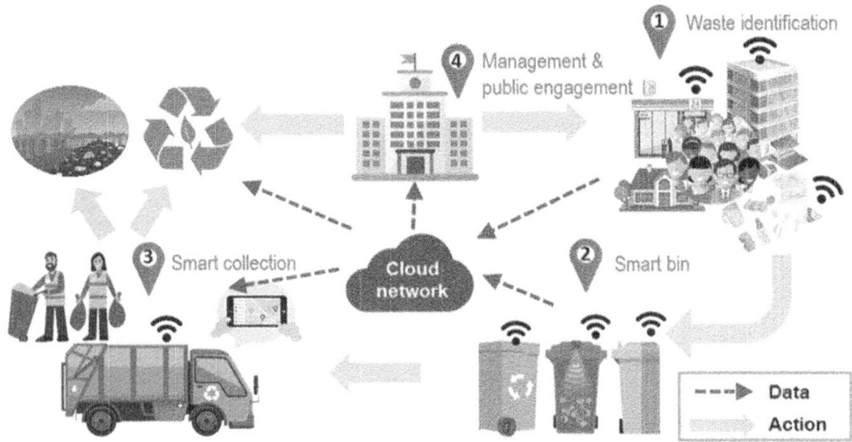

Figure 11.15 Smart waste management based on IoT.

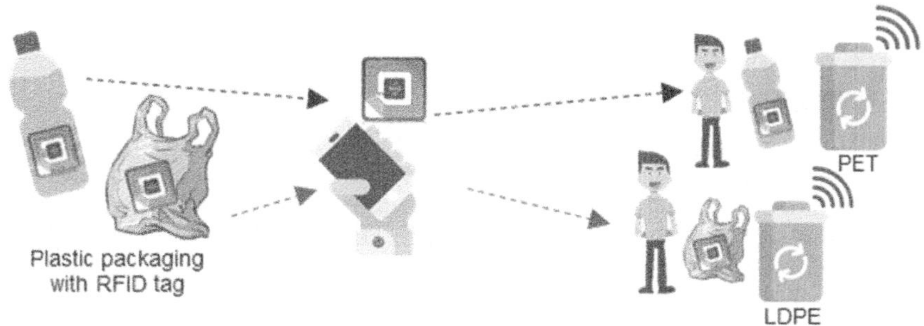

Figure 11.16 Application of IoT for efficient plastic waste identification and sorting at the customer end.

at the forefront in the application of IoT because of its maturity, low cost and wide acceptance by the business communities (Atzori *et al.*, 2010). As shown in Figure 11.16, RFID tags which contain information can be embedded in any plastic products especially recyclable plastic packages. Users will access the information such as plastic types by scanning the RFID tags using smart phones and will be instructed with proper sorting and disposal methods. RFID technology aids waste sorting at its source and avoids unsegregated plastic waste. It also helps to reduce the burden of waste segregation and enhance the opportunity for plastic wastes recycling based on its type.

Except for RFID tags, several other innovative tags to help with efficient plastic waste sorting and recycling have been developed and utilized as well. For instance, the HolyGrail Project developed digital watermarks and chemical tracers (Figure 11.17) to help in facilitating the high-quality recycling of plastic packages (New Plastics Economy, 2017). Both digital watermarks and chemical tracers are machine-readable codes or identifiers that facilitate the sorting of products by pointing the sorting system to a database where information on how the product can be recycled can be found. This technology in turn

Figure 11.17 Digital tagging technologies applied for plastic waste identification (*Sources*: Amritkar, 2017; New Plastics Economy, 2017).

makes the sorting process more reliable and efficient at the customer end, and increases the amount of plastic that is efficiently recycled.

- **Smart trash bin**

Smart trash bins have been developed with the help of the recent available digital technologies especially IoT and its associated cloud computing and data analytics. The main functions which are currently developed and incorporated in smart bin technologies are summarized in Figure 11.18 (Angin *et al.*, 2018; Balamurugan *et al.*, 2017; Huh *et al.*, 2021; Kumar *et al.*, 2017; Noiki *et al.*, 2021; Sohail *et al.*, 2019; Wahab *et al.*, 2014). Three most notable features of smart bins are waste monitoring, communication and waste segregation (Soni & Kandasamy, 2018). To monitor the waste amount in trash bins, statuses such as the trash level, trash weight, and trash volume are usually measured using proper sensors. This function is usually associated with the dynamic waste collection system. Another attractive function of smart trash bins is to separate different types of waste. With properly selected sensing devices and designed mechanical mechanisms, a smart bin is possible to separate dry and wet waste, organic and inorganic waste, or even separate types of recyclable wastes such as metals, glass, paper and plastic. Moreover, safety and hygiene management is also a factor in the smart trash bin design. Many smart bin products carry the flame alert function and/or odor alert function in order to maintain good environmental hygiene conditions during use. Communication is another common function that may be incorporated while designing a smart bin. Some examples of typical communication functions are short messages sent to remind the users, connection to smart phone apps or other online platforms

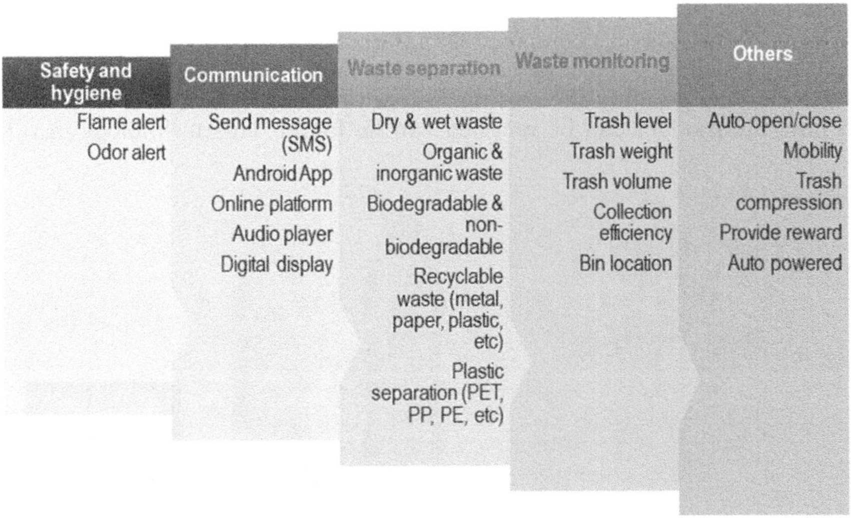

Safety and hygiene	Communication	Waste separation	Waste monitoring	Others
Flame alert	Send message (SMS)	Dry & wet waste	Trash level	Auto-open/close
Odor alert	Android App	Organic & inorganic waste	Trash weight	Mobility
	Online platform	Biodegradable & non-biodegradable	Trash volume	Trash compression
	Audio player	Recyclable waste (metal, paper, plastic, etc)	Collection efficiency	Provide reward
	Digital display	Plastic separation (PET, PP, PE, etc)	Bin location	Auto powered

Figure 11.18 Key functions of smart trash bins.

for real-time monitoring, and voice or video playback functions. Several other typical designed functions for smart bins include auto-open/close, mobility, trash compression, and auto power. To encourage proper sorting of recyclable waste, several innovative smart bins incorporate the function to provide rewards to the disposers.

To accomplish the diverse designed functions of smart bins, a number of sensors have been installed and tested. An ultrasonic sensor is one of the most commonly used trash level sensors in smart bins, owing to its easy access, simple installation, and low cost. Meanwhile, visual-based sensors such as optical camera are widely used in combination of machine learning algorithms in order to identify and segregate multiple types of recyclable wastes. In particular, spectroscopy sensors such as VIS, NIR, and FIR can be utilized alone or in combinations to separate different types of plastic polymers, which will potentially enhance the efficiency of plastic recycling depending on their polymer types. Table 11.4 lists several most frequently applied sensors that assist the smart bins to achieve the desired features.

Singapore has installed thousands of smart bins around the city to battle overflowed waste (Teng, 2018). These smart bins, also called 'Bigbelly' bins are equipped with ultrasonic sensors to measure the level of trash inside. Once the bin is full, the sensor is able to detect and send an alert through 3G, email or text messages to the mobile phones of cleaners or scavengers. Waste compaction up

Table 11.4 Sensors used in smart trash bins.

Waste Monitoring		References
Trash level	Ultrasonic sensor Optical camera	Huh *et al.* (2021); Kumar *et al.* (2017); Noiki *et al.* (2021); Sohail *et al.* (2019); Wahab *et al.* (2014)
Trash weight	Digital weight scale Force sensor	Balamurugan *et al.* (2017)
Location/Live map	GPS	Kumar *et al.* (2017); Sohail *et al.* (2019)
Waste separation		
Dry and wet waste	Moisture sensor	Jaikumar *et al.* (2020); Jayson *et al.* (2018); Singh *et al.* (2019)
Organic and inorganic waste	Optical camera	Salimi *et al.* (2019)
Metallic and non-metallic waste	Metal sensor	Jayson *et al.* (2018); Soni and Kandasamy (2018)
Multiple recyclable waste	Optical camera Infrared sensor LED screen	Costa *et al.* (2018); Satvilkar (2018); Vo *et al.* (2019); Yang *et al.* (2020)
Plastic types	Visible sensor (VIS) far-infrared laser diodes (FIR) Near-infrared laser diodes (NIR)	Thakker and Narayanamoorthi (2015)

to eight times of the capacity of normal trash bins is possible, which maximizes the efficiency of the bins. In addition, the smart bin is empowered by solar energy with an installed solar panel on the bin cover. The solar panels power an internal battery that drives the compaction mechanism, internal sensors and the communication module. To ensure the safety and hygiene need while using the bins, installation of fire alert and a closed design to prevent odor emission are also present (see Figure 11.19).

Angin *et al.* (2018) designed a smart trash splitter which carried multiple sensors and was able to sort mixed garbage into metals, organics, paper and plastics. Three sensors – metal sensor, infrared sensor, and LED light sensor were used in sequence in order to conduct automatic on-site sorting of domestic wastes. The prototype was able to achieve accuracy in sorting different types of garbage at 95% for metals, 28% for organic wastes, 12% for paper, and 41% for plastics (Figure 11.20).

- **Smart waste collection and dynamic logistics**

IoT system incorporating smart bins and smart routing technologies can be adopted for waste collection and logistics management. An application concept of such smart waste collection systems is illustrated in Figure 11.21. By integrating sensor systems in garbage bins, it is possible to measure waste levels or waste volumes in the bins real time. The monitoring data from garbage bins will be transmitted automatically to a server for storage and processing in real time. When waste levels reach the threshold, alerts will be sent to waste collectors notifying them of the collection time for necessary locations only. Once the bins are emptied again, the sensors will also notify and send related data back to the control system. For the waste collection process, the system

Figure 11.19 The smart 'BigBelly' bins used in Singapore (YTL Community, 2016).

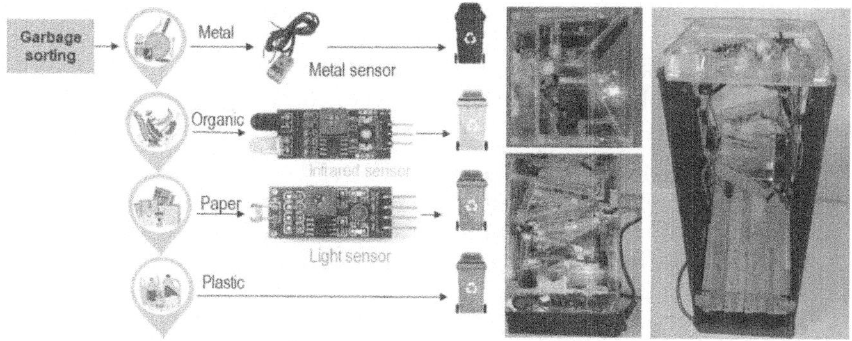

Figure 11.20 A shematic diagram of the smart trash splitter (*Source*: modified from Angin *et al.*, 2018).

can optimize routes for waste collection which will be displayed to waste collectors via smartphone or navigation devices. Moreover, as data have been gathered from sensors over time, AI can predict the waste generation rate of each bin. Such systems can also estimate the size of vehicles, routes for waste collection and other related information associated with the collection process such as traffic congestion or the weather. With the optimized routing for waste collection, it is possible to save costs and time as the vehicles will only target the identified locations. In addition, fuel and energy consumption will be reduced together with the increase in working efficiency.

A combination of new smart logistic strategies and smart bin technologies is typically the main feature of such smart waste collection systems. It ensures that cities are able to categorically sort, recycle and dispose of their waste in a smooth and efficient manner. The integrated process from using sensors to

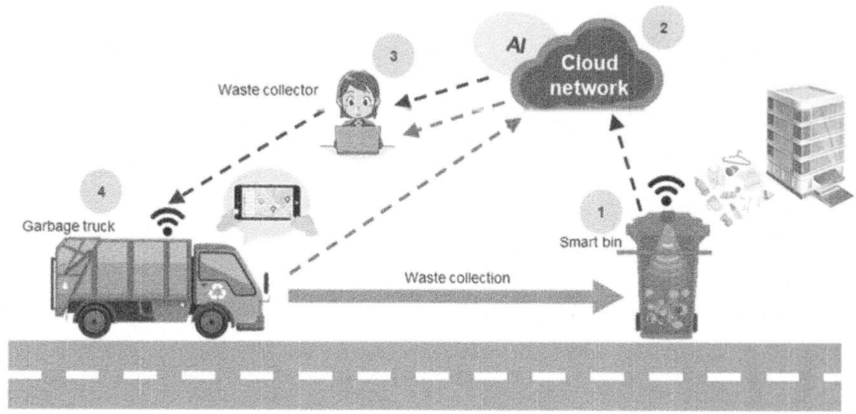

Figure 11.21 A conceptual flowchart of the IoT-based smart waste collection system.

determine the trash level in the bins to providing garbage truck drivers optimized route suggestions via predictive analytics is expected to efficiently save costs, time and energy. Taking the Fasterstream Technologies' smart waste collection system as an example (Faststream Technologies, 2021), five core technologies/ solutions constitute the smart waste management system (Figure 11.22).

- **Trashfill sensing technology**: A combo reading from ultrasonic sensors placed on the lid of the trash bin and a weight sensor under the bottom panel is used to get an accurate measurement of the weight of the trash stored. This transformation of the trash bins allows live notifications of filled bins and their locations to be easily accessed.
- **Route optimization**: The system determines the best routes for the garbage trucks to take when collecting only filled containers.
- **TuckFill sensing**: A smart waste management to take better care of the truck assets with sensor technology that notifies the operator and the driver once the truck has reached its maximum load capacity.
- Data analytics help to learn residents' waste disposal trends and habits, which can be used to further enhance the operational efficiency of the system. It reduces time consumption for waste pick-up and provides the ability to predict filling trends. This helps minimize bin overfill and maintain healthy conditions around the garbage bins.
- **SmartFetch interface**: The interface presents all the pertinent information including the number of empty and full bins, their locations together with the weight and fill level in a management dashboard. It performs route optimization and sends the route to the truck driver's tablet.

- **Multiple sensors-based plastic waste sorting**

Segregating commingled plastic waste is a challenging task. The efficiency and accuracy of separating the mixed plastics into their respective categories are essential in determining the capacity of plastic recycling. One potential solution

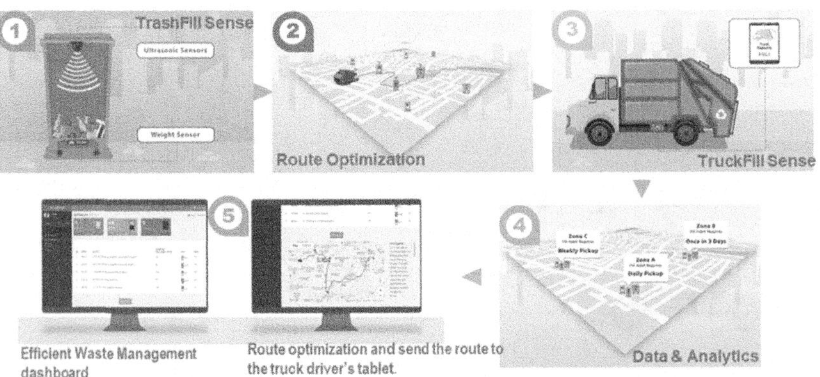

Figure 11.22 An example of commercialized smart Bin and smart waste collection system (*Source*: modified from Faststream Technologies, 2021).

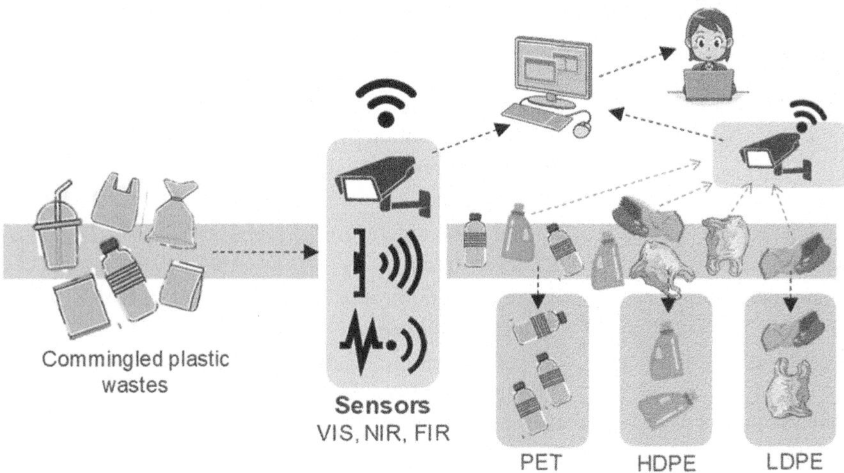

Figure 11.23 Plastic waste identification and segregation using multiple sensors-based smart system (*Source*: Chidepatil *et al.*, 2020).

is to adopt multiple sensors incorporated with AI technology to automatically identify and separate the plastics based on their characteristics such as color, transmittance and chemical composition. Figure 11.23 illustrates a conceptual process flow of the plastic waste segregation system that is empowered by multiple sensors and AI (Chidepatil *et al.*, 2020). Several types of sensors such as visual sensors (VIS), near-infrared (NIR) laser sensors and far-infrared (FIR) laser sensors can be applied to accomplish the mission. Optical sensors can retrieve the visual information such as shape, color and texture from an objective. Via proper AI trainings, the system can identify and separate certain shaped or colored plastic bottles and flakes. It has been proven that a high accuracy up to 99% color-based segregation can be achieved in the industrial scale applications. Laser diodes (e.g., NIR and FIR) have also been applied in differentiating various plastic polymer types such as PET, HDPE, LDPE, PP, PS and PVC. These laser diodes can scan the plastic wastes over a designed wavelength range (typically within 1–1000 μm) and retrieve the absorption spectrum. Depending on their specific resonant frequencies of different plastic polymers, the sensors can identify plastic types with a high accuracy up to 97% (Zhu *et al.*, 2019).

11.2.1.2 Artificial intelligence

AI is a digital solution characterized by their ability to solve problems independently and efficiently (Mainzer, 2016). It deals with the design of computer systems and programs that are capable of simulating human intelligence such as learning, perceiving, understanding, problem-solving and decision-making through different algorithm models. Various models including artificial neural network (ANN), genetic algorithm (GA) and linear regression (LG) have been developed and applied to fulfill different desired functions in

Table 11.5 AI models applied for various waste management objectives.

Application Domain		Applied AI Model
Waste characteristics	Waste generation	ANN, SVM, LR, DT, GA
	Waste classification	ANN, SVM, DT, GA
	Waste compression	DT
	Heating value	ANN, SVM, GA
	Co-melting temperature	ANN
Smart waste bin	Bin level status	ANN, SVM
Vehicle routing	Waste collection route	ANN, LR, GA
	Collection frequency	LR
Recycling/disposal process	Biogas generation, leachate formation; energy recovery	ANN, SVM, LR, GA
Waste management planning	Illegal dump sites and waste accumulation	DT, GA
	Waste facility siting	GA
	Management costs and environmental impacts	GA

Note: ANN, artificial neural network; SVM, support vector machine; LR, linear regression; DT, decision trees; GA, genetic algorithm.
Source: Modified from Abdallah *et al.* (2020).

the application of AI technology today. In the field of solid waste management, AI is extensively applied to forecast and characterize waste generation, optimize waste collection truck routes, locate waste management facilities, and simulate waste treatment and recycling processes (Abdallah *et al.*, 2020). Table 11.5 summarizes the research and application domains of AI in solid waste management with different AI models.

Applications of AI in plastic waste management and pollution control has raised increasing attention as it helps in addressing multiple issues such as to enhance operational efficiency during plastic waste sorting and recycling, and to improve environmental management and citizen engagement to mitigate plastic litter pollution. In combination with digital technologies such as IoT, robotic and automation, AI technology has been developed and applied in customer services such as voice/image recognition to support waste identification and proper disposal at the source, smart bins for waste level detection or on-site waste segregation, autonomous vehicles for collection and transportation, AI chatbots, waste sorting robots, and citizen information (Ahmed & Asadullah, 2020; Gibson, 2020; Wilson *et al.*, 2021). Furthermore, AI in combination with remote sensing technologies such as cameras, unmanned aerial vehicles (UAV) and satellites have been applied to automatically identify and count plastic debris in rivers and coastal areas. This information is used to support decision making on plastic mitigation measures and to improve public awareness and engagement (Acuña-Ruz *et al.*, 2018; Andriolo *et al.*, 2020; Gonçalves *et al.*, 2020; Jakovljevic *et al.*, 2020; van Lieshout *et al.*, 2020).

11.2.1.3 Robotics

The Robot Institute of America defines robotics as 'reprogrammable, multifunctional manipulator designed to perform variety of tasks through variable programmed motions' (Considine & Considine, 1986). In the context of plastic waste management, robotics is a two-faced technology. On the one hand, it is used in the production process starting from resource extraction up until packaging and shipping of the final goods. This results in increased productivity and resource efficiency and prevents the waste generation in production and service sections. On the other, automation and AI in combination with robotics allows segregation of the contaminated waste, resulting in high purity which is mandatory for efficient recycling process. It meanwhile helps to avoid potential contamination and reduce the safety risk to the waste collectors, segregators and recyclers through direct contact while handling the wastes. Particularly in the circumstance of global pandemic of COVID-19, robotics-based waste sorting and recycling provides a highly secured and efficient solution in managing infectious plastic waste such as used surgical masks and PPEs. Moreover, utilization of robotics is likely expanded to waste collection process as well as the logistic management in the future.

11.2.1.4 Blockchain

Blockchains are a virtual distributed ledger on which data can be permanently stored such that it is verifiable and auditable (Swan, 2015). Blockchains store data in a series of connected blocks, such that any change in one block is easily detectable by computing its cryptographic hash and comparing it with the hash stored in the following block. Any difference between the generated and stored hashes is evidence of a change in the first block. Furthermore, updating the hash stored in the second block is detectable by computing and comparing its hash with the one stored in the subsequent block. As a result, any change to a historical block requires all subsequent blocks to be updated. Distributed consensus algorithms such as proof of work, stake or space ensure it is computationally or financially infeasible to do so. Blockchain is a revolutionary technology as it helps reduce the risk and clears out any fraudulent activity in a mappable way. As a result, it is attracting increasing applications in business management such as supply chain management and smart contract.

The application of blockchains within the waste management sector is of increasing attention (Kouhizadeh & Sarkis, 2018; Saberi *et al.*, 2018). Current applications of blockchains may contribute to plastic waste management in two ways: (1) facilitating the incentive payment or rewards that promote plastic recycling (Agora Tech Lab, 2018; Plastic Bank, 2020); and (2) monitoring and track the plastic waste stream for better management (Taylor *et al.*, 2020). In the first scenario, an entity depositing waste is rewarded or paid with a blockchain-secured digital token, which can be redeemed for goods or exchanged for other currencies. One example is the Plastic Bank Project. The project aims to establish an ecosystem for plastic recycling that is empowered by blockchain technology. It uses such blockchain rewards to incentivize individuals to become plastic waste collectors, particularly in developing countries, with an aim of reducing the amount of plastic that ends up in the oceans. The gathered waste is brought to collection locations where the waste is weighed before a payment is made to the collector through a blockchain-based banking application. The immutability and transparency of blockchains prevent fraudulent and corrupt practices. The checking of waste is often performed by humans, but could be automated in some cases such as a reverse vending machine. The plastic bank that is empowered by the IBM blockchain has been developed and trialed in Haiti, the Philippines and Indonesia to turn plastic waste into currency. In addition, a company in the UK delivered an application, 'Reward4Waste', which utilizes blockchain technology to deliver solutions that revolutionize the way items are tracked through a Deposit Return Scheme for plastic bottles (Business Green, 2018). These examples of blockchain payment or rewards facilitation are typically designed and implemented with social and environmental aims.

In the second case, data on the type of waste collected and waste transfer is recorded on the blockchain. An example is the 'Waste Transport on Blockchain' Project which was adopted by the Dutch Ministry for Infrastructure (LTO NETWORK, 2019). The project utilizes blockchain technology to streamline and automate waste transportation between the Netherlands and Belgium. It is expected to enhance the efficiency and transparency in the cross-border waste transportation process through adopting blockchain technology. By monitoring the waste amount and type data, it enables all stakeholders including engaged government bodies and companies to get real-time verification of each step in the management process.

11.2.1.5 *Cloud computing and data analytics*

Cloud computing is an on-demand availability of computer system resources, especially data storage and computing power, without direct active management by the user. The term is generally used to describe data centers available to many users over the internet. Cloud computing can facilitate the network connection between the stakeholders involved in the plastic waste management as shown in Figure 11.24. Storage and processing of sensor data or software solutions for the management, collection, administration and documentation can be made possible through cloud computing.

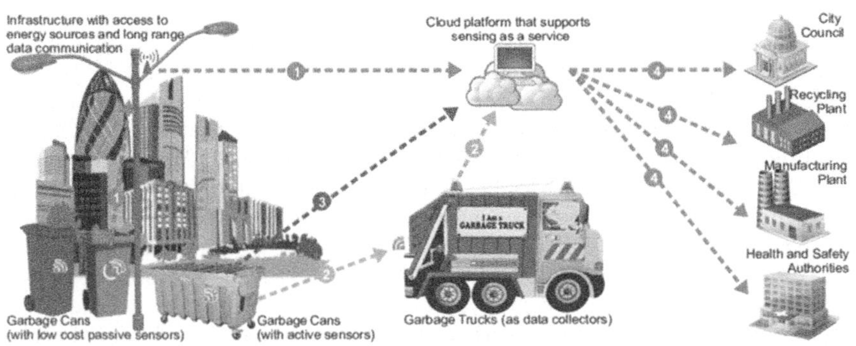

Figure 11.24 Cloud computing for waste management (*Source*: Perera *et al.*, 2017).

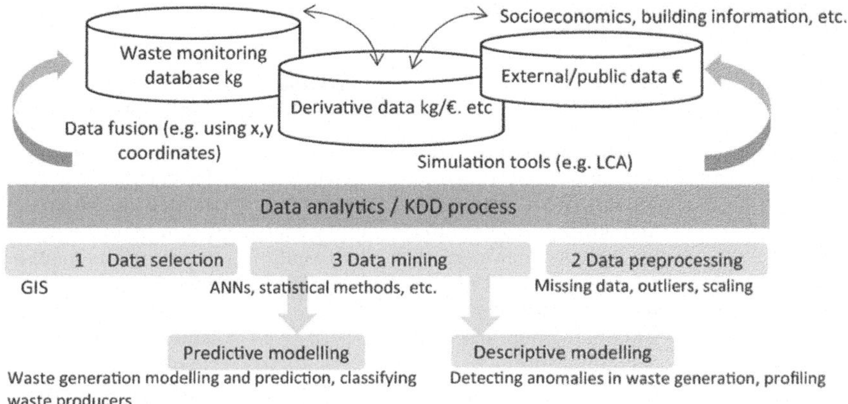

Figure 11.25 Data analytics approach to create waste generation profiles for waste management and collection (*Source*: Niska & Serkkola, 2018).

Data analysis is the task of processing and analyzing data in order to identify patterns, extract information, discover trends or calibrate models. Data analytics is important as large amount of data are logged digitally (e.g., in IoT and blockchain systems), and these need to be analyzed in order to make the data trend understandable and generate the information for communication and visualization. Data analytics are used in waste management such as electronically supported disposition of plastic waste, funding or conducting plastic cleanup projects, automated sorting plants in recycling facilities, plastic waste identification and control of incineration plants as shown in Figure 11.25. Material flow analysis of plastics in countries, regions and communities is also possible by data analytics.

11.2.2 Benefits of digitalization for plastic waste management

Despite variable technological options, and diversified structures and scales to adopt them, the digital technologies for plastic waste management are typically deployed to achieve one or more of the following objectives: (1) to improve communication and citizen engagement, (2) to advance operational efficiency, or (3) to increase management and business transparency and efficiency (Wilson *et al.*, 2021). Table 11.6 summarizes the key benefits of different application objectives of digital technologies and the available technologies for plastic waste management.

11.2.2.1 Digitalization to improve communication and citizen engagement

Communication or citizen engagement technology makes up the biggest part of digital solutions already in use in the waste sector. Here the transfer from other sectors is easy because the necessary investment in hardware is relatively low. For municipal waste disposers, communication is important to stay in touch with citizens. For commercial waste businesses, communication with

Table 11.6 Application objectives of digital technologies in plastic waste management and their key benefits.

Application Objective	Key Benefits	Available Technologies	Digital Applications
To improve communication and citizen engagement	• Driving the public and institutional outreach • Raising public awareness and willingness in better plastic waste management • Incentive/reward public recycling actions • Enhancing public service accountability	Digital interfaces and soft applications, data analytic, and so on.	Various mobile app for plastic waste-related campaign, promotion, games, tracking and mapping, identification, payment/rewarding, and education, and so on.
To advance operational efficiency	• More efficient plastic waste collection and sorting • More efficient logistic management • More efficient plastic recycling	IoT, AI and machine learning, robotics, could computing, data analytic, and so on.	Tagged plastic packages, smart trash bin, smart waste collection and transportation management system, AI and robotic waste sorting, and so on.
To increase management and business transparency and efficiency	• Better integration of multiple sources information • Access to market information for fairly managing the plastic supply chain • Ensure recordable and traceable transaction in plastic-related business • Help to implement EPR • Better measuring environmental and social impact	Blockchain, remote sensing, AI, could computing, data analytic, and so on.	Plastic bank, plastic supply chain management software, deposit return scheme and system, coastal plastic debris monitoring and mapping system (e.g., UVA + AI + mobile app), waste generation and characterization models, and so on.

customers is a mandatory and important part especially when tasks such as billing and documentation are included. One aspect is communication from the waste disposal company to customers or citizens. It includes information on pick-up dates, information on waste separation into different types, bills, and reports on disposed waste. Communication from the citizen or customer to the waste disposal company is needed to schedule pick-ups, change data for ongoing waste contracts, inquire on prices and services, and handle any other type of customer contact initiated by the customer. Typical applications to enhance waste management communication through digital solutions include various websites, mobile applications, or integrating information/functions into the proprietary communication apps or other third-party services.

Websites serve above all to display general information on the offered service and prices among others. When designed as a web-portal, it facilitates the process of contacting the service or changing data needed for service fulfilment by providing appropriate input forms. Municipal services often show information on waste separation, provide information on public collection stations and containers, offer reminder services if a routine pick-up service is scheduled, give information on special waste such as hazardous waste and provide the possibility to report illegal littering or other issues.

Mobile applications allow all the information presented on a website to be easily accessed via a mobile device as well. Additionally, the possibility of individualized apps exists. It allows for an easy personalization to deliver individual calendars, push-messages for news and important events. Furthermore, the built-in sensors of mobile handheld devices allow integration of data such as positions or the use of functionalities such as QR-Codes, images or NFC Tags. Dedicated software for business-to-business communication allows for easy creation or modification of jobs, such as collection requests, and facilitates the subsequent documentation and billing tasks if the required information is passed along in or between the digital systems. The application of 'County Waste & Recycling' is designed for residential customers to properly manage their household wastes (County Waste & Recycling, n.d.). Users can easily access their normal and holiday collection schedule, set up reminders, pay their bill and look up where to safely recycle or dispose of materials. Customers also find it easy to connect with customer service through the app.

If there is a communication solution for other public services available, the waste-related information for public services can be integrated in these existing apps or websites. This reduces the number of different information access points and increases convenience for citizens. Furthermore, several proprietary communication apps operated by third parties such as WhatsApp, Facebook, Twitter or Instagram are used by some waste companies for public relations purposes or even to communicate with customers. In a ranking of technologies that have the most impact on solid waste management, social media has been ranked in fifth place (ISWA, 2017).

11.2.2.2 Digitalization to advance operational efficiency

Several digital solutions have been introduced to the key operational processes of plastic waste management such as collection, transportation and sorting, and have been developed to meet various requirements such as higher efficiency, lower cost and carbon emission, and increased recycling rate and efficiency. Waste collection and transportation is an important section of the waste business, and the digitalization process is already taking place. Several parts of the waste collection process can be digitalized such as automated documentation, use of sensors in the collection process, routing and the inventory tracking and resource planning.

Route planning is an integral part of the waste collection process and benefits from digital tools based on geographic information systems (GIS) to optimize routes and save operation time. The routing task was handled manually in the past, nowadays digital tools allow for the optimization of fixed routes or can include real-time data to optimize routes for the current traffic situation. Additional boundary conditions and inputs such as avoidance of schools in opening times or ignoring containers that are serviced less frequently can be taken into account. Individual routes can then be transmitted to the vehicle's navigation hardware to communicate them to the drivers. These routing solutions are applied in the waste collection industry already and are integrated in various software applications. Due to the high costs for running, maintaining and operating the crewed collection vehicles, even minor improvements in the routing can have a fast return on investment (ISWA, 2017).

Efficient plastic waste sorting either from the customer end or at the waste sorting facilities is essential for a high quality and quantity waste plastic recycling. Thanks to the rapid development of sensing devices and associated application technologies (e.g., IoT, AI, and robotics), smart sorting of waste and plastic wastes is becoming possible. The applications of smart waste sorting bins and automatic waste sorting robotics have largely reduced the requirement on labor cost and increased the accuracy and efficiency of waste sorting process.

11.2.2.3 *Digitalization to enable transparent and efficient management and business*

Integration of digital technology in waste management is not a new thing, however, various promising technologies are on the rise to enhance the effectiveness and the ease of the plastic waste management. It can also help in the internal process like billing, accounting, controlling, processing and documentation, covering the whole life cycle of plastic waste management, from collection, to sorting, to recycling, to end-of-life management. The design and application of plastic supply chain platforms, enables efficient management decision and rapid response to incidents through obtaining the 'up-to-date' information during the process. Plastic organizations, formal or informal, are able to access to vast amounts of data including the location, volumes and types of plastic materials being collected, prices being paid for raw materials, payment methods, aggregator stock levels and recycling rates (Wilson *et al.*, 2021). The creative application of blockchain technology in plastic waste management further increase the reliability and transparency of information and allows for longitudinal analysis to identify the market trends and precisely predict the waste flows in the supply chain, which is an important requirement.

11.2.3 Future projection of the digitization in plastic waste management

With the digitization being embraced in every part in today's age and time, it would only be logical to infer that digitization can help evolve the waste management technology and procedures. The results of ISWA's recent global survey revealed the future interests in developing and investing the digital solutions in waste management and recycling sectors (ISWA, 2017). It is believed that except for seeking for alternative materials, innovative digital technologies such as advanced sensors, robots, mobile apps and IoTs are ranking as the top solutions which will bring the most significant impact to address the plastic

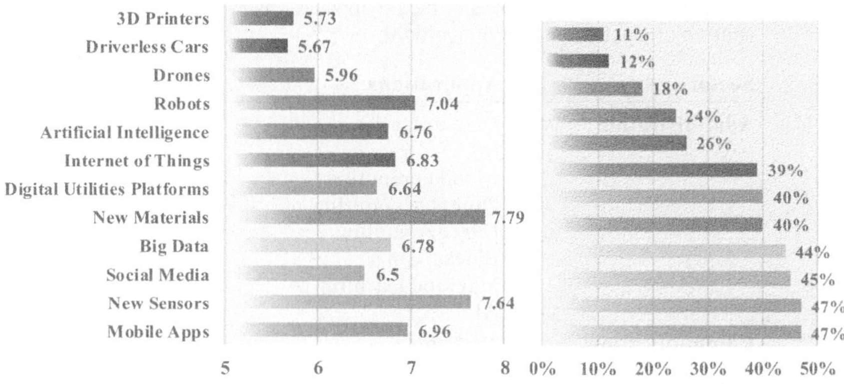

Figure 11.26 Expected impact of advanced technologies to waste management sector in 15 years (a) and preferred areas of digital technologies for future investment for waste management sector (b) (*Source*: modified from ISWA, 2017).

waste management and recycling in the next 15 years (see Figure 11.26 left). Meanwhile, mobile apps, advances in sensors, social media and big data will become the most attractive technologies for the market due to their relatively low cost and high popularity (see Figure 11.26 right).

Further innovations in applying digital solutions to address the plastic waste issues are expected such as fully robotic waste sorting and recycling facilities, digital consulting and engineering in waste management and plastic circularity, chatbots guiding citizens for proper waste prevention and recycling and driverless garbage trucks. Data collection and analysis followed by deep insights can also help in executing policy interventions and identifying enablers and disablers for plastic waste management, prospering plastic recycling business and standardized plastic circular economy. Table 11.7 summarizes several future directions of how digital technologies can improve plastic waste management.

11.2.4 Limitation/barriers for adoption of digitization in plastic waste management

Although, digitization of waste management is favorable, there are still factors that slow down its speed or prevent the application. Some of the reasons are listed below.

- **Digital literacy:** Installation and maintenance of digital tools require skilled workers with digital technology and knowledge. It might potentially replace the existing workforce which include waste pickers, waste sorters, and local hawkers. This may create unemployment of the people involved at the grassroot level of waste management. It may create conflict between waste regulators and market drivers for material recovery and recycling.

Table 11.7 Future directions of digital applications in various sectors of plastic waste management.

Sector	Applications
Administration	Data anlaytics
	IoT
	Cloud computing
	Digital accounting
	Interoperability
	Blockchain
	Machine learning
	AI
Communication	Nudging
	Augmented reality
	AI chatbots
	Interactive gamings for awareness
	Voice recognitions
	Digitally readable QR codes, labelling
	Satellite for plastic waste mapping
Logistics	Robotics
	Autonomous vehicles
	Pneumatic waste collection
	Smart bins
Sorting	Digital labelling
	Tracking
	Tracing
	Advanced data analytics
	Plastic mapping
	Identifying hotspots
Recycling	Quality assurance systems
	Smart bins
	Automated data transfer
	Transparency
	Data analytics

- **Costs:** Digitization has high upfront investment and relies on high operational and secure software and hardware for its main tasks. Shift to digitization may be resisted by actors due to easy availability of traditional waste management practices with lower prices.
- **Security:** Effective and integrated waste management requires data collection, survelliance, IoT and cloud computing whereas there is also risk of data breaching, unauthorized access and data manipulation. It can pose risks such as complete monitoring of citizens, not only by the state but by marketers and criminals through illegal data breaching/hacking. Digital data are high-priority targets for foreign offensive IT-Warfare Operations, digital attacks and hackers. Hence, this requires well designed and thoroughly managed software, which typically extends the technology development cycle and increased the cost.

- **Digital ecosystem:** Effective application of digital tools and technologies demands a cohesive and legalized digital framework. Digitization requires broadband access, digital data symbiosis, network coverage, standards for data exchange, interfaces and public digital literacy. All of these are not possible without a symbiotic digital ecosystem.

11.3 INNOVATIONS TO COMBAT MARINE PLASTIC POLLUTION

Plastics have been widely used in numerous applications due to its countless advantages, such as durability and low cost. However, mismanaged or littered plastic waste can cause severe harm to wildlife and human beings as it takes hundreds of years to break down. Additionally, increased single-use plastic such as take-away containers and surgical masks has also worsened the plastic pollution problem. Every year, an estimated 8 million metric tons of plastic ends up in the oceans, which is equivalent in terms of individual pieces to 150 million tons. By 2050, it is predicted that there will be more plastic in the oceans than fish.

Not only big fragments of plastics, microplastics and nanoplastics have also been found in the ocean as a result of plastic fragmentation into smaller pieces because of environmental factors, referred to as secondary microplastics. Additionally, they can also be released into the environment in the form of microbeads from personal care products, nurdles from plastic production and other forms of primary microplastics. As these microplastics integrate the food chain, toxins leach, bioaccumulate and are stored in the fatty tissues of animals later consumed by humans. In light of the growing concern about the adverse impacts of plastics on environmental and human health, more and more governments have been increasingly responding to this problem through improved policy implementation and environmental governance at the local, national and international levels.

Despite great efforts on improving environmental governance, the materialized impact may not be sufficient. More active engagement by public and private sectors with technology innovations is necessary in combatting marine plastic pollution considering the global nature of the problem and the extent of stakeholders involved.

Emerging technologies and strategies have been developed and invented around the world to identify, prevent, remediate and recycle the waste plastics in the environment. These innovative technologies can focus on different life cycle stages of plastics, including production, consumption and end-of-life management. The majority of these technologies target riverine debris as approximately 80% of marine plastic debris is land-sourced, and finds their way to the ocean via rivers. Technologies addressing these issues are mainly geared into four stages including (i) identifying the sources and occurrence of waste plastics and assessing their associated impact in the environment; (ii) preventing plastic leakage into main waterways (iii) collecting/removing existing plastic debris from the rivers or oceans; and (iv) recycling the collected plastic wastes. Various technological innovations further boost the incubation and growth of a number of startups and social entrepreneurs

	Objectives	Strategy & Technology Innovation	Business Startup Opportunity	Measurement
Observation & Assessment	• Source identification & characterization • Hotspot/hot moment identification • Impact assessment	• Ground-based monitoring • Citizen science • Remote sensing • Dynamic modeling	• Monitoring services • Smart phone applications • Awareness, outreach and knowledge	
Prevention	• Environmental leakage minimization	• Source prevention via alternative materials • Leakage prevention	• Marine degradable products • Leakage Prevention	• Number of technology patents and scientific report • Number of startups and entrepreneurs
Clean-up	• Reduce the accumulation of environmental plastic pollution	• Macroplastic clean-up (Barrier, receptance, and vessel) • Microplastic clean-up (Filter and nanotechnology)	• Marine litter collection • Rivers/ waterways clean-up • Fishing nets and gear collection	
Recycling	• Recycle the collected marine plastic litter	• Waste to material • Waste to energy	• Plastic waste upcycling • Mixed plastic waste recycling • Energy recovery	

Figure 11.27 A strategic framework for combating marine plastic pollution via technology innovation.

(Dijkstra *et al.*, 2021). Figure 11.27 briefly illustrates a strategic framework to combat marine plastic pollution through technological innovation. This section will discuss state-of-the-art innovations in marine plastic monitoring, prevention and collection aiming at combatting marine plastic pollution.

11.3.1 Innovations in environmental monitoring of plastics
11.3.1.1 Introduction of environmental monitoring
Environmental monitoring can be understood as systematic sampling of air, water, soil, sediment and biota in order to observe and study the environment and to derive knowledge from these processes (Artiola *et al.*, 2004). Scientists need to observe changes to study the dynamics of not only natural processes and cycles but also anthropogenic impacts (Artiola & Brusseau, 2019). Proper environmental monitoring provides valuable information about pollution levels and environmental changes, as well as the effectiveness of the implemented mitigation measures, public policies and pollution control technologies. Beyond assessment, the purpose of environmental monitoring is to learn from past experiences. It is, therefore, essential to define clear scopes, adopt consistent approaches, select proper indicators, acquire knowledge, improve analytical tools and methods, and reduce uncertainties. Thanks to development of environmental monitoring and analysis approaches, management interventions can be undertaken to improve decision making and the effectiveness of management responses.

Depending on the questions to be addressed, four types of monitoring, namely surveillance monitoring, implementation monitoring, effectiveness monitoring, and ecological effects monitoring, can be selected and designed to meet different goals (Hutto & Belote, 2013). Among these four types of monitoring, surveillance monitoring occurs independently of particular management activity while it can also inform management if response variables are linked to conditions on the ground. The other three types of monitoring are tightly linked with management activity and are designed to answer very different types of questions as indicated below.

- Surveillance monitoring – to uncover changes in target variables over space and time.
- Implementation monitoring – to record whether management actions have been applied as prescribed.
- Effectiveness monitoring – to evaluate whether a given management action has been effective in meeting a stated management objective.
- Ecological effects monitoring – to explore unintended ecological consequences of management actions.

The sampling strategies, protocols and used indicators are important aspects to take into consideration when designing a monitoring program. In addition, assessment feedbacks are also used critically to design suitable mitigation measures and public policies. Any monitoring scheme starts by defining several main features such as monitoring objectives, study area, protocols and methods, sampling strategy, and time period for monitoring. Additionally, for representative measurements in any environmental monitoring program, four key aspects should be taken into consideration at the design stage (Hanke, 2013): (1) spatial and temporal scales should be defined; (2) rigorous and repeatable sampling and analytical procedure protocols should be used; (3) suitable mapping and dissemination tools to show the environmental status of the different indicators should be developed; and (4) the sampling scales and indicators to management issues (e.g. mitigation measures), as well as resource considerations, should be linked.

11.3.1.2 *Monitoring marine plastics and microplastics*

Monitoring the environment for the presence and abundance of plastic litter is of significance to evaluate the accumulation extent and potential impact of plastic pollution, to formulate mitigation plans to reduce the inputs, and to assess the effectiveness of adopted mitigation measures. Since it was first reported in 1972, global concerns on marine plastic litter especially the microparticles have grown rapidly especially in the recent 15 years. A number of sampling and measurement methods have been developed for both macroplastics and microplatics in the marine ecosystem. Macroplastic refers to the relatively large plastic debris with the size typically larger than 5 mm. Such plastic debris is visible and can be detected and measured using various approaches including direct observation and remote sensing. In addition to the traditional monitoring of marine macroplastics by scientists, a number of innovations and practices have appeared to advance the knowledge regarding marine macroplastics, including involving citizen scientists in data collection and reporting, adopting smart monitoring based on remote sensing technologies, and developing plastic hydrodynamic models. On the other hand, microplastic is defined as the plastic fragments in the marine environment smaller than 5 mm in diameter (Lavender & Thompson, 2014). Microplastics presenting in the environment can be

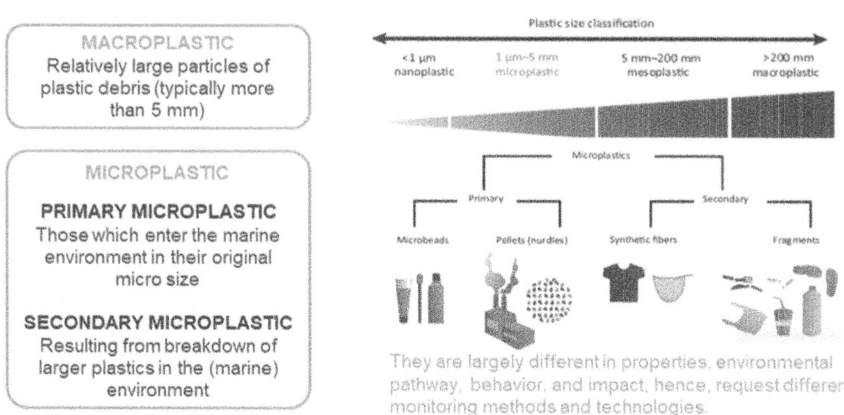

Figure 11.28 Definitions and classifications of macroplastic and microplastic.

sourced from two routes, primary microplastics and secondary microplastics, with the latter being considered as the majority in the marine environment (see Figure 11.28). The early concern on microplastics can be traced back to 2004. Since the subject on microplastic monitoring is still in its infancy, ground-based sampling and measurement is currently the major available method.

11.3.1.3 An integrated system for monitoring marine macroplastics

Maximenko *et al.* (2019) have proposed an integrated system to monitor and report the marine plastics debris. Such system is composed of three layers of monitoring approaches, ground-based observation, remote sensing and dynamic models (see Figure 11.29).

• **Platforms and facilities for ground-based observation**

The ground-based observation provides the 'ground-truth' information in terms of plastic abundance, type and characteristics. Various in-situ facilities and platforms have been or can be applied for ground-based observation of marine plastic debris (Figure 11.30). First, ships are traditionally used to collect data on marine debris floating on or near the surface. They provide a platform from which a broad variety of sensors and samplers can be used for comprehensive study of the entire water column, from seabed to the surface. Second, autonomous platforms such as floats, gliders (both seagliders and wavegliders), and autonomous surface and underwater vehicles (ASVs and AUVs), equipped with adequate sensors can provide unique measurements in hard-to-reach parts of the ocean. Another platform is fixed point observatories. These are important and efficient platforms to monitor temporal variability and, in particular, long-term trends of the problem. Also, benthic landers and crawlers which represent a range of devices have been developed. They remain on the seabed for protracted periods of time, some of which photograph the seabed repeatedly or can take time-series sediment samples when crawling over

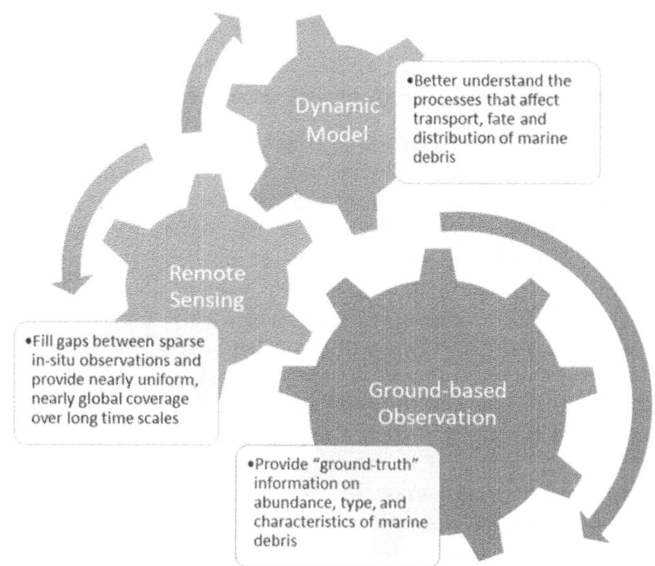

Figure 11.29 Structure of an integrated marine debris observation platform. (*Source*: Maximenko *et al.*, 2019)

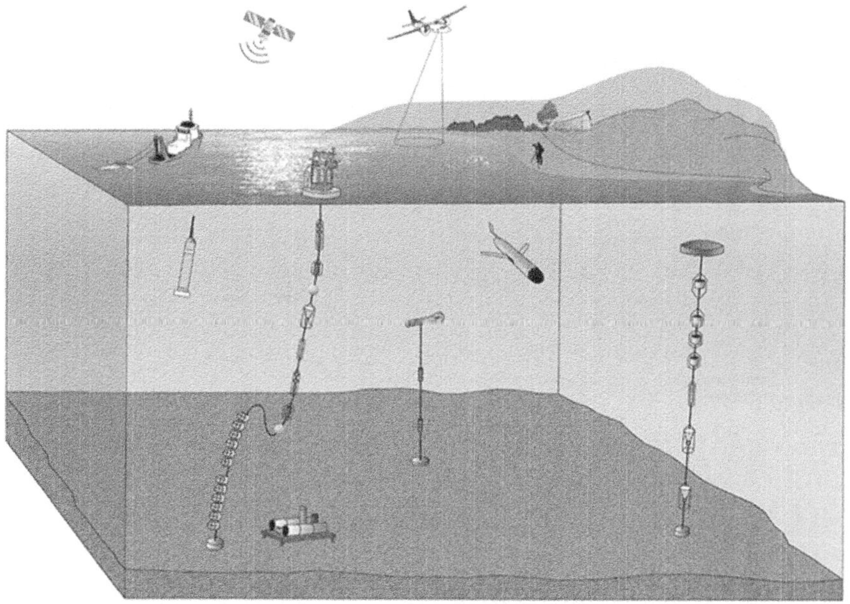

Figure 11.30 Platforms and facilities for marine plastic observation (*Source*: Maximenko *et al.*, 2019).

the seabed. Finally, shoreline monitoring and beachcombing are also commonly used (Maximenko *et al.*, 2019).

- **Citizen science and its role in plastic debris monitoring**

Environmental monitoring is traditionally an activity conducted by scientists and researchers, while, citizen science has been recognized as a robust and important approach to collecting more information and knowledge about the environment. Citizen scientists with proper education and training can play valuable roles in monitoring, communicating, and reporting the abundance and characteristics regarding marine litters. The earliest citizen volunteers conducting marine litter research appeared in early 1990s (Hidalgo-Ruz & Thiel, 2015; Zettler *et al.*, 2017). Such volunteer activities become more and more popular in recent years due to the increasing awareness and concerns on marine plastic pollution among the public.

Citizen science plays an important role as it allows for easy large-scale data collection by the larger public. Citizen scientists provide valuable scientific data to professional scientists and may influence the general public by reporting about their experience, for example through social media platforms. This, in turn, encourages the general public and eventually decision makers to act on the environmental problem in question (GESAMP, 2019). Figure 11.31 shows the role of citizen science in plastic litter monitoring.

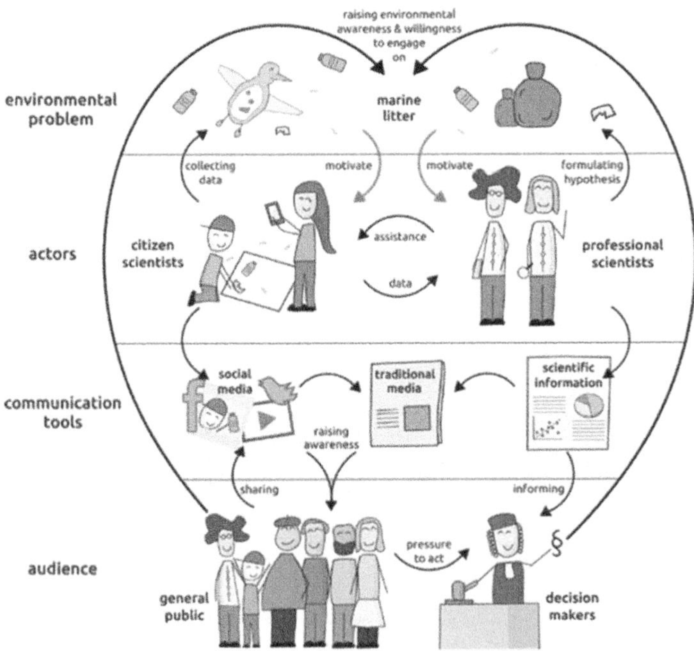

Figure 11.31 Roles of citizen science in plastic monitoring (*Source*: GESAMP, 2019).

Citizen science data collection can be done either visually or through simple tools and/or technology, such as smartphone apps. Different types of samples or data can be gathered by citizen science projects such as observational data of litter impacts, collection of specific litter items, bulk estimates of gross amounts of litter, frequency data on litter types, and quantitative data on litter densities. Figure 11.32 shows the level of participation of citizen scientists in litter research of marine areas. Top panels illustrate methods and types of data gathered by citizen scientists. Bottom panels show the degree of involvement by citizen scientists (turquoise area) in the scientific process. A number of citizen science projects and activities focusing on marine plastic litter have been carried out globally. They are motivated by various reasons, target different goals, and are performed at either individual or organizational levels. Table 11.8 describes several examples of marine debris monitoring and reporting projects based on citizen science.

- **Remote sensing-based plastic monitoring**

Remote sensing can help to fill the gaps between sparse in-situ observations and provide data with nearly global coverage over long time scales. Despite the great potential of remote sensing, applications of such technologies in marine plastics debris observation is still in its infancy (Garaba & Dierssen, 2018; Maximenko *et al.*, 2019).

Currently, unmanned aerial vehicles (UAV) have been applied to assess plastic debris in rivers, on beaches, coastal areas and shallow marine zones

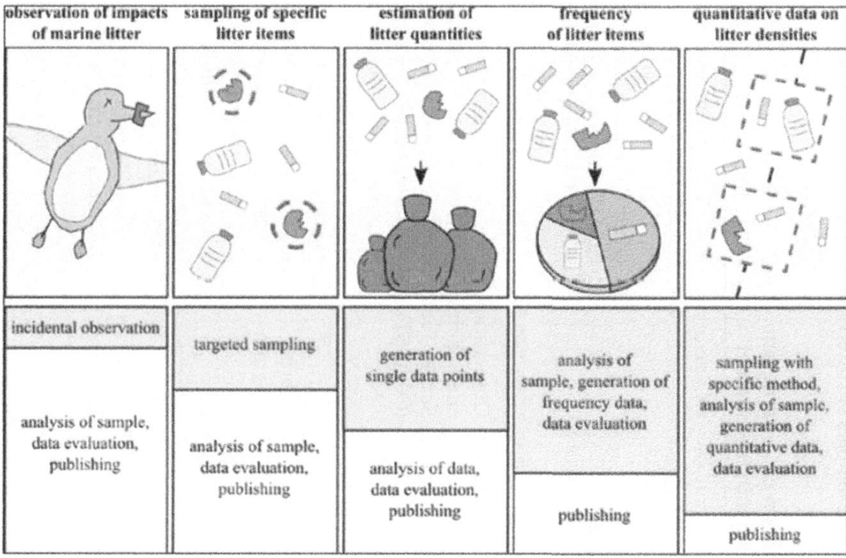

Figure 11.32 Levels of participation of citizen scientists in marine litter research (*Source*: GESAMP, 2019).

Table 11.8 Examples of citizen science projects for marine litter monitoring and management.

Level of Participation of Citizen Scientists	Program/ Project	Description	References
Data collection – active sampling	eXXpedition	The eXXpedition, led by an ocean advocate Emily Penn, was founded in 2014 to shift the way people feel, think and act by building a global network of multidisciplinary women who can contribute to world-class scientific studies, explore solutions, and use their unique skills to tackle the ocean plastic problem from all angles. The eXXpedition crew has conducted a number of expeditions to inform the global knowledge and provide better understanding on the causes and solution to plastic in the oceans.	Exxpedition (2014)
Data collection – active sampling	Coastal Observation and Seabird Survey Team (COASST)	Established in 1999, the citizen science team that monitors marine resources and ecosystem health at more than 350 beaches from northern California to Alaska. Although COASST received funding from the National Science Foundation, had long focused on collecting data on beach-cast seabird carcasses as an indicator of coastal health, the team has been planning to focus on collecting data on beached marine debris. Resulting data could be used to help support efforts to reduce the impacts of marine debris on coastal wildlife.	COASST (1999)
Data collection	Marine Debris Program	The Marine Debris Monitoring and Assessment Project (MDMAP) is a citizen science initiative that engages NOAA partners and volunteers across the nation to survey and record the amount and types of marine debris on shorelines. Each partner in the network selects a nearby shoreline monitoring site that they return to monthly to conduct surveys and submit meaningful data to NOAA's MDMAP Database. This project continues adding new resources to the toolbox, including FAQs, training videos, a database user guide, and a GPS point guide.	NOAA (n.d.)
Application	Debris Tracker	The smart phone application is designed to help citizen scientists make a difference by contributing data on plastic pollution in the community. Scientific organizations and citizen scientists from all around the world record data on inland and marine debris with the easy-to-use application, contributing to the open data platform and scientific research.	Stanley (2010)

(Continued)

Machine learning	The Zooniverse Plastic Tide citizen science project	By registering at the zooniverse web site, citizen scientists can use their computer or mobile device to analyze images in the plastic tide's database for plastics and litter, tag each piece of plastic spotted by drawing a rectangle around it on your screen and identify it as fragments, fishing line, drink bottles or some other type of plastic waste. By tagging plastics and litter in the images taken with the drone, citizen scientists directly teach the computer program to autodetect, measure and monitor plastics to help researchers answer how much of the missing 99% ends up on our beaches.	Greenemeier (2017)
Data Collection – Surveys	The Ocean Cleanup	This non-profit-organization investigates plastic transport through rivers and plastic accumulation in the oceans. There are currently two different ways, that is, riverways and ocean survey to be involved in this survey as a citizen scientist.	
Data Collection – Surveys	International Pellet Watch	IPW is a volunteer-based global monitoring program designed to monitor the pollution status of the oceans. This focuses on persistent organic pollutants (POPs) such as PCBs and organochlorine pesticides. IPW is based on the fact that POPs are accumulated in resin pellets (plastic raw material) from the surrounding seawater by a factor of millions. Similar accumulation occurs with broken plastic fragments in the ocean. Pellets are surrogates for all marine plastics. IPW reveals the magnitude and spatial variability of POPs in marine plastics for risk assessment. It has demonstrated that marine plastics transport POPs in marine environments, even to remote areas.	International Pellet Watch (2005)
Scientific literacy	Following the Pathways of Plastic Litter	This is an international citizen science project for promoting K12 students' scientific literacy. The project collected scientific data, understanding of the scientific process was promoted among the participants and a rethinking of their own actions was anticipated. Guided by an accompanying book, the students dealt with important scientific principles on marine litter, which was required for the subsequent sampling on the beach and the evaluation of the data. Since the project was carried out in Germany and Chile at the same time, the participants were able to exchange information about their research on a project website and thus gained insight into the conditions in the partner country.	Kruse *et al.* (2020)

(Andriolo *et al.*, 2020; Escobar-Sánchez *et al.*, 2021; Jakovljevic *et al.*, 2020; van Lieshout *et al.*, 2020). In combination with machine learning technology, UAVs equipped with RGB cameras can recognize and characterize large plastic debris accumulating on the water surface and ground surface with acceptable accuracy. In addition to its low cost, it allows the collection of imagery data with high resolution over a spatial scale of hundreds of hectares, access to remote areas, flexibility in the frequency of data collection, and, under certain conditions, the use of satellites (e.g., in cloudy weather). However, UAVs have high requirements on the weather conditions, such as calm waters, low winds, and minimal sun glint, which present limitations on further application in plastic debris observation especially in offshore areas. Satellite observation opens a new window for the monitoring of marine plastic debris and in support of the implementation of effective remedial actions and policies. Recent research has demonstrated the possibility to combine satellite imageries and optical spectro-radiometric reflectance or hyperspectral information to detect large floating plastic debris and/or the cumulative plastic litter on the beaches (Acuña-Ruz *et al.*, 2018; Arii *et al.*, 2012; Garaba & Dierssen, 2018; Kikaki *et al.*, 2020; Topouzelis *et al.*, 2019). Despite a number of encouraging research outputs, the overall marine debris monitoring using remote sensing is still facing technological challenges, such as the availability of proper sensors, insufficient spectral resolution, effects of weather conditions, high costs and lack of standardization.

- **Dynamic modeling of marine debris**

Numerical models are essential tools with which we can get more insight into the complex processes that affect transport, fate and distribution of marine debris. For example, the hypothesized sinking and rising of plastic particles due to biofouling and possible subsequent remineralization at depth (Kooi *et al.*, 2017) will be extremely difficult to observe, but its large-scale effect on the horizontal distribution of marine plastic can be modeled. Furthermore, numerical models are not constrained to the particle size of plastic debris. This means, they can be utilized to study the distribution and fate of microplastics in a global perspective (Hardesty *et al.*, 2017; Peeken *et al.*, 2018). A number of different models have been developed over the past decades. These range from statistical observation-driven Markov models (Maximenko *et al.*, 2012; van Sebille *et al.*, 2012), to highly idealized mathematical models (Koelmans *et al.*, 2017), to full-blown two- and three-dimensional particle tracking models (Lebreton *et al.*, 2012; Robinson *et al.*, 2017).

11.3.1.4 Ground-based monitoring of marine microplastics

- **Sampling of microplastics**

Active sampling methods are typically used for microplastic monitoring. Microplastic samples may be collected from water, sediment, and biota from the marine ecosystem. Microplastic sampling from water can be distinguished into bulk sampling and volume reduction sampling. Bulk sampling refers to the collection of entire volume of aquatic samples, while, volume-reduced sampling means that the volume of sample is reduced during the sampling process with only the fraction containing target substance for further laboratory processing. This method is preferable especially for sampling microplastics from relatively clean media such as drinking water and surface water, to avoid extremely large sampling volume. Depending on the selected devices, it is possible to collect samples from different levels of the water column from the surface to the bottom. Sampling devices used for sampling microplastics from aquatic environment can be subdivided into three types: (1) non-discrete sampling devices such as nets and pumping systems; (2) discrete sampling devices including rosette, integrating water samplers, buckets, bottle and steel sampler; and (3) sampling devices with surface microlayer like sieves and rotating drum samplers (Cutroneo *et al.*, 2020; Prata *et al.*, 2019; Wu *et al.*, 2020).

Sediment samplings are conducted to investigate ambient conditions and the influence of potential pathways for microplastic transport. The studied location should be chosen carefully as sediment areas can be depositional, or erosional in addition to other unique attributes that impact microplastics accumulation. In general, studied sediments are depositional and located in the nearshore bay margins most likely affected by stormwater runoff and wastewater effluents. Sediment samples can be collected using a variety of devices such as shovels, grab samplers, and sediment corers. The depth of the sediment to be sampled is an important consideration, with shallow samples more likely to represent the recent conditions. In contrast, sediment cores can give an indication of cumulation trends over time, provided the area has remained depositional over time and the sediment has not been significantly disturbed. The European Union Marine Strategy Framework Directive recommends that surface sediment samples be collected from the top 5 cm of sediment because this layer is affected by seasonal erosion and deposition and thus likely to represent recent conditions, which is as a result the most frequently studied in the literature (Miller *et al.*, 2021; Prata *et al.*, 2019; Stock *et al.*, 2019).

Microplastics enter the food chain through ingestion and cause serious physical and toxicological effects due to bioaccumulation to various marine organisms. Sampling and measuring the microplastics present in different aquatic species are of importance to better understand the hazards associated with microplastics to aquatic organisms. A variety of techniques can be used to sample aquatic organisms. Depending on the target species and their habitat, organisms can be collected in grasps, traps, creels or bottom crawling (benthic invertebrates), by manta or bongo nets (planktonic and nektonic invertebrates), by trawls in different water levels (fish), by hand (e.g., bivalves or crustaceans) or by electrofishing (Stock *et al.*, 2019). Table 11.9 summarizes the most frequently used sampling methods and devices for microplastics from water, sediment and biota, and analyzes the pros and cons of each method.

- **Laboratory analysis of microplastics**

Laboratory separation and analysis are carried out following the sampling procedures. Figure 11.33 summarizes the primary procedures and methods that are used for microplastics analysis. Separation processes such as sieving,

Table 11.9 Sampling methods and available devices for marine microplastics.

Sample Type	Sampling Method	Pros	Cons
Water	Neuston and Manta trawl	Easy to use; Sample large volumes of water; Largely used (good to compare between locations); Produces large numbers of microplastics for further testing.	Expensive equipment; Requires boat; Time consuming; Potential contamination by vessel and tow ropes; Lower limit of detection is 333 μm.
	Plankton net	Easy to use; Lowest limit of detection 100 μm; Quick to use; Samples medium volumes of water.	Expensive equipment; Requires boat; Static sampling requires water flow; May become clogged or break; Sampling of lower volumes of water than Manta trawl.
	Sieving	Does not require specialized equipment nor boat; Easy to collect samples.	Laborious and time consuming; Samples medium volumes; Manual transfer of water with buckets
	Pump	Samples large volumes of water; Effortless; Allows choice of mesh size.	Requires equipment; Requires energy to work; Potential contamination by the apparatus; May be difficult to carry between sampling locations.
	Bulk sampling	Easy to collect samples; Known volume of water; Allows choice of mesh size.	Sampling of low volumes; Transportation of water samples to the lab; Potential contamination by the apparatus; Time consuming depending on mesh size.

(Continued)

Sediment	Corer	Collect sediment core; Easy to use; Allow duplication	Maybe expensive; Need boat; Sampling may disturb sediment.
	Grab sampler	Collect surface sediment; Easy to use; Allow duplication	Expensive equipment; Requires boat; Variation with sampled area and depth; Sampling may disturb sediment surface.
	Shovel	Does not require specialized equipment; Beach and shallow seabed sampling	Laborious and time consuming; driving equipment; Sample medium volume; Manual transfer of sample
Biota	Manual	Does not require specialized equipment nor boat; Easy to collect some species of biota samples; Sample individual biota sample.	Laborious and time consuming.
	Trawl and net	Easy to use; Quick to use; Sample various species; Sample medium volumes of biota.	Requires boat.
	Electrofishing	Quick to use; Sample biota species such as fishes; Sample individual biota sample.	Need skilled labor; Expensive equipment.

Sources: Prata *et al.* (2019) and Stock *et al.* (2019).

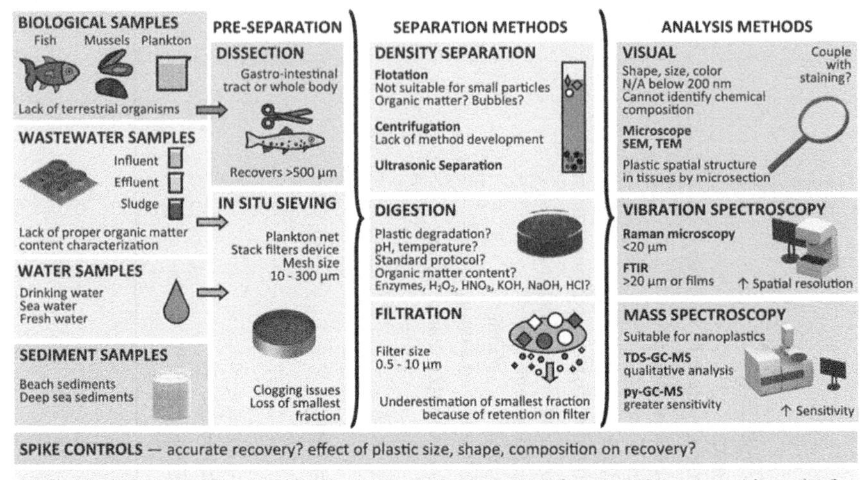

Figure 11.33 An overview of microplastic separation and analysis methods in different environmental samples (*Source*: Nguyen *et al.*, 2019).

density separation (NaCl, ZnCl$_2$, NaI, and sodium lauryl sulfate (SLS), and so on), filtration (polycarbonate, polyamide, glass, cellulose, ANOPORE inorganic membrane filters), and digestion methods (acidic, enzymatic, alkaline, and oxidative, etc.) are typically used to separate microplastics from various environmental samples. Analysis of microplastics typically includes physical, chemical and biological characterizations. Physical characterization identifies the abundance, shape, color, size, aging and surface texture of the microplastics. Direct visual observation, optical microscopes and electrical microscopes are used to accomplish these measurements. Chemical characterization adopts advanced analytic equipment such as Fourier-transform infrared spectroscopy (FTIR), Raman microscopy and mass spectroscopy to identify the polymer type, weathering status, additives and adsorbed chemicals associated with microplastics. Biological analysis may be applied to identify the microbial species attached on marine microplastics or to study the toxicity of microplastics through bioassay (Campanale *et al.*, 2020; GESAMP, 2019; Hermsen *et al.*, 2018; Nguyen *et al.*, 2019).

11.3.1.5 Impact assessment of plastic litter

The impact assessment of marine plastic litter is still at its beginning stage. Several recent studies have proposed different approaches to identify and quantify the potential impact and/or risks caused by marine plastic litter (Beaumont *et al.*, 2019; Compa *et al.*, 2019; Saling *et al.*, 2020; Woods *et al.*, 2021). Among the proposed assessment strategies, life cycle impact assessment provides a systematic and quantifiable approach to assist the decision making with regard to marine plastic abatement (Saling *et al.*, 2020). Woods *et al.* (2019)

proposed a comprehensive framework to assess the impact of marine plastic litter in life cycle impact assessment. The framework links inventory data with respect to mass of plastic leakage into a specified environmental compartment (i.e., air, terrestrial, freshwater, and marine) to six areas of protections: ecosystem quality, human health, socio-economic assets, ecosystem services, natural heritage and cultural heritage. The key fates of plastic litter include transportation, fragmentation, and degradation were taken into consideration in the pathway modelling. Exposure and effect modelling steps differentiate among six routes including inhalation, ingestion, entanglement, invasive species rafting, accumulation, and smothering, that potentially compromise sensitive receptors, such as ecosystems, humans and manmade structures. The framework includes both existing, for example, human toxicity and ecotoxicity, and proposed new impact categories, for example, physical effect to biota. Figure 11.34 illustrates a proposed life cycle impact assessment model framework for comprehensively assessing the impact of marine plastic litter.

11.3.2 Innovations in marine plastic prevention

Technologies and innovations to prevent the leakages of plastics into the marine ecosystem include source prevention (e.g., replacement by alternative materials) and key leakage point prevention (e.g., household leakage, stormwater leakage, and wastewater leakage).

Advancing in biodegradable plastic suggests a growing market and demand for sustainable alternatives to conventional plastic products and provides a potential approach to prevent marine plastic leakage from the source. In particular, marine degradable plastics which can benignly degrade when ending up in the marine environment attracts industrial invention and business

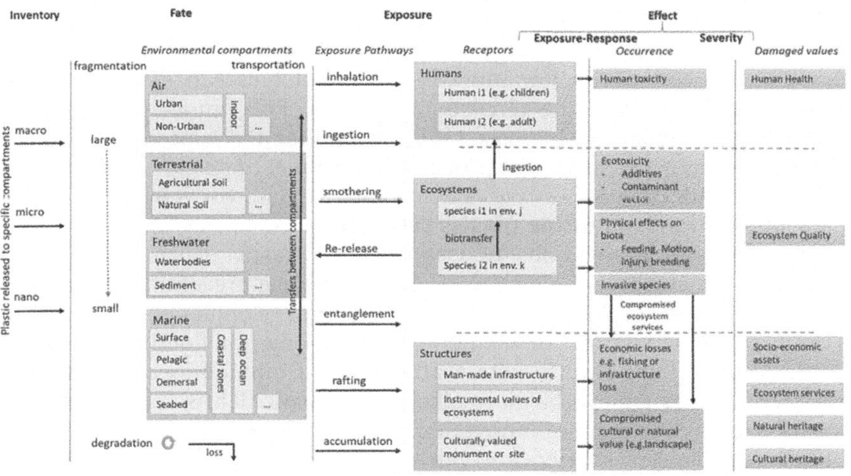

Figure 11.34 A proposed framework for marine plastic litter life cycle impact assessment (*Source*: Woods *et al.*, 2021).

investment. These materials can be made from renewable feedstock including marine products such as seaweed, microalgae and seafood wastes, as well as other agricultural byproducts and organic waste. They have been designed to partially replace plastic packages and many single-used plastic products like straws, food containers, garbage bags and some small plastic parts that inevitably leaked into the environment. A recent review study has identified a number of startups and entrepreneurs that engaged in marine degradable materials to overcome marine plastic pollution (Dijkstra *et al.*, 2021). Most of these companies have been founded in recent years and attracting increasing investment although some of them are still at conceptual and prototyping phases. Notably, attempts have been made to have biodegradable materials replace primary microplastic products such as microbeads (Nam & Park, 2020; Naturbeads, 2018). However, solutions to microplastic pollution at the source are remaining underdeveloped.

Urban waterways are identified as the most important routes for waste plastics entering the rivers and oceans. Emerging technologies have been developed to prevent leakages of macro and microplastics from various key urban water leakage points such as the households, stormwater and wastewater facilities. Table 11.10 summarizes different technologies being developed and applied to achieve such objectives. Emerging technologies and devices for household leakage prevention are mainly focusing on microplastics and microfibers from household laundries and showering. Such filter devices are placed at the household drainages and able to capture microparticles especially the microfibers that shed when synthetic fibers are washed, and prevent them from being discharged into the municipal sewage systems.

Technologies preventing leakages of plastics debris from stormwater and wastewater discharges are typically composed of mesh nets or metal rags/traps which are installed at the wastewater treatment plants, stormwater outfalls and downstream of urban drainage systems to prevent their leakage into surface water systems like canals and rivers. Such devices are usually targeting macroplastics, whereas, a few innovative technologies can meanwhile filter out microplastics such as the jellyfish mucus filter. This innovative technology applies jellyfish mucus as a bio-based solution to filter and remove microplastics during the wastewater treatment process (Freeman *et al.*, 2020).

11.3.3 Innovations in marine plastic clean-up

Technologies and innovations to clean-up the plastic debris from aquatic ecosystems are much variable compared with leakage prevention. Most of the plastic clean-up technologies in use and/or under development today are based on three fundamental types of technologies: (1) booms and floating barriers, (2) receptacles and traps, and (3) watercraft vehicles. Combinations of these three basic technologies tend to be more efficient in plastic debris collection, hence becoming the mainstream of innovation.

Booms and floating barriers are widely used to capture and guide the floating plastics and other debris in the waterways and prevent their flow from entering the oceans. Booms can be manufactured into different floatation structures using light materials such as recycled PE and foams, as well as waste plastic bottles and

Table 11.10 Emerging technologies for plastic debris leakage prevention.

Technology	Description	Target Particulates	Similar Projects and/ or Products	References
Household leakage				
Cora Ball	Balls placed in the laundry machine capture microfibers shed when washing synthetic fibers	Micro	FiberFree	Cora Ball (2020); FibreFree (2018)
Lint LUV-R	Water filter on laundry machines captures microfibers when water is drained through the machine	Micro	Showerloop	Environmental Enhancements (n.d.); Shower Loop (n.d.)
Stormwater & Wastewater leakage				
PumpGuard	Mesh net system uses water flow to capture and remove trash, floatables, and solids from stormwater and wastewater	Macro	StormX Netting Trash Trap	Storm Water Systems (n.d.); StormTrap (2016)
Inline litter separator	Trap attached to the drainage system downstream of shopping areas removes litter from passing stormwater	Macro	Osprey Litter Interceptor; StormTrap Trash Trap	Osprey (2021a); Phillips (1999); StormTrap (2020)
Jellyfish mucus filter	Jellyfish mucus extraction made into filters to capture and bind to nano-sized particles, and remove microplastics from wastewater	Micro		Freeman *et al.* (2020); GoJelly (2018)

Source: Modified from Schmaltz *et al.* (2020).

Figure 11.35 Marine litter containment floating boom developed by the CLAIM Project. (*Source*: Credit: CLAIM)

containers (Figure 11.35). They are often attached with curtains or skirts suspended into the shallow water column to capture buoyant waste. Barriers work with the same mechanism as booms but are designed in various shapes and structures such as chain-link fencing, reinforcing bar, and light mesh and net (Figure 11.36). These devices are typically anchored to the shoreline and the riverbed downstream of hotspot trash outfalls, and function as a durable or semi-permanent solution. Sizes of such devices are easily customized depending on the size of the waterway and the load of floating debris in the waterway. Trashes captured in a boom or floating barrier can be removed manually, or by skimmers and excavators. As passive collection devices, they usually do not require any energy consumption. In addition, relatively low requirement

Figure 11.36 Blue barriers by the Sea Defense Solution (SEADS). (*Source*: Credit: SEADS)

on maintenance and low cost makes this technology an attractive option for local applications in small rivers and canals, or projects with limited funding resources. However, several limitations such as the low collection efficiency and instability under complex environmental conditions need to be considered in technology development and selection in the future.

In addition to being anchored on the riverbanks for independent application, booms and floating barriers can be used as a component of most other plastic debris collection system, such as receptacles and watercrafts, to funnel the plastic debris into the designated collection devices. Large-size boom systems are also applied in the open oceans to cumulate and remove marine plastic debris directly. The Ocean Cleanup System 001 was a 62-mile-long floating barrier that targets plastic debris and ghost nets from the gyres, also known as the Great Pacific Garbage Patch (The Ocean Cleanup, 2018). The unit has an arch design that moves slowly with the ocean currents to collect ocean plastic and contain debris. Once full, the accumulated plastic will be removed for recycling by a vessel. This innovative design has cleaned up 80 000 tons of garbage within a four-month period (Figure 11.37).

Receptacles typically refer to floating containers that actively or passively trap the debris and hold them in a confined space. These devices are typically made of aluminum material or HDPE and are installed in hotspot areas for debris outfalls to capture litter as it flows downstream. They are often combined with booms and floating barriers to increase the trash collection efficiency in a large waterway. Collection of plastic debris can be in a passive mode meaning

Figure 11.37 The Ocean Cleanup System 001 for marine plastic debris collection (The Ocean Cleanup, 2018).

driven by water flow and current, or in an active mode meaning driven by motors and pumps. Trash removal can be done with lift-out baskets, conveyor belts or boats. To increase mobility and ease of maintenance and emptying, the device can be designed with a detachable metal collection trap.

Several notable innovative plastic receptacle systems have been developed and applied in recent years. The trash wheel technology developed and commercialized by Clearwater Mills, LLC, is an innovative water wheel that captures trash in rivers, streams and harbors, and then turns it into electricity (Helinski *et al.*, 2021; Schmaltz *et al.*, 2020; Waterfront Partnership of Baltimore, 2019). It can clean up to 38 000 pounds of trash daily and has prevented 1.6 million pounds of debris from entering the ocean thus far. It uses a conveyor belt powered by a unique blend of solar and hydro power to move the trash into a floating barge. The collected trash is then used for electricity generation (see Figure 11.38).

The Intercepter™ designed by the Ocean Cleanup is an integrated floating plastic extraction system that consists of a solid barrier to catch and concentrate debris towards a conveyor belt fitted onto a floating pontoon (Schmaltz *et al.*, 2020; The Ocean Cleanup, 2016). The debris flows through the central channel of the catamaran toward an extraction conveyor belt. A shuttle belt automatically distributes the debris across six dumpsters. Using sensor data, the containers are filled equally until they reach full capacity. Up to 50 m³ of trash can be contained in the barge of the Intercepter™. When the barge is completely full, an alert is automatically sent to the local operators to come and remove the barge, bring it to the side of the river, and empty the dumpsters. All electronics on the Interceptor, including the conveyor belt, shuttle, lights, sensors and data transmission are solar-powered (see Figure 11.39).

A passive receptacle does not request energy supply but may collect debris at a low rate especially when water movement is slow. The Great Bubble Barrier is designed with an energy-efficient configuration and utilize air bubbles to

Figure 11.38 Mr. Trash Wheel (cnet.com, 2021).

Figure 11.39 The Interceptor™ by the Ocean Cleanup (Ocean Cleanup, 2016).

collect and remove plastics from waterways (The Great Babble Barrier, n.d.). A bubble curtain is generated by pumping air through a perforated tube on the bottom of the waterway. The bubble curtain creates an upward current which directs the plastics to the surface. By placing the Bubble Barrier diagonally in the river, the natural flow will push the plastic waste to the side and into the catchment system (see Figure 11.40).

In addition to capturing macroplastics, several innovative receptacle systems have been designed with potential functions to capture and remove microplastics from water. The Seabin V, created by the Seabin Project, is aimed at gathering floating trash located in areas with calm water, like marinas and harbors (Seabin Project for Cleaner Oceans, n.d.). Made from recyclable materials, this vacuum-like design pumps water into the filter bag located inside the device. Water is then pumped back into the surrounding area, leaving the trash trapped inside. With proper filter material, the device can remove not only macroplastics, but also oil and fuel, and microplastics from the ocean (see Figure 11.41).

Watercraft vehicle-based technologies are primarily composed of a buoyant structure which can travel on or in the water for debris clean-up, such as boats, interceptors, vessels, and robots. Skimmers, conveyer belts, manipulator arms and nets are typically used to collect the buoyant debris from the water. Such

Figure 11.40 The Bubble Barrier system (The Great Babble Barriar, n.d.).

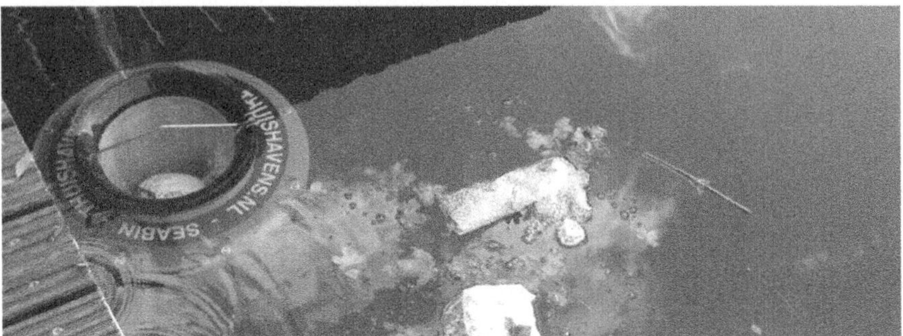

Figure 11.41 Seabin V by the Seabin project (Seabin Project, n.d.).

devices are usually equipped with receptables or containers for temporally storing the collected debris. They can also be used in combination with floating barriers to guide the trash into the catchment system. Large scales of watercraft vehicles are usually human operated, and used primarily in big lakes, rivers, harbors and bays that have navigable conditions. Such watercraft vehicles are typically powered by conventional energy such as fuel and electricity to deal with heavy debris loads. In recent years, small-scale watercrafts have been invented and favored for waterway clean-up due to their higher flexibility. They are typically designed for manual or AI remote control. Such technologies operate on fully autonomous energy harvesting (e.g., solar, wave) while others require external energy sources (e.g., batteries, fuel).

In addition to various debris collection devices for waterway clean-up, emerging sand cleaners have been developed and applied to remove macro and microplastics on the beach. The Surf Rake and Sand Man manufactured by H. Barber & Sons, Inc. are able to clean large and small beaches using sand shifting machines, respectively (Barber, 2021a, 2021b). Such technology has been applied in over 90 countries as an efficient solution for plastic pollution on the beach.

The Hoola One vacuum machine uses the principle of buoyancy to separate small plastic litter from the beach (Hoola One, 2020). All the materials, for example, plastics, sands, wood and debris, are vacuumed into a container full of water where microplastics will float, leaving the sand sinking and resting at the bottom. It then removes the microplastics and puts the clean sand back on the beach. The technology is believed more efficient compared to existing technologies as it recovers plastic particles as small as 50 μm.

Nowadays, nanotechnologies also find their ways to contribute to marine plastic cleaning, especially targeting at microplastics. Ferreira used ferrofluids, a combination of oil and magnetite powder, and magnets to extract microplastics from water (EcoWatch, 2019). Kang *et al.* (2019) developed a carbon nanotube catalyst which can boost the removal of microplastics and generate stable and environmentally friendly salt compounds, carbon dioxide and water in approximately 8 h. Uheida *et al.* (2021) demonstrated a novel strategy for the elimination of microplastics using photocatalytic degradation which was enabled by zinc oxide nanorods (ZnO NRs) immobilized onto glass fiber substrates. An average reduction of PP particle volume by 65% was obtained within two weeks under visible light irradiation. These innovative nanotechnologies make in-situ remediation of microplastics possible and bring great potential to solve the marine plastic issue in the future. Table 11.11 summarizes currently reported technologies for marine plastic clean-up.

11.3.4 Marine plastic debris recycling

Improved collection of marine plastic debris alone is not sufficient to deal with the problem; development of environmentally ethical techniques to recycle those recovered plastic wastes is essential to provide a comprehensive solution to the marine plastic pollution. Marine debris is usually collected as mix of contaminated and weathered materials that requires intensive separation, cleaning and preprocessing before being recycled. There are typically limited recycling options for such plastic debris. Hence, it poses great challenge for technology innovation to recycle marine plastic debris in an efficient and financially sound way. Dijkstra *et al.* (2021) reviewed the currently applied marine plastic recycling methods and subdivided them into four categories: (1) recycling collected fishing nets and gears; (2) recycling specific polymer; (3) recycling mixed plastic debris, and (4) plastic debris to energy. Several types of marine plastic debris recycling have been developed and even industrialized in recent years.

Recycling collected fishing nets and gear is the largest subset of current practice with regard to marine plastic recycling, due to the abundance of lost and discarded fish gears into the oceans every year, accounting for 10–11% of overall marine litter (Peña-Rodriguez *et al.*, 2021). Different materials are used for manufacturing fishing nets and gears, with the most commonly used polymers being polyamides, nylon, PE, PP and polyesters. There are a number of inventions developed to recycle the collected fishing nets and gears into plastic pellets, flakes, filaments, and/or other plastic products. Mechanical recycling is usually used to recycle such type of marine waste. Typically, the collected fishing nets are washed, chopped, milled and dried, then undergo extrusion processing to produce filament or pellet feedstocks. Manufacturers use the filaments and pellets from marine fishing nets to produce various plastic products. Notably, depending on the material types and conditions, a loss of properties in the final products after mechanical recycling typically occurs.

Specific polymers (such as PET, PP, PE) may be recycled after proper segregation and cleaning processes of recovered marine plastic debris. The OceanWorks recycles polymer pellets and flakes of PET, HDPE and PP from a variety of sources including the collected marine debris (OceanWorks, 2021). Mechanical and chemical recycling processes are considered for this type of marine plastic recycling. In the case of mechanical recycling, to avoid significant polymer degradation due to high processing temperature, mixing in raw materials or utilizing additives are usually carried out, as most of the collected marine litter has been weathered by UV radiation and other marine environment at different levels. Chemical recycling may be a good

Table 11.11 Emerging technologies for plastic debris clean-up.

Technology	Description	Target Particulates	Similar Projects and/or Products	References
Boom and barriers				
PermaFence	Durable floating booms installed across waterways, acting as a barrier to collect or deflect debris.	Macro	B.o.B Litter Trap; Bandalong Boom System™; The LitterBoom Floating Barrier; AlphaMERS Floating Barrier; Plastic Fischer TrashBoom; Osprey Litter Boom	AlphaMERS Ltd, (n.d.); Bandalong Iternational (2020); ELASTEC (n.d.); Mareaverde (2018); Osprey (2021b); Plastic Fischer (2021); The Great Babble Barriar (n.d.); The Litterboom Project (n.d.)
Blue Barrier	Floating fences to capture and divert floating debris towards the riverbanks.		Azure system; Nash Run Trash Trap; Chemolex Fence	Chemolex (2021); Ichthion (2020); Plastic Soup Foundation (n.d.); SEADS (2019)
Ocean Cleanup System 001	An artificial mobile coastline which can concentrate plastics in the ocean. It is composed of a 600-m-long floater and a 3-m-deep skirt attached below, and driven by natural oceanic forces including wind, waves and current.	Macro		The Ocean Cleanup (2018)
Receptacle				
Seabin V5	A floating garbage bin skimming the surface of the water by pumping water into the device and filtering out the marine litters	Macro and micro	Marina Trash Skimmer	Seabin Project for Cleaner Oceans (n.d.); Clean Ocean Access (2016)
The Great Bubble Barrier	Bubbles are created by pumping air through a perforated pipe which is placed at the bottom of a waterway; plastic debris is pushed to the side and guided into the catchment system	Macro and micro		The Great Babble Barriar (n.d.)

(Continued)

Technology	Description	Size	Examples	References
SDG litter trap	Floating litter trap that uses the water flow and pressure to capture trash with a bypass flap in the rivers	Macro		All Around Plasctics (2020)
Receptacle with boom				
Brute Bin	Trapping basket/chamber that utilizes booms to guide trash, capturing floatables from stormwater conveyances and outfalls before they reach the waterways	Macro	Clear River Litter Trap; Osprey Litter Gitters; Allseas Plastic Catcher; Shoreliner; Bandalong Litter Trap™	AllSeas (n.d.); Bandalong Iternational (2020); ClearRivers (n.d.); ELASTEC (n.d.); Osprey (2021a); TAUW (n.d.)
Mr. Trash Wheel™	Solar/hydro-powered anchored/floating device that uses a rake and conveyor belt system to lift debris into holding container. Floating debris is guided by booms connecting across the river mouth	Macro	The Interceptor™	The Ocean Cleanup (2016); Waterfront Partnership of Baltimore (2019)
CirCleaner	Running water sets a paddle wheel in motion, plastic debris from the upper water are guided by floating booms, captured by the moving paddles and placed into a receptacle	Macro		Plastic Soup Foundation (2019)
Watercraft vehicles-based technology				
Trashbot	Small scale, remotely controlled robot that is suited for a variety of applications; can be equipped with manipulator arm, nets, or baskets for removal of suspending and submerged trash	Macro	Hector the Collector; ORCA-SMURF Water Cleaning Robot; WasteShark™; Jellyfishbot; BluePhin	BluePhin Technologies, (2020); Interactive Autonomous DYnamic Systems (2020); ORCA-TECH (2020); Rozalia Project (2017); Urban Rivers (n.d.); WasteShark (2020)

(Continued)

Table 11.11 Emerging technologies for plastic debris clean-up. (*Continued*)

Technology	Description	Target Particulates	Similar Projects and/or Products	References
Ocean Cleanup System 002	Large floating boom/collection module driven by vessels which can collect plastic debris in the open oceans and rivers	Macro	Holy Turtle; SeeKuh II	Monllos (2018); One Earth One Ocean (2019); The Ocean Cleanup (2021)
RiverVax™	Large-scale trash skimming watercrafts to remove floating debris in the large rivers and coastal areas	Macro	SeaVax; Clewat Cleansweep Vessel; Trash Hunters; Marina Cleaner/Trash Skimmer; Havester; Omni Catamaran; Veri-Cat Trash Skimmer Boat; Water Witch Workboats; MariClean; FRED (Floating Robot for Eliminating Debris); SeeHamster; Manta; OC-Tech Horizon; ERVIS	Aquarius Systems, (2010); Clean Ocean Access (2016); Clear Blue Sea (2019); Clewat (2021); Echavez (2020); ELASTEC (n.d.); ERVIS Foundation (n.d.); Ocean Cleaner Technology (n.d.); One Earth One Ocean (2021); The Sea Cleaners (n.d.); Thepbamrung (2013); Water Witch (2021a, 2021b)
Others (Sand cleaner)				
Barber Surf Rake	Tractor-towed machine removes waste on beaches	Macro		Barber (2021a)
Barber Sand Man	Walk-behind sand shifting machine uses a vibrating screen to shift debris from sand and soil on beaches	Macro		Barber (2021b)

(*Continued*)

Hoola One Vacuum	Vacuum sand and debris into a tank that separates particles by buoyancy, allowing for plastic separation and removal	Macro and micro	Hoola One (2020)
Marine Microplastic Removal Tool	Sand is piled on a sheet of fine mesh stretched between two long poles, and the mesh catches plastic and other foreign materials while allowing sand to fall through	Macro and micro	Ward (2015)
Others (nanotechnology)			
Ferrofluid	Combined ferrofluids and microplastics being extracted by the magnetite powder and removed from water.	Micro	EcoWatch (2019)
Magnetic coil	Carbon nanotube catalysts laced with nitrogen which can boost the removal of microplastics and generate stable and environmentally friendly salt compounds, carbon dioxide and water in a short time.	Micro	Cell Press (2019); Kang *et al.* (2019)
Nano-coating for plastic	Zinc oxide nanorods (ZnO NRs) immobilized onto glass fiber substrates to make the nano coating material for plastics. It causes the photocatalytic reaction to sunlight in a way that weakens the plastic, causing it to crack and break apart.	Micro	Uheida *et al.* (2021)

Sources: Modified from Helinski *et al.* (2021) and Schmaltz *et al.* (2020).

option for marine polymers with low purity. This process breaks down waste polymers into its monomers, which are then reassembled into plastic polymers. Depolymerization of PET using processes such as methanolysis, hydrolysis and glycolysis have been tested and proven for their less sensitivity to the polymer degradation in marine environment (Barboza *et al.*, 2009; Carta *et al.*, 2003; Myren *et al.*, 2020).

Technical options for recycling mixed and/or contaminated marine plastics are relatively limited. Mechanical recycling is typically used to convert the mixed marine plastic waste to building blocks, briskets, fences or other construction materials (Dijkstra *et al.*, 2021). Energy recovery is another suitable option for these mixed and low quality marine plastics. However, mechanical recycling to high-quality plastic products and polymer recycling are difficult to be implemented with such plastic waste.

11.3.5 Challenges of technological innovation combatting marine plastic pollution

Despite the growing global concern and robust development on innovations to fight against marine plastic pollutions, there are still several challenges, and further efforts are required.

11.3.5.1 Challenges of and future directions for plastic debris monitoring and assessment

Monitoring and assessment of the sources, presence, distribution and impact of plastic debris paves the foundation for deploying any policies, strategies and technologies to defeat the marine plastic pollution. Several limitations on current available marine plastic monitoring technologies are listed below.

- **Ensure technological development and global application for marine plastic debris monitoring**: Although an integrated technical structure for marine debris observation have been proposed by scientists, many of the technologies are at their infant stage, meaning great research efforts are needed to enhance the capacity and reduce the technical barriers to adopt these technologies. For example, to build the ground-based observation systems requires variable platforms and facilities, which is typically associated with high capital and O&M cost. To enable such systems for low-income countries, development of cost-effective options as well as enhancement of international collaboration are needed. Furthermore, very few research studies have successfully used remote sensing technologies especially the public satellites to observe and report the distribution of plastic debris, despite advantages of the technology itself. On the other hand, advanced technologies for in-situ microplastic identification and characterization under different environmental media remain to be explored.
- **Harmonize standard methodology and data reporting format**: A number of global and regional organizations such as NOAA, UNEP, WESTPAC, MSFD and WIOMSA have published guidelines for monitoring, sampling and analyzing marine plastic litter. However, one major remaining

obstacle is the compatibility of these methods. This obstacle results in the inability to compare results due to variations in sampling procedures (sampling protocols, dealing with contamination, reporting units and analysis), which eventually impacts the reliability of the results. It is, therefore, critical to develop standard methods for environmental monitoring, aiming at harmonization of the plastic and microplastic monitoring and analytical methodologies.

- **Build up global monitoring networks and platform to enhance data sharing**: Currently, monitoring and data collection for marine plastic litter has been performed by national and regional programs as well as several non-governmental organizations. These observation programs are organized either on a regular basis or occasionally in the form of short-term projects, hence, generating huge heterogenicity in terms of the data coverage and quality. With the fast-growing number of local/regional initiatives and datasets, their assembling into larger products and databases for joint analyses is of essential values to benefit all key stakeholders in defeating marine plastic pollutions. However, a successful data platform and network for global marine plastic pollution requires unified definitions, standards and formats, complemented by well-developed infrastructure for data flow and storage. Meanwhile, the global governance for marine debris data acquisition, streaming, quality control and distribution to users is crucial but has yet to be designed and built. An example of good practice in building such regional partnership information infrastructure is the European Marine Observation and Data Network (EMODnet). Future efforts are needed to expand such practices at a global scale that inclusively covers the hotspots of global plastic leakages.

- **Establish scientific impact assessment methodology for marine plastic debris**: Despite a comprehensive life cycle impact assessment framework being integrated and proposed, a number of knowledge and methodology gaps remain, which constraints data coverage, indicator selection and effect quantification. For example, current research indicates that the elements of fate modelling, particularly transport processes in marine compartment, together with effect modelling relative to microplastic ingestion and microplastic entanglement have, to date, received most research attention (Woods *et al.*, 2019). Hence, impact assessment regarding plastic litter on ecosystem quality is currently the most feasible direction to be assessed. However, there is still a long way to go in order to fulfill the lack of data and expertise on the other potential concerned areas such as human health, social-economic assets, ecosystem services, and natural and cultural heritage.

11.3.5.2 Challenges of and future directions for marine plastic debris prevention and collection

- **Expensive cost:** The cost of different prevention and collection technologies vary from the relative low-cost booms and barriers (~100–300 USD/m) to the extremely advanced devices such as trash wheel and interceptor (up to 8 00 000–1 200 000 USD/device). In addition, most of the active technologies require intensive operation (e.g., labor cost and energy cost) and maintenance. As such, the cost of these technologies becomes one of the most important obstacles hindering the application and selection of different technologies. These problems can be alleviated by improving policy support, reducing cost through technology development, ensuring project funds security and enhancing international collaboration.

- **Potential environmental impact**: Marine wildlife entanglement may occur due to the installation of plastic clean-up devices. Although most of the technologies claim their low impact on underwater life during operation, concerns have been raised on several types of technologies which apply suction mechanism and possibly entangle small organisms in the surroundings. The possible environmental impact may also come from the energy consumption of active technologies such as wheels and watercraft. Nonetheless, several advanced technologies utilize the natural renewal energy sources and declaim their carbon neutrality as their highlight (e.g., the trash wheel and interceptor, etc.). However, technologies such as the bubble barrier and various watercraft vehicles typically consume fossil fuel and hence raise concerns about greenhouse gas (GHG) emission. Using cleaner and renewable energy to power the plastic clean-up devices hence will be a future trend for their development.

- **Microplastic prevention and collection technologies**: Macroplastics continuously fragment in the environment and eventually end up as microplastics. Current available prevention and collection technics are primarily targeting large particulates of plastic debris. Technologies designed for microplastic clean-up are still in their infancy; hence, limited options are available to accomplish such objectives. It is exciting that several emerging solutions such as advanced filtering equipment and nanotechnologies have been proposed and explored for their potential to clean up microplastics (Kang *et al.*, 2019; Schmaltz *et al.*, 2020). However, most of these technologies are still in the research and development stage. There are still great uncertainties for future large-scale applications.

- **Enhance the recycling of collected plastic debris**: It is difficult to recycle the collected plastic debris from the ocean since most of it has been contaminated by various chemical and biological pollutants. In addition, weathering by various processes under marine environment including wave fragmentation, UV radiation, and biochemical oxidation causes damage to its structure. The mixed and contaminated marine debris requires intensive separation, cleaning and preprocessing before it can be recycled. Also, limited recycling options and technologies are currently available. For these reasons, collected plastic debris is often landfilled, which leads to harmful CO_2 emissions. Due to these constraints, there is still large room for innovation and development of marine plastic waste recycling technologies.

Variable technology innovations and inventions provide stakeholders options to prevent plastic leakage into the waterways and to clean up plastic pollution accumulating in rivers and oceans. However, there are no simple solutions to such a complex and extensive problem as marine plastic pollution. Technological innovation and development cannot work effectively without proper policy and public effort. Only through continuous efforts to find creative solutions across technology, policy and advocacy can we possibly stop plastic leakage into the oceans and mitigate their impact.

REFERENCES

Abdallah M., Abu Talib M., Feroz S., Nasir Q., Abdalla H. and Mahfood B. (2020). Artificial intelligence applications in solid waste management: a systematic research review. *Waste Management*, **109**, 231–246. https://doi.org/10.1016/j.wasman.2020.04.057

Acuña-Ruz T., Uribe D., Taylor R., Amézquita L., Guzmán M. C., Merrill J., Martínez P., Voisin L. and Mattar B. C. (2018). Anthropogenic marine debris over beaches: spectral characterization for remote sensing applications. *Remote Sensing of Environment*, **217**, 309–322, https://doi.org/10.1016/J.RSE.2018.08.008

Agora Tech Lab. (2018). Waste Management Fueled by Blockchain [WWW Document]. https://www.agoratechlab.com/ (accessed 21 September 21).

Ahmed A. A. A. and Asadullah A. (2020). Artificial intelligence and machine learning in waste management and recycling. *Engineering International*, **8**, 43–52, https://doi.org/10.18034/ei.v8i1.498

All Around Plasctics. (2020). SCG delivering 20 SCG-DMCR Litter Traps and launched the 'SCG Smart Litter Trap 4.0' Prototype [WWW Document]. https://www.allaroundplastics.com/en/article/news-en/2513 (accessed 20 October 21).

AllSeas. (n.d.). River plastics removal [WWW Document]. https://allseas.com/allseas-river-plastics-removal-project/ (accessed 20 October 21).

AlphaMERS Ltd. (n.d.). Floating trash barrier [WWW Document]. https://sites.google.com/a/alphamers.com/alphamers/home/floating-trash-barrier?authuser=0 (accessed 20 October 21).

Amritkar M. V. (2017). Automatic waste management system with RFID and ultrasonic sensors. *International Journal of Computational Science and Engineering*, **5**, 240–242.

Andriolo U., Gonçalves G., Bessa F. and Sobral P. (2020). Mapping marine litter on coastal dunes with unmanned aerial systems: a showcase on the Atlantic Coast. *Science of the Total Environment*, **736**, 139632, https://doi.org/10.1016/J.SCITOTENV.2020.139632

Angin D. P., Siagian H., Suryanto D. E. and Sashanti R. M. (2018). Design and development of the trash spliter with three different sensors. *Journal of Physics Conference Series*, **1007**, 012057, https://doi.org/10.1088/1742-6596/1007/1/012057

Aquarius Systems. (2010). Trash hunters [WWW Document]. http://www.aquarius-systems.com/pages/54/trash_hunters.aspx (accessed 20 October 21).

Arii M., Aoki Y. and Koiwa M. (2012). Effective monitoring for marine debris after the Great East Japan Earthquake by using spaceborne synthetic aperture radar. Proc. SPIE Asia-Pacific Remote Sensing, Kyoto, Japan. https://doi.org/10.1117/12.975938

Artiola J. F. and Brusseau M. L. (2019). The role of environmental monitoring in pollution science. *Environmental and Pollution Science*, 149–162, https://doi.org/10.1016/B978-0-12-814719-1.00010-0

Artiola J. F., Brusseau M. L. and Pepper I. L. (2004). Environmental Monitoring and Characterization. Elsevier Academic Press, Burlington, MA, USA.

Atzori L., Iera A. and Morabito G. (2010). The internet of things: a survey. *Computer Networks*, **54**, 2787–2805, https://doi.org/10.1016/J.COMNET.2010.05.010

Balamurugan S., Ajithx A., Ratnakaran S., Balaji S. and Marimuthu R. (2017). Design of smart waste management system. 2017 International Conference on Microelectronic Devices, Circuits and Systems, Vellore, India, pp. 1–4, https://doi.org/10.1109/ICMDCS.2017.8211709

Bandalong Iternational. (2020). Bandalong Cleaning your Waterways [WWW Document]. https://www.bandalong.com.au/bandalong-boom-system.html (accessed 20 October 21).

Barber. (2021a). Surf rake – Tractor-towed beach cleaner machines [WWW Document]. http://www.hbarber.com/Cleaners/SurfRake/Default.html (accessed 20 October 21).

Barber. (2021b). Walk behind sand cleaner: The Barber Sand Man™ 850 [WWW Document]. http://www.hbarber.com/Cleaners/SandMan/Default.html (accessed 20 October 21).

Barboza E. S., Lopez D. R., Amico S. C. and Ferreira C. A. (2009). Determination of a recyclability index for the PET glycolysis. *Resources, Conservation & Recycling*, **53**, 122–128, https://doi.org/10.1016/j.resconrec.2008.10.002

BAX & Company. (2021). Waste Management Hierarchy [WWW Document]. https://baxcompany.com/insights/circularity-of-polymer-composites/waste-management-hierarchy/ (accessed 29 October 21).

Beaumont N. J., Aanesen M., Austen M. C., Börger T., Clark J. R., Cole M., Hooper T., Lindeque P. K., Pascoe C. and Wyles K. J. (2019). Global ecological, social and economic impacts of marine plastic. *Marine Pollution Bulletin*, **142**, 189–195, https://doi.org/10.1016/j.marpolbul.2019.03.022

Berg H., Sebestyén J., Bendix P., Blevennec K. Le and Vrancken K. (2020). Digital waste management.

BluePhin Technologies. (2020). Meet BluePhin [WWW Document]. https://bluephin.io/

Business Green. (2018). Cryptocycle Limited. Bus. Green.

Campanale C., Savino I., Pojar I., Massarelli C. and Uricchio V. F. (2020). A practical overview of methodologies for sampling and analysis of microplastics in riverine environments. *Sustainability*, **12**(17), 6755, https://doi.org/10.3390/su12176755

Carta D., Cao G. and D'Angeli C. (2003). Chemical recycling of poly(ethylene terephthalate) (pet) by hydrolysis and glycolysis. *Environmental Science and Pollution Research*, **10**, 390–394, https://doi.org/10.1065/espr2001.12.104.8

Cell Press. (2019). Magnetic 'springs' break down marine microplastic pollution. ScienceDaily.

Chemolex. (2021). Plastic Capture Device [WWW Document]. https://chemolex.co.ke/service/plastic-capture-device/ (accessed 21 October 21).

Chidepatil A., Bindra P., Kulkarni D., Qazi M., Kshirsagar M. and Sankaran K. (2020). From trash to cash: how blockchain and multi-sensor-driven artificial intelligence can transform circular economy of plastic waste? *Administrative Sciences*, **10**(2), 23, https://doi.org/10.3390/admsci10020023

Clean Ocean Access. (2016). Marina Trash Skimmers [WWW Document].

Clear Blue Sea. (2019). Meet Fred [WWW Document]. https://www.clearbluesea.org/meet-fred/ (accessed 20 October 21).

ClearRivers. (n.d.). Litter Traps [WWW Document]. https://www.clearrivers.eu/litter-traps (accessed 20 October 21).

Clewat. (2021). Collection of plastic waste [WWW Document]. https://clewat.com/en/collection-of-plastic-waste/ (accessed 20 October 21).

COASST. (1999). COASST [WWW Document]. https://coasst.org/ (accessed 24 January 22).

Coliba. (2021). App Mobile Coliba [WWW Document]. http://coliba.ci/ (accessed 29 October 21).

Compa M., Alomar C., Wilcox C., van Sebille E., Lebreton L., Hardesty B. D. and Deudero S. (2019). Risk assessment of plastic pollution on marine diversity in the Mediterranean Sea. *Science of the Total Environment*, **678**, 188–196, https://doi.org/10.1016/J.SCITOTENV.2019.04.355

Considine D. M. and Considine G. D. (1986). Robot technology fundamentals. In: Standard Handbook of Industrial Automation. Springer, Boston, MA, USA, pp. 262.

Cora Ball. (2020). Cora ball [WWW Document]. https://www.coraball.com/ (accessed 21 October 21).

Costa B. S., Bernardes A. C. S., Pereira J. V. A., Zampa V. H., Pereira V. A., Matos G. F., Soares E. A., Soares C. L. and Silva A. F. (2018). Artificial intelligence in automated sorting in trash recycling. Anais Do XV Encontro Nacional de Inteligência Artificial e Computacional (ENIAC), Sao Paulo, Brazil. pp. 198–205, https://doi.org/10.5753/eniac.2018.4416

CounterMEASURE. (2021). Plastic-free rivers through science [WWW Document]. https://countermeasure.asia/ (accessed 21 January 22).

County Waste & Recycling. (n.d.). County Waste [WWW Document]. https://www.county-waste.com/ (accessed 24 January 22).

Cutroneo L., Reboa A., Besio G., Borgogno F., Canesi L., Canuto S., Dara M., Enrile F., Forioso I., Greco G., Lenoble V., Malatesta A., Mounier S., Petrillo M., Rovetta R., Stocchino A., Tesan J., Vagge G. and Capello M. (2020). Microplastics in seawater: sampling strategies, laboratory methodologies, and identification techniques applied to port environment. *Environmental Science and Pollution Research*, **27**, 8938–8952, https://doi.org/10.1007/s11356-020-07783-8

Devasahayam S., Raman R. K. S., Chennakesavulu K. and Bhattacharya S. (2019). Plastics – Villain or Hero? Polymers and recycled polymers in mineral and metallurgical processing – A review. *Materials*, **12**(4), 655, https://doi.org/10.3390/ma12040655

Dijkstra H., van Beukering P. and Brouwer R. (2021). In the business of dirty oceans: overview of startups and entrepreneurs managing marine plastic. *Marine Pollution Bulletin*, **162**, 111880, https://doi.org/10.1016/J.MARPOLBUL.2020.111880

Echavez J. (2020). MariClean: A Sustainable Marine Litter Clean-up Vessel. United Nations, New York City, USA.

EcoWatch. (2019). Irish teen wins $50 000 for his project to extract microplastics from water. BRIGHTVIBES.

ELASTEC. (n.d.). PermaFence [WWW Document]. https://www.elastec.com/products/floating-boom-barriers/oil-containment-boom/permafence/ (accessed 20 October 21a).

ELASTEC. (n.d.). Brute bin trash collector [WWW Document]. https://www.elastec.com/products/floating-boom-barriers/trash-debris-boom/brute-bin/ (accessed 20 October 21b).

ELASTEC. (n.d.). Omni catamaran [WWW Document]. https://www.elastec.com/products/work-boats/omni-cat-waterway-maintenance/ (accessed 20 October 21c).

Ellen Macarthur Foundation. (2020). Upstream Innovation: a guide to packaging solutions [WWW Document]. https://plastics.ellenmacarthurfoundation.org/upstream (accessed 29 October 21).

Environmental Enhancements. (n.d.). About the Lint LUV-R Filter [WWW Document]. https://environmentalenhancements.com/store/index.php/products/products-lint-filter (accessed 21 October 21).

ERVIS Foundation. (n.d.). Ervis – The Ship [WWW Document]. http://www.ervisfoundation.org/ervis-the-ship/ (accessed 20 October 21).

Escobar-Sánchez G., Haseler M., Oppelt N. and Schernewski G. (2021). Efficiency of aerial drones for macrolitter monitoring on Baltic Sea Beaches. *Frontiers in Environmental Science*, **8**, 283, https://doi.org/10.3389/fenvs.2020.560237

European Commission. (2018). Digital transformation scoreboard 2018 – EU businesses go digital: opportunities, outcomes and uptake.

Exxpedition. (2014). ABOUT EXXPEDITION [WWW Document]. https://exxpedition.com/ (accessed 24 January 22).

Faststream Technologies. (2021). Smart Waste Management Solutions [WWW Document]. https://www.faststreamtech.com/solutions/smart-waste-management/ (accessed 22 September 21).

FibreFree. (2018). FibreFree [WWW Document]. https://www.livefibrefree.com/ (accessed 21 October 21).

Foster C. and Heeks R. (2013). Conceptualising inclusive innovation: modifying systems of innovation frameworks to understand diffusion of new technology to low-income consumers. *The European Journal of Development Research*, **25**, 333–355, https://doi.org/10.1057/ejdr.2013.7

Freeman S., Booth A. M., Sabbah I., Tiller R., Dierking J., Klun K., Rotter A., Ben-David E., Javidpour J. and Angel D. L. (2020). Between source and sea: the role of wastewater treatment in reducing marine microplastics. *Journal of Environmental Management*, **266**, 110642, https://doi.org/10.1016/J.JENVMAN.2020.110642

Fressoli M., Arond E., Abrol D., Smith A., Ely A. and Dias R. (2014). When grassroots innovation movements encounter mainstream institutions: implications for models of inclusive innovation. *Innovative Development*, **4**, 277–292, https://doi.org/10.1080/2157930X.2014.921354

Garaba S. P. and Dierssen H. M. (2018). An airborne remote sensing case study of synthetic hydrocarbon detection using short wave infrared absorption features identified from marine-harvested macro- and microplastics. *Remote Sensing of Environment*, **205**, 224–235, https://doi.org/10.1016/J.RSE.2017.11.023

GESAMP. (2019). Guidelines for the monitoring and assessment of plastic litter in the ocean.

Gibson T. (2020). Recycling robots. *Mechanical Engineering*, **142**, 32–37, https://doi.org/10.1115/1.2020-JAN2

Girlfriend Collective. (2021). Welcome to Girlfriend Collective [WWW Document]. https://girlfriend.com/pages/about (accessed 29 October 21).

GoJelly. (2018). Jellyfish mucus efficiently captures nanoparticles [WWW Document]. https://gojelly.eu/ (accessed 21 October 21).

Gonçalves G., Andriolo U., Pinto L. and Duarte D. (2020). Mapping marine litter with unmanned aerial systems: a showcase comparison among manual image screening and machine learning techniques. *Marine Pollution Bulletin*, **155**, 111158, https://doi.org/10.1016/J.MARPOLBUL.2020.111158

Greenemeier L. (2017). The Plastic Tide.

Hahladakis J. N., Iacovidou E. and Gerassimidou S. (2020). Plastic waste in a circular economy. In: Plastic Waste and Recycling, T. M. B. T.-P. W. Letcher (ed.), Elsevier Academic Press, Burlington, MA, USA, pp. 481–512, https://doi.org/10.1016/B978-0-12-817880-5.00019-0

Hanke G. (2013). Guidance on monitoring of marine litter in European seas.

Hardesty B. D., Harari J., Isobe A., Lebreton L., Maximenko N., Potemra J., van Sebille E., Vethaak A. D. and Wilcox C. (2017). Using numerical model simulations to improve the understanding of micro-plastic distribution and pathways in the marine environment. *Frontiers in Marine Science*, **4**, 30, https://doi.org/10.3389/fmars.2017.00030

Helinski O. K., Poor C. J. and Wolfand J. M. (2021). Ridding our rivers of plastic: a framework for plastic pollution capture device selection. *Marine Pollution Bulletin*, **165**, 112095, https://doi.org/10.1016/J.MARPOLBUL.2021.112095

Henderson R. M. and Clark K. B. (1990). Architectural innovation: the reconfiguration of existing product technologies and the failure of established firms. *Administrative Science Quarterly*, **35**, 9–30, https://doi.org/10.2307/2393549

Heriawan A. (2020). Upcycling Plastic Waste for Rural Road Construction in India: An Alternative Solution to Technical Challenges.

Hermsen E., Mintenig S. M., Besseling E. and Koelmans A. A. (2018). Quality criteria for the analysis of microplastic in biota samples: a critical review. *Environmental Science & Technology*, **52**, 10230–10240, https://doi.org/10.1021/acs.est.8b01611

Hidalgo-Ruz V. and Thiel M. (2015). The contribution of citizen scientists to the monitoring of marine litter. *Marine Anthropogenic Litter*, **16**, 429–447.

Hoola One. (2020). Plastic removal device [WWW Document]. https://hoolaone.com/home/ (accessed 20 October 21).

Huh J. H., Choi J. H. and Seo K. (2021). Smart trash bin model design and future for smart city. *Applied Sciences*, **11**, 4810, https://doi.org/10.3390/app11114810

Hutto R. L. and Belote R. T. (2013). Distinguishing four types of monitoring based on the questions they address. *Forest Ecology and Management*, **289**, 183–189, https://doi.org/10.1016/J.FORECO.2012.10.005

Ichthion. (2020). Azure – macro plastics in rivers [WWW Document]. https://ichthion.com/technology/ (accessed 21 October 21).

Interactive Autonomous DYnamic Systems. (2020). The Jellyfishbot [WWW Document]. https://www.iadys.com/en/jellyfishbot-2/ (accessed 20 October 21).

International Pellet Watch. (2005). What's International Pellet Watch (IPW) [WWW Document]. http://pelletwatch.org/ (accessed 24 January 22).

ISWA. (2017). The Impact of the 4th Industrial Revolution on the Waste Management Sector.

Jaikumar K., Brindha T., Deepalakshmi T. K. and Gomathi S. (2020). IOT Assisted MQTT for Segregation and Monitoring of Waste for Smart Cities. 6th International Conference on Advanced Computing and Communication Systems (ICACCS), Coimbatore, India. https://doi.org/10.1109/ICACCS48705.2020.9074399

Jakovljevic G., Govedarica M. and Alvarez-Taboada F. (2020). A deep learning model for automatic plastic mapping using unmanned aerial vehicle (UAV) data. *Remote Sensing*, **12**(9), 1515, https://doi.org/10.3390/RS12091515

Jayson M., Hiremath S. and Lakshmi H. R. (2018). SmartBin-Automatic waste segregation and collection. SmartBin-Automatic waste segregation and collection, 2018 Second International Conference on Advances in Electronics, Computers and Communications (ICAECC), Bangalore, India.

Kang J., Zhou L., Duan X., Sun H., Ao Z. and Wang S. (2019). Degradation of cosmetic microplastics via functionalized carbon nanosprings. *Matter*, **1**, 745–758, https://doi.org/10.1016/J.MATT.2019.06.004

Kikaki A., Karantzalos K., Power C. A. and Raitsos D. E. (2020). Remotely sensing the source and transport of marine plastic debris in Bay Islands of Honduras (Caribbean Sea). *Remote Sensing*, **12**(11), 1727, https://doi.org/10.3390/rs12111727

Kiran D. R. (2019). Chapter 35 – internet of things. In: Production Planning and Control, D. R. B. T. -P. P. Kiran (ed.), Butterworth-Heinemann, Oxford, United Kingdom, pp. 495–513, https://doi.org/10.1016/B978-0-12-818364-9.00035-4

Koelmans A. A., Kooi M., Law K. L. and van Sebille E. (2017). All is not lost: deriving a top-down mass budget of plastic at sea. *Environmental Research Letters*, **12**, 114028, https://doi.org/10.1088/1748-9326/aa9500

Kogabayev T. and Maziliauskas A. (2017). The definition and classification of innovation. *Holistica – Journal of Business and Public Administration*, **8**, 59–72, https://doi.org/10.1515/hjbpa-2017-0005

Kooi M., Nes E. H. van., Scheffer M. and Koelmans A. A. (2017). Ups and downs in the ocean: effects of biofouling on vertical transport of microplastics. *Environmental Science & Technology*, **51**, 7963–7971, https://doi.org/10.1021/acs.est.6b04702

Kouhizadeh M. and Sarkis J. (2018). Blockchain practices, potentials, and perspectives in greening supply chains. *Sustainability*, **10**(10), 3652, https://doi.org/10.3390/su10103652

Kruse K., Knickmeier K., Honorato-Zimmer D., Gatta-Rosemary M., Weinmann A., Kiessling T., Schöps K., Thiel M. and Parchmann I. (2020). Dem plastikmüll auf der spur – Ein internationales citizen-science-projekt zur förderung der naturwissenschaftlichen grundbildung von schülerinnen und schülern. *CHEMKON*, **27**, 328–336, https://doi.org/10.1002/ckon.201800093

Kumar S. V., Kumaran T. S., Kumar A. K. and Mathapati M. (2017). Smart garbage monitoring and clearance system using internet of things. 2017 IEEE International Conference on Smart Technologies and Management for Computing, Chennai, India, pp. 184–189, https://doi.org/10.1109/ICSTM.2017.8089148

Kumar R., Verma A., Shome A., Sinha R., Sinha S., Jha P. K., Kumar R., Kumar P., Shubham Das S., Sharma P. and Vara Prasad P. V. (2021). Impacts of plastic pollution on ecosystem services, sustainable development goals, and need to focus on circular economy and policy interventions. *Sustainability*, **13**(17), 9963, https://doi.org/10.3390/su13179963

Lavender L. K. and Thompson R. C. (2014). Microplastics in the seas. *Science (80-.)*, **345**, 144–145, https://doi.org/10.1126/science.1254065

Lebreton L. C. M., Greer S. D. and Borrero J. C. (2012). Numerical modelling of floating debris in the world's oceans. *Marine Pollution Bulletin*, **64**, 653–661, https://doi.org/10.1016/J.MARPOLBUL.2011.10.027

Loop. (2021). How It Works [WWW Document]. https://loopstore.com/ (accessed 29 October 21).

Lopez J. (2015). Types of Innovation [WWW Document]. Constant Contact. https://techblog.constantcontact.com/software-development/types-of-innovation/ (accessed 29 October 21).

LTO NETWORK. (2019). Waste transport on Blockchain [WWW Document]. https://www.ltonetwork.com/ (accessed 21 September 21).

Lufadeju Y. and Chavanel S. (2019). UNICEF Breaks Ground on Africa's First-of-its-Kind Recycled Plastic Brick Factory in Côte D'Ivoire. UNICEF, New York, USA.

Mainzer K. (2016). Toward a theory of intelligent complex systems: from symbolic AI to embodied and evolutionary AI. In: Fundamental Issues of Artificial Intelligence. Springer, Cham, Switzerland, pp. 241–259.

Mareaverde. (2018). B.o.B. Litter Trap [WWW Document].

Maximenko N., Hafner J. and Niiler P. (2012). Pathways of marine debris derived from trajectories of Lagrangian drifters. *Marine Pollution Bulletin*, **65**, 51–62, https://doi.org/10.1016/J.MARPOLBUL.2011.04.016

Maximenko N., Corradi P., Law K. L., Van Sebille E., Garaba S. P., Lampitt R. S., Galgani F., Martinez-Vicente V., Goddijn-Murphy L., Veiga J. M., Thompson R. C., Maes C., Moller D., Löscher C. R., Addamo A. M., Lamson M. R., Centurioni L. R., Posth N. R., Lumpkin R., Vinci M., Martins A. M., Pieper C. D., Isobe A., Hanke G., Edwards M., Chubarenko I. P., Rodriguez E., Aliani S., Arias M., Asner G. P., Brosich A., Carlton J. T., Chao Y., Cook A.-M., Cundy A. B., Galloway T. S., Giorgetti A., Goni G. J., Guichoux Y., Haram L. E., Hardesty B. D., Holdsworth N., Lebreton L.,

Leslie H. A., Macadam-Somer I., Mace T., Manuel M., Marsh R., Martinez E., Mayor D. J., Le Moigne M., Molina Jack M. E., Mowlem M. C., Obbard R. W., Pabortsava K., Robberson B., Rotaru A.-E., Ruiz G. M., Spedicato M. T., Thiel M., Turra A. and Wilcox C. (2019). Toward the integrated marine debris observing system. *Frontiers in Marine Science*, **6**, 447, https://doi.org/10.3389/fmars.2019.00447

McCarthy S. (2019). WWF Launches Activation Hub to Help Prevent 10 Million Metric Tons of Global Plastic Waste.

Miller E., Sedlak M., Lin D., Box C., Holleman C., Rochman C. M. and Sutton R. (2021). Recommended best practices for collecting, analyzing, and reporting microplastics in environmental media: lessons learned from comprehensive monitoring of San Francisco Bay. *Journal of Hazardous Materials*, **409**, 124770, https://doi.org/10.1016/J.JHAZMAT.2020.124770

Monllos K. (2018). SodaStream Created a New Device, the 'Holy Turtle,' to Remove Plastic in the Oceans [WWW Document]. https://www.adweek.com/brand-marketing/sodastream-created-a-new-device-the-holy-turtle-to-remove-plastic-in-the-oceans/#:~:text=The Holy Turtle was 'designed, %2C' according to a release (accessed 21 October 21).

Myren T. H. T., Stinson T. A., Mast Z. J., Huntzinger C. G. and Luca O. R. (2020). Chemical and electrochemical recycling of end-use poly(ethylene terephthalate) (PET) plastics in batch, microwave and electrochemical reactors. *Molecules*, **25**(12), 2742, https://doi.org/10.3390/molecules25122742

Nam H. C. and Park W. H. (2020). Aliphatic polyester-based biodegradable microbeads for sustainable cosmetics. *ACS Biomaterials Science & Engineering*, **6**, 2440–2449, https://doi.org/10.1021/acsbiomaterials.0c00017

Naturbeads. (2018). About NatureBeads [WWW Document]. http://www.naturbeads.com/ (accessed 21 October 21).

New Plastics Economy. (2017). HolyGrail: tagging packaging for accurate sorting and high-quality recycling [WWW Document]. https://www.newplasticseconomy.org/ (accessed 21 September 21).

Nguyen B., Claveau-Mallet D., Hernandez L. M., Xu E. G., Farner J. M. and Tufenkji N. (2019). Separation and analysis of microplastics and nanoplastics in complex environmental samples. *Accounts of Chemical Research*, **52**, 858–866, https://doi.org/10.1021/acs.accounts.8b00602

Niska H. and Serkkola A. (2018). Data analytics approach to create waste generation profiles for waste management and collection. *Waste Management*, **77**, 477–485, https://doi.org/10.1016/j.wasman.2018.04.033

NOAA. (n.d.). Marine Debris Program [WWW Document] https://marinedebris.noaa.gov/ (accessed 24 January 22).

Noiki A., Afolalu S. A., Abioye A. A., Bolu C. A. and Moses E. E. (2021). Smart waste bin system: a review. IOP Conference Series: Earth and Environmental Science. p. 655, https://doi.org/10.1088/1755-1315/655/1/012036

Ocean Cleaner Technology. (n.d.). OC-Tech Vessels [WWW Document]. https://www.oceancleaner.es/products/ (accessed 20 October 21).

OceanWorks. (2021). Recycled Plastic Solutions [WWW Document]. https://oceanworks.co/ (accessed 28 October 21).

One Earth One Ocean. (2019). SeeKuh II [WWW Document]. https://oneearth-oneocean.com/en/the-seekuh/ (accessed 20 October 21).

One Earth One Ocean. (2021). The SeeHamster [WWW Document].

OpenMap Development Tanzania. (2017). OpenMap Development Tanzania [WWW Document]. https://www.omdtz.or.tz/ (accessed 29 October 21).

ORCA-TECH. (2020). ORCA-SMURF water cleaning robot [WWW Document]. https://www.orca-tech.cn/EN/projectIntro.html (accessed 20 October 21).

Osprey. (2021a). Litter Collection Devices [WWW Document]. https://osprey.world/services (accessed 20 October 21).

Osprey. (2021b). Litter Interceptor [WWW Document]. https://osprey.world/litter-interceptor (accessed 21 October 21).

Patel K. K. and Patel S. M. (2016). Internet of things-IOT: definition, characteristics, architecture, enabling technologies, application & future challenges. *International Journal of Engineering Science and Computing*, **6**, 6122–6131.

Peeken I., Primpke S., Beyer B., Gütermann J., Katlein C., Krumpen T., Bergmann M., Hehemann L. and Gerdts G. (2018). Arctic Sea ice is an important temporal sink and means of transport for microplastic. *Nature Communications*, **9**, 1505, https://doi.org/10.1038/s41467-018-03825-5

Peña-Rodriguez C., Mondragon G., Mendoza A., Mendiburu-Valor E., Eceiza A. and Kortaberria G. (2021). Recycling of marine plastic debris. In: Recent Developments in Plastic Recycling. Composites Science and Technology, J. Parameswaranpillai, S. Mavinkere Rangappa, A. G. Rajkumar and S. Siengchin (eds.), Springer Singapore, Gateway East, Singapore. pp 121–141, https://doi.org/10.1007/978-981-16-3627-1_6

Perera C., Qin Y., Estrella J. C., Reiff-Marganiec S. and Vasilakos A. V. (2017). Fog computing for sustainable smart cities: a survey. *ACM Computing Surveys (CSUR)*, **50**(3), 1–43, https://doi.org/10.1145/3057266

Phillips D. I. (1999). A new litter trap for urban drainage systems. *Water Science and Technology*, **39**, 85–92, https://doi.org/10.1016/S0273-1223(99)00011-6

Plastic Bank. (2020). Empowering the World to Stop Ocean Plastic [WWW Document]. https://plasticbank.com/ (accessed 21 September 21).

Plastic Fischer. (2021). Plastic Fischer TrashBoom [WWW Document]. https://plasticfischer.com/projects (accessed 20 October 21).

Plastic Soup Foundation. (2019). New Dutch plastic catcher from Noria [WWW Document]. https://www.plasticsoupfoundation.org/en/2019/11/new-dutch-plastic-catcher-from-noria/ (accessed 20 October 21).

Plastic Soup Foundation. (n.d.). Nash run trash trap [WWW Document]. https://www.plasticsoupfoundation.org/en/solutions/nash-run-trash-trap/#content (accessed 21 October 21).

Postecapes. (2015). What Is The 'Internet of Things'? [WWW Document]. https://www.postscapes.com/what-exactly-is-the-internet-of-things-infographic/ (accessed 21 September 21).

Prata J. C., da Costa J. P., Duarte A. C. and Rocha-Santos T. (2019). Methods for sampling and detection of microplastics in water and sediment: a critical review. *Trends in Analytical Chemistry*, **110**, 150–159, https://doi.org/10.1016/J.TRAC.2018.10.029

Procto. (2019). What is Architectural Innovation & 3 Examples for it [WWW Document]. https://www.procto.biz/architectural-innovation (accessed 29 October 21).

Regenize. (2021). Recycling collections that reward you [WWW Document]. https://www.regenize.co.za/ (accessed 29 October 21).

Robinson J., New A. L., Popova E. E., Srokosz M. A. and Yool A. (2017). Far-field connectivity of the UK's four largest marine protected areas: four of a kind? *Earth's Future*, **5**, 475–494, https://doi.org/10.1002/2016EF000516

Rothy's. (2021). Rothy's [WWW Document]. https://rothys.com/ (accessed 29 October 21).

Rozalia Project. (2017). Technology: Hector the Collector, seeing through the dark and knowing where we are! [WWW Document]. http://rozaliaproject.org/about/technology/ (accessed 20 October 21).

Saberi S., Kouhizadeh M. and Sarkis J. (2018). Blockchain technology: a panacea or pariah for resources conservation and recycling? *Resources, Conservation & Recycling*, **130**, 80–81, https://doi.org/10.1016/J.RESCONREC.2017.11.020

Salimi I., Bayu Dewantara B. S. and Wibowo I. K. (2019). Visual-based trash detection and classification system for smart trash bin robot. International Electronics Symposium on Knowledge Creation and Intelligent Computing, Bali, Indonesia, IES-KCIC 2018 – Proceedings. pp. 378–383, https://doi.org/10.1109/KCIC.2018.8628499

Saling P., Gyuzeleva L., Wittstock K., Wessolowski V. and Griesshammer R. (2020). Life cycle impact assessment of microplastics as one component of marine plastic debris. *International Journal of Life Cycle Assessment*, **25**, 2008–2026, https://doi.org/10.1007/s11367-020-01802-z

Satvilkar M. (2018). Image Based Trash Classification Using Machine Learning Algorithms for Recyclability Status. National College of Ireland, Dublin 1, Ireland.

Schmaltz E., Melvin E. C., Diana Z., Gunady E. F., Rittschof D., Somarelli J. A., Virdin J. and Dunphy-Daly M. M. (2020). Plastic pollution solutions: emerging technologies to prevent and collect marine plastic pollution. *Environment International*, **144**, 106067, https://doi.org/10.1016/J.ENVINT.2020.106067

Seabin Project for Cleaner Oceans. (n.d.). SEABIN V5 [WWW Document]. https://seabinproject.com/the-seabin-v5/ (accessed 20 October 21).

SEADS. (2019). Blue barriers technology [WWW Document]. https://www.seadefencesolutions.com/project/blue-barriers-technology/ (accessed 21 October 21).

Shower Loop. (n.d.). Shower Loop [WWW Document]. https://showerloop.org/ (accessed 21 October 21).

Singh C. P., Manisha M., Hsiung P.-A. and Malhotra S. (2019). Automatic Waste Segregator as an integral part of Smart Bin for waste management system in a Smart City. 5th International Conference On Computing, Communication, Control and Automation (ICCUBEA), Pune, India.

Sohail H., Ullah S., Khan A., Samin O. B. and Omar M. (2019). Intelligent trash bin (ITB) with trash collection efficiency optimization using IOT sensing. 2019 8th International Conference on Information and Communication Technologies, Karachi, Pakistan, pp. 48–53, https://doi.org/10.1109/ICICT47744.2019.9001982

Soni G. and Kandasamy S. (2018). Smart garbage bin systems – a comprehensive survey BT – smart secure systems – IoT and analytics perspective. In: Smart Secure Systems – IoT and Analytics Perspective, G. P. Venkataramani, K. Sankaranarayanan, S. Mukherjee, K. Arputharaj and S. Sankara Narayanan (eds.), Springer Singapore, Singapore, pp. 194–206.

Stanley M. (2010). Debris Tracker [WWW Document]. https://debristracker.org/ (accessed 24 January 22).

Steffen A. D. (2019). Check Out These Easy Refill Machines For Cleaning Products in Colombia. Intell. Living.

Stock F., Kochleus C., Bänsch-Baltruschat B., Brennholt N. and Reifferscheid G. (2019). Sampling techniques and preparation methods for microplastic analyses in the aquatic environment – a review. *Trends in Analytical Chemistry*, **113**, 84–92, https://doi.org/10.1016/J.TRAC.2019.01.014

Storm Water Systems. (n.d.). StormX Netting Trash Trap [WWW Document]. https://stormwatersystems.com/stormx-netting-trash-trap/ (accessed 21 October 21).

StormTrap. (2016). PumpGuard [WWW Document]. https://stormtrap.com/products/pumpguard/ (accessed 10.21.21).

StormTrap. (2020). TRASHTRAP [WWW Document]. https://stormtrap.com/products/trashtrap/ (accessed 21 October 21).

Swan M. B. (2015). Blockchain: Blueprint for a New Economy, O'Reilly Media, Inc., Sebastopol, CA.

TAUW. (n.d.). Shoreliner – Vermindering aanvoer plastic naar de oceaan [WWW Document]. https://www.tauw.nl/op-welk-gebied/circulaire-economie/shoreliner-plastic-soup-oplossing.html (accessed 20 October 21).

Taylor P., Steenmans K. and Steenmans I. (2020). Blockchain technology for sustainable waste management. *Frontiers in Political Science*, **2**, 15, https://doi.org/10.3389/fpos.2020.590923

Teng A. (2018). JTC puts up more 'smart' waste bins that alert cleaners via SMS when full. Today.

Thakker S. and Narayanamoorthi R. (2015). Smart and Wireless Waste Management: An innovative way to manage waste and also produce energy, https://doi.org/10.1109/ICIIECS.2015.7193141

The Great Babble Barriar. (n.d.). The Bubble Barrier [WWW Document]. https://thegreatbubblebarrier.com/technology/ (accessed 20 October 21).

The Litterboom Project. (n.d.). The Litterboom Project [WWW Document]. https://www.thelitterboomproject.com/ (accessed 20 October 21).

The Ocean Cleanup. (2016). THE INTERCEPTORTM [WWW Document]. https://theoceancleanup.com/rivers/ (accessed 20 October 21).

The Ocean Cleanup. (2018). SYSTEM 001 [WWW Document]. https://theoceancleanup.com/milestones/system001/ (accessed 20 October 21).

The Ocean Cleanup. (2021). SYSTEM 002 [WWW Document]. https://theoceancleanup.com/milestones/system-002/ (accessed 20 October 21).

The Sea Cleaners. (n.d.). The Manta [WWW Document]. https://www.theseacleaners.org/en/the-manta-a-revolutionary-vessel/ (accessed 20 October 21).

Thepbamrung N. (2013). New river fleet helps weeding go with the flow. Bangkok Post.

Topouzelis K., Papakonstantinou A. and Garaba S. P. (2019). Detection of floating plastics from satellite and unmanned aerial systems (plastic litter project 2018). *International Journal of Applied Earth Observation and Geoinformation*, **79**, 175–183, https://doi.org/10.1016/J.JAG.2019.03.011

Uheida A., Mejía H. G., Abdel-Rehim M., Hamd W. and Dutta J. (2021). Visible light photocatalytic degradation of polypropylene microplastics in a continuous water flow system. *Journal of Hazardous Materials*, **406**, 124299, https://doi.org/10.1016/J.JHAZMAT.2020.124299

UNCTAD. (2017). New innovation approaches to support the implementation of the Sustainable Development Goals.

Urabe K., Child J. and Kagono T. (eds.). (2018). Innovation and Management: International Comparisons. De Gruyter, Berlin, Germany, https://doi.org/10.1515/9783110864519

Urban Rivers. (n.d.). Trashbot [WWW Document]. https://www.urbanriv.org/innovations (accessed 20 October 21).

van der Have R. P. and Rubalcaba L. (2016). Social innovation research: an emerging area of innovation studies? *Research Policy*, **45**, 1923–1935, https://doi.org/10.1016/j.respol.2016.06.010

van Lieshout C., van Oeveren K., van Emmerik T. and Postma E. (2020). Automated river plastic monitoring using deep learning and cameras. *Earth and Space Science*, **7**(8), e2019EA000960, https://doi.org/10.1029/2019EA000960

van Sebille E., England M. H. and Froyland G. (2012). Origin, dynamics and evolution of ocean garbage patches from observed surface drifters. *Environmental Research Letters*, **7**, 44040, https://doi.org/10.1088/1748-9326/7/4/044040

Vo A. H., Hoang Son L. E., Vo M. T. and Le T. (2019). A novel framework for trash classification using deep transfer learning. *IEEE Access*, **7**, 178631–178639, https://doi.org/10.1109/ACCESS.2019.2959033

Wahab M. H. A., Kadir A. A., Tomari M. R. and Jabbar M. H. (2014). Smart recycle bin: a conceptual approach of smart waste management with integrated web based system. 2014 International Conference on IT Convergence and Security, Beijing, China, pp. 1–4, https://doi.org/10.1109/ICITCS.2014.7021812

Ward M. (2015). Marine microplastic removal tool. US8944253B2.

WasteShark. (2020). RanMarine Technology [WWW Document]. https://www.wasteshark.com/ (accessed 20 October 21).

Water Witch. (2021a). Water Witch Workboats Mk. 2/Mk. 3 [WWW Document]. https://waterwitch.com/ (accessed 20 October 21).

Water Witch. (2021b). Versi-Cat Trash Skimmer Boat [WWW Document]. https://waterwitch.com/ww/versi-cat-2/ (accessed 20 October 21).

Waterfront Partnership of Baltimore. (2019). Mr. Trash Wheel®: A proven solution to ocean plastics [WWW Document]. https://www.mrtrashwheel.com/ (accessed 20 October 21).

Wilson M., Kitson N., Beavor A., Ali M., Palfreman J. and Makarem N. (2021). Digital Dividends in Plastic Recycling.

Woods J. S., Rødder G. and Verones F. (2019). An effect factor approach for quantifying the entanglement impact on marine species of macroplastic debris within life cycle impact assessment. *Ecological Indicators*, **99**, 61–66, https://doi.org/10.1016/J.ECOLIND.2018.12.018

Woods J. S., Verones F., Jolliet O., Vázquez-Rowe I. and Boulay A. M. (2021). A framework for the assessment of marine litter impacts in life cycle impact assessment. *Ecological Indicators*, **129**, 107918, https://doi.org/10.1016/J.ECOLIND.2021.107918

World Economic Forum. (2019). The Loop Alliance plans to eliminate plastic waste.

Wu M., Yang C., Du C. and Liu H. (2020). Microplastics in waters and soils: occurrence, analytical methods and ecotoxicological effects. *Ecotoxicology and Environmental Safety*, **202**, 110910, https://doi.org/10.1016/J.ECOENV.2020.110910

Yang H.-Q., Lu T., Guo W.-J., Chang S. and Song J.-F. (2020). RecycleTrashNet: Strengthening Training Efficiency for Trash Classification Via Composite Pooling. Proceedings of the 2020 2nd International Conference on Advanced Control, Automation and Artificial Intelligence, Shanghai, China, pp. 77–82.

YTL Community. (2016). Smart bins with Free WiFi to Compact Trash on Orchard Road from November 2016. https://www.ytlcommunity.com/printnews.asp?newsID=427 (accessed 15 May 2021).

Zettler E. R., Takada H., Monteleone B., Mallos N., Eriksen M. and Amaral-Zettler L. A. (2017). Incorporating citizen science to study plastics in the environment. *Analytical Methods*, **9**, 1392–1403, https://doi.org/10.1039/C6AY02716D

Zhu S., Chen H., Wang M., Guo X., Lei Y. and Jin G. (2019). Plastic solid waste identification system based on near infrared spectroscopy in combination with support vector machine. *Advanced Industrial and Engineering Polymer Research*, **2**, 77–81, https://doi.org/10.1016/J.AIEPR.2019.04.001

IWA Publishing's authorised EU representative for General Product Safety Regulations is Diane D'Arras, 15 rue Duret, 75116 Paris, France, e-mail: safety@iwap.co.uk.

Printed and bound by CPI Group (UK) Ltd, Croydon, CR0 4YY
16/06/2026
02140367-0001